Group Theory in Non-Linear Problems

NATO ADVANCED STUDY INSTITUTES SERIES

Proceedings of the Advanced Study Institute Programme, which aims at the dissemination of advanced knowledge and the formation of contacts among scientists from different countries

The series is published by an international board of publishers in conjunction with NATO Scientific Affairs Division

A	Life Sciences	Plenum Publishing Corporation
B	Physics	London and New York
C	Mathematical and Physical Sciences	D. Reidel Publishing Company Dordrecht and Boston
D	Behavioral and Social Sciences	Sijthoff International Publishing Company Leiden
E	Applied Sciences	Noordhoff International Publishing Leiden

Series C – Mathematical and Physical Sciences

Volume 7 – Group Theory in Non-Linear Problems

Group Theory in Non-Linear Problems

Lectures Presented at the NATO Advanced Study Institute on Mathematical Physics, held in Istanbul, Turkey, August 7–18, 1972

edited by

A. O. BARUT

International Centre for Theoretical Physics, Trieste, Italy and University of Colorado, Boulder, Colo., U.S.A.

D. Reidel Publishing Company

Dordrecht-Holland / Boston-U.S.A.

Published in cooperation with NATO Scientific Affairs Division

Library of Congress Catalog Card Number 73–91202

ISBN-13: 978-94-010-2146-3 e-ISBN-13: 978-94-010-2144-9
DOI: 10.1007/978-94-010-2144-9

Published by D. Reidel Publishing Company
P.O. Box 17, Dordrecht, Holland

Sold and distributed in the U.S.A., Canada, and Mexico
by D. Reidel Publishing Company, Inc.
306 Dartmouth Street, Boston, Mass. 02116, U.S.A.

Softcover reprint of the hardcover 1st edition 1974

TABLE OF CONTENTS

INTRODUCTION

This is the second volume of a series of books in various aspects of Mathematical Physics. Mathematical Physics has made great strides in recent years, and is rapidly becoming an important discipline in its own right. The fact that physical ideas can help create new mathematical theories, and rigorous mathematical theorems can help to push the limits of physical theories and solve problems is generally acknowledged. We believe that continuous contacts between mathematicians and physicists and the resulting dialogue and the cross fertilization of ideas is a good thing. This series of studies is published with this goal in mind.

The present volume contains contributions which were originally presented at the Second NATO Advanced Study Institute on Mathematical Physics held in Istanbul in the Summer of 1972. The main theme was the application of group theoretical methods in general relativity and in particle physics. Modern group theory, in particular, the theory of unitary irreducible infinite-dimensional representations of Lie groups is being increasingly important in the formulation and solution of dynamical problems in various branches of physics. There is moreover a general trend of <u>approchement</u> of the methods of general relativity and elementary particle physics. We hope it will be useful to present these investigations to a larger audience.

A.O. BARUT

RELATIVISTIC SYMMETRY GROUPS

Roger Penrose
Department of Mathematics
Birkbeck College, London

1. ORTHOGONAL AND CONFORMAL GROUPS

A mathematical fact of very great significance for relativity theory is the existence of the familiar homomorphism* between the group SL(2,C) of complex unimodular (2 x 2) matrices and the Lorentz group O(1,3). This homomorphism

$$SL(2,C) \to O(1,3) \tag{1.1}$$

is a local isomorphism and maps SL(2,C) onto the identity-connected component of O(1,3), in an essentially (2-1) fashion. The term "essentially" here refers to the fact that SL(2,C) is connected. If A and B are the two elements of SL(2,C) which map to some given element Q of O(1,3) (actually B = -A), then A may be connected to B by some curve in SL(2,C). The image of this curve in O(1,3) is a closed curve κ through Q (topologically equivalent to a continuous rotation through 2π). Neither A nor B can be preferred as *the* SL(2,C) image of the Lorentz transformation Q. For as we pass from Q back to Q along the curve κ in O(1,3), the inverse image in

* A homomorphism between continuous groups is simply a continuous mapping from the first to the second which preserves the group operations. A local isomorphism is such a mapping which is 1-1 in the neighbourhood of the identity elements; then there is induced an isomorphism between the corresponding Lie algebras of infinitesimal group elements. The identity-connected component of a continuous group is its maximal connected subgroup (i.e. consisting of elements continuously connected with the identity element). For a discussion of the classical groups SL(2,C), O(1,3), etc. see references [1,2].

SL(2,C) must pass continuously from A to B, or else back from B to A. The ambiguity between A and B is therefore essential.

There is a higher-dimensional analogue of (1.1) which also has considerable importance for relativity theory, namely the homomorphism

$$SU(2,2) \rightarrow O(2,4) \tag{1.2}$$

which is also a local isomorphism, and which similarly maps SU(2,2) onto the identity-connected component of O(2,4) in an essentially (2-1) fashion. The group SU(2,2) of unimodular pseudo-unitary (++--) (4 x 4)-matrices gives rise to the algebra of twistors, analogously to the way that SL(2,C) gives rise to the algebra of 2-component spinors. Twistors will be discussed in Section 4. The significance of the pseudo-orthogonal group O(2,4), for relativity theory, lies in its relation to the 15-parameter conformal group of Minkowski space-time. I shall denote this conformal group by C(1,3) and give its precise definition shortly. We have, in fact, a homomorphism

$$O(2,4) \rightarrow C(1,3) \tag{1.3}$$

which is again a local isomorphism, mapping O(2,4) onto C(1,3) in an essentially (2-1) fashion. The homomorphism which is the composite of (1.2) with (1.3)

$$SU(2,2) \rightarrow C(1,3) \tag{1.4}$$

is thus a local isomorphism which maps SU(2,2) onto the identity-connected component of C(1,3) in an essentially (4-1) fashion.

Similar to (1.3) is a homomorphism

$$O(1,3) \rightarrow C(2) \tag{1.5}$$

where C(2) denotes a 6-parameter conformal group for the Euclidean plane analogous to C(1,3). (I shall be more precise shortly.) The homomorphism (1.5) is a local isomorphism which is (2-1), but it is not essentially (2-1). The map from the identity-connected component of O(1,3) onto the identity-connected component of C(2) is actually (1-1). The homomorphisms (1.3), (1.5) are part of a more general pattern of local isomorphisms:

$$O(p+1,\ q+1) \rightarrow C(p,q) \tag{1.6}$$

The local isomorphisms (1.1) and (1.2) are, on the other hand, special features of the low dimensionalities involved. (We may note, in passing, there is the "non-relativistic" essentially (2-1) local isomorphism SU(2) → O(3), which is closely related to these, to quaternions and to non-relativistic spinors.) For the remainder of this section I shall be primarily concerned with (1.6) and its particular instances (1.3), (1.5). The special isomorphism (1.4) will play a basic role in Section 4.

First, consider (1.5). I have yet to define what I mean by the conformal group C(2). The orientation-preserving local conformal maps of the plane to itself may be conveniently represented by

$$\zeta \to \tilde{\zeta} = f(\zeta), \tag{1.7}$$

where f is a holomorphic (i.e. complex-analytic) function and where

$$\zeta = x + iy,$$

x and y being standard Cartesian coordinates for the plane. We have, for the line-element,

$$d\sigma^2 = dx^2 + dy^2 = d\zeta d\bar{\zeta} \tag{1.8}$$

so

$$d\sigma^2 \to d\tilde{\sigma}^2 = |f'(\zeta)|^2 \, d\sigma^2$$

illustrating the conformal nature of (1.7). Since f is arbitrary holomorphic, the local conformal maps of the Euclidean plane to itself constitute an infinite-parameter system. For a global map, we would require that both f and its inverse map be non-singular over the whole plane. This restricts f to be a linear function

$$f(\zeta) = \alpha\zeta + \beta$$

showing that the group of (orientation-preserving) conformal maps of the plane to itself is described by four real parameters (real and imaginary parts of α, β). These maps may be generated by the Euclidean motions ($|\alpha|=1$) and the dilations (α real, $\beta=0$).

This group is not what I mean by C(2), however. For that, we require to compactify the plane by the addition of a point at infinity. This is a standard procedure in complex variable theory, and is most graphically illustrated by means of a stereographic projection of the unit sphere S^2 to the plane (Figure 1). Let S^2 be given by the equation

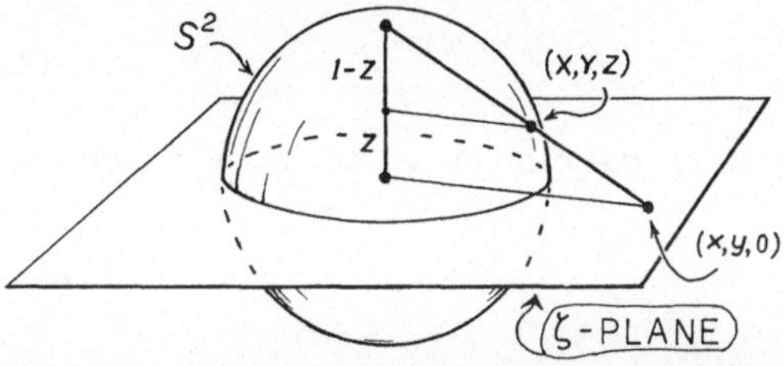

Fig. 1. The unit sphere S^2 is projected stereographically from the north pole (0,0,1) to the plane Z = 0, this plane being regarded as the Argand plane of the complex number ζ = x+iy. The first formula (1.9) is readily obtained from the geometry of the picture.

$$X^2 + Y^2 + Z^2 = 1,$$

where X,Y,Z are standard Cartesian coordinates for Euclidean 3-space. We project the point (X,Y,Z) on S^2, from the north pole (0,0,1), into the point (x,y,0) on the equatorial plane Z=0, where

$$\zeta = x + iy = \frac{X + iY}{1 - Z}; \quad X + iY = \frac{2\zeta}{1 + \zeta\bar{\zeta}}, \quad Z = \frac{\zeta\bar{\zeta} - 1}{1 + \zeta\bar{\zeta}} \quad . \tag{1.9}$$

The complex number ζ, which previously labelled a point in the plane, may now be used to label the corresponding point (X,Y,Z) of the sphere S^2. We have one further point, however, namely the north pole of S^2 (corresponding to $\zeta = \infty$). The stereographic projection, defined by (1.9), of S^2 to the plane, is in fact a conformal map. One way of seeing this is to re-express ζ in terms of standard spherical polar coordinates θ,ϕ for the sphere

$$\zeta = e^{i\phi} \cot\frac{\theta}{2} \tag{1.10}$$

and then to observe that the metric $d\sigma^2$ for S^2 is given by

$$d\ell^2 = d\theta^2 + \sin^2\theta d\phi^2 = \frac{4d\zeta d\bar{\zeta}}{(1+\zeta\bar{\zeta})^2} = \frac{4d\sigma^2}{(1+\zeta\bar{\zeta})^2} \tag{1.11}$$

$d\sigma^2$ being the metric for the plane Z = 0, as in (1.8). Thus, S^2 with the north pole removed, is conformally identical with the Euclidean plane. The addition of the north pole provides the required <u>conformal compacitification</u> of the plane.

The group C(2) may now be defined as the group of all conformal maps of the compactified plane (i.e. of S^2) to itself. The connected component of the identity in C(2) consists of the orientation preserving conformal maps of S^2. These are given by (1.7) where f has the form

$$\zeta \to f(\zeta) = \frac{\alpha\zeta + \beta}{\gamma\zeta + \delta} \quad , \tag{1.12}$$

these being regular at $\zeta = \infty$. The three complex ratios $\alpha:\beta:\gamma:\delta$ serve to define f, so we have a <u>six</u>-real-parameter group. If desired, we can normalize $\alpha,\beta,\gamma,\delta$ by

$$\alpha\delta - \beta\gamma = 1. \tag{1.13}$$

Then the unimodular complex (2 x 2) matrix (i.e. SL(2,C) element)

$$\begin{pmatrix} \alpha & \beta \\ \gamma & \delta \end{pmatrix} \tag{1.14}$$

may be used to represent our transformation (with only an overall sign ambiguity for (1.14)) the composition of two transformations (1.12) being now represented by matrix multiplication. (This can be used to derive the local isomorphism SL(2,C) → C(2).) The remaining (i.e. orientation-reversing) elements of $\underline{C}(2)$ are the compositions of (1.12) with complex conjugation: $\zeta \to \bar{\zeta}$. The elements of C(2) are generated by the Euclidean motions, dilations, and the inversion

$$\zeta \to \tilde{\zeta} = \bar{\zeta}^{-1} . \tag{1.15}$$

(This inversion is disconnected from the identity.) The group C(2) is transitive over the whole of S^2, showing that the adjoined point at infinity is (conformally) on an equal footing with all the other points. The inversion (1.15) actually interchanges the origin $\zeta = 0$ with the point at infinity $\zeta = \infty$ (i.e. it interchanges the north and south poles of S^2, being a reflection in the equatorial plane).

To establish the relation of C(2) with O(1,3), consider Minkowski 4-space with standard coordinates T,X,Y,Z, the metric being given by

$$ds^2 = dT^2 - dX^2 - dY^2 - dZ^2 .$$

I shall be concerned primarily with the null cone N of the origin, its equation being

$$T^2 - X^2 - Y^2 - Z^2 = 0. \tag{1.16}$$

The generators of N are the null rays through the origin, given by

$$T:X:Y:Z = \text{const.}$$

with T,X,Y,Z satisfying (1.16). Let us consider S^2 to be the section of N by the spacelike 3-plane T = 1. Then there is a (1-1)-correspondence between the generators of N and the points of S^2 (namely that given by the intersections of the generators with T = 1). We may regard S^2 as a realization of the space of generators of N. But equally well we could have used any other cross-section $\hat{S}^2$ of N to represent this space. However, the important point to realize is that the map which carries any one such cross-section into another, with points on the same generator of N corresponding to one another, is a conformal map (Figure 2). Thus,

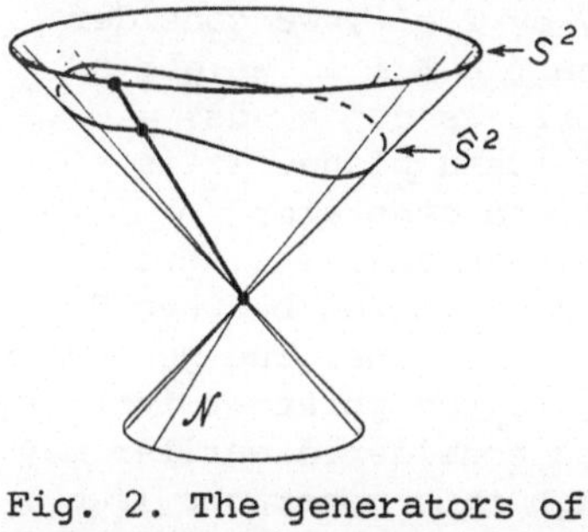

Fig. 2. The generators of the null cone N establish a 1-1 map between any two cross-sections of N. This map is conformal, so the space of generators of N may itself be assigned a conformal structure, namely that of any one of these sections.

the conformal structure of S^2 - or, equivalently, of $\hat{S}^2$ - reflects an intrinsic conformal structure on the space of generators of N.

To see that this generator map is a conformal map we may re-express the metric on N in a form

$$ds^2 = -r^2\gamma_{\alpha\beta}dx^\alpha dx^\beta + 0.dr^2 \tag{1.17}$$

where x^α and r are coordinates on N, the generators being given by the coordinate lines x^α = const. The quantities $\gamma_{\alpha\beta}$ are to be independent of r. (There are clearly many ways of attaining the form (1.17); one is to use ordinary spherical polar coordinates for Minkowski space, giving $ds^2 = dT^2 - dr^2 - r^2(d\theta^2 + \sin^2\theta d\phi^2)$, but with T = r on N). A cross-section of N is given by specifying r to be some function of the x^α. It is obvious from the form of (1.17) that any two cross-sections give conformally related metrics, being mapped to one another by the generators of N (i.e. keeping x^α constant). It is clear, moreover, that many other cone-like null surfaces will share this property of N, provided their metrics can be put in the form (1.17).

Let us now consider the effect of an (active) Lorentz transformation L on our configuration. This will send N into itself and send generators of N into other generators of N. It will send the cross-section S^2 into some other cross-section (say $\hat{S}^2$) whose intrinsic metric is the same as that of S^2. Since the map from $\hat{S}^2$ back to S^2 along generators of N is conformal, it follows that L induces a transformation on the space of generators of N which amounts to a conformal map of S^2 to itself. This establishes a homomorphism $O(1,3) \to C(2)$. Since the dimensionality of each group is the same, namely six, the inverse image of each element of C(2) must be a <u>discrete</u> set of elements of O(1,3). From this it follows that the mapping is a local isomorphism. Indeed, it is intuitively clear that L, up to a space-time reflection in the origin, is <u>defined</u> by its effect on N. The homomorphism $O(1,3) \to C(2)$ is thus (2-1). But it is only inessentially (2-1). For restriction to orthochronous O(1,3) transformations would yield a global isomorphism between the two groups.

If, instead of the cross-section of N by T = 1, we consider the intersection of N with the null 3-plane T = Z + 1, we get a "parabolic" section E^2 whose intrinsic metric is $d\sigma^2 = -ds^2 = dX^2 + dY^2$. Thus, E^2 is intrinsically a Euclidean plane. It is not quite a cross-section of N because the one generator T - Z = X = Y = 0, which is parallel to the 3-plane, fails to give rise to a point on E^2. Nevertheless the relation between E^2 and S^2, for all other generators, is a conformal one. The generator parallel to the 3-plane simply corresponds to the point at infinity on E^2. In fact, the sterographic projection considered earlier may be established by examining 2-planes through this generator (see Figure 3). Such a 2-plane meets S^2, E^2, and the plane Z = 0 = T - 1 in corresponding points. The details are left as an exercise.

Let us next consider $O(2,4) \to C(1,3)$. This will also serve to

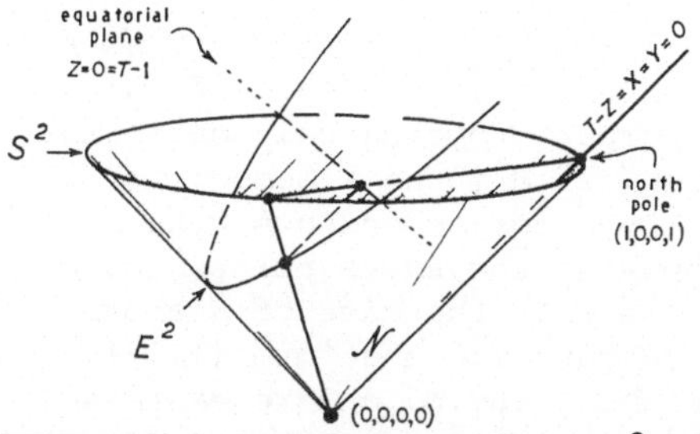

Fig. 3. The Euclidean plane E^2 may be imbedded as a "parabolic" section of N. Its conformal relation to the unit sphere S^2 is established via generators of N, or via 2-planes through the "parallel" generator $T - Z = X = Y = 0$.

illustrate the general case $O(p + 1, q + 1) \to C(p,q)$, which is basically no more complicated. In the first place, it will be necessary to define the group $C(1,3)$ (or, more generally, $C(p,q)$) precisely. As in the case of $C(2)$, we may regard $C(1,3)$ as the group of conformal self-transformations of the appropriate pseudo-Euclidean space (in this case, Minkowski 4-space) which has been compactified [3,4,5,6] in a suitable conformal way, by the addition of extra "points at inifinity". In the present context, the most rapid way of achieving this conformal compactification of Minkowski space is to go directly to the null cone of the origin in a six-dimensional pseudo-Euclidean (++----) space. Let us choose coordinates T, V, W, X, Y, Z for this space, the metric being given by

$$ds^2 = dT^2 + dV^2 - dW^2 - dX^2 - dY^2 - dZ^2 , \qquad (1.18)$$

the null cone K of the origin having the equation

$$T^2 + V^2 - W^2 - X^2 - Y^2 - Z^2 = 0 . \qquad (1.19)$$

By analogy with E^2 above we can consider the intersection M^4 of K with the null hyperplane $V - W = 1$. The intrinsic metric of M^4 is

$$ds^2 = dT^2 - dX^2 - dY^2 - dZ^2 .$$

The coordinates T, X, Y, Z suffice for M^4 and are unrestricted in range. Thus, M^4 is intrinsically identical with Minkowski space-time. In the 6-space, however, M^4 has the form of a "paraboloid", the remaining coordinates being defined in terms of T, X, Y, Z by

$$V = \tfrac{1}{2} (1 - T^2 + X^2 + Y^2 + Z^2) = W + 1 .$$

Every generator of K (set of points for which T:V:W:X:Y:Z are constant and for which (1.19) is satisfied), except for those lying in the null hyperplane $V = W$, will meet M^4 in a unique point.

We may think of those generators of K which do lie in V + W as representing "points at infinity" for M^4. Thus, the set G of (unoriented) generators of K (being a compact topological space) may be interpreted as a "compactification" of M^4, those members of K which do not lie in V = W being in (1-1)-correspondence with M^4, and those which do lie in V = W supplying the extra points necessary for the compactification. Now, as with the case of N above, the metric of K can (in many ways) be put into the form (1.17). (For example, we may select any variable, say W, and re-express (1.18) as $ds^2 = W^2\{d(T/W)^2 + d(V/W)^2 - d(X/W)^2 - d(Y/W)^2 - d(Z/W)^2\}$, using (1.19); and then eliminate one of the redundant variables T/W, ---, Z/W by expressing it in terms of the others, again using (1.19).) Thus, as with N, the generators of K establish conformal maps between any two (local) cross-sections of K. This gives G the structure of a conformal manifold, which we can identify as the conformal compactification of M^4. The group C(1,3) is now defined as the group of self-transformations of G which preserve its conformal structure. The definition of C(p,q) is similar.

There are certain differences arising in the case of C(1,3) from the earlier situation for C(2). In the first place, the infinite parameter set of "local" conformal transformations of the plane given by (1.7) has no analogue for dimension higher than two, and so does not occur here. On the other hand, the global conformal self-transformations of M^4 are like those for E^2. They may be generated by (pseudo-) Euclidean motions (here constituting the Poincaré group) and dilations. The 11-parameter (orthochronous) group thus arising is sometimes known as the causal group. In "compactifying" M^4 we must now adjoin not just a point but an entire null cone. One way of seeing this is to examine the "inversion" of M^4, which is the analogue of (1.15). This corresponds to the reflection W → -W in 6-space and can be expressed

$$(T, X, Y, Z) \to -\{T^2 - X^2 - Y^2 - Z^2\}^{-1} (T, X, Y, Z) \qquad (1.20)$$

in terms of Minkowski coordinates. The transformation (1.20) is not well-defined on the null cone $T^2 - X^2 - Y^2 - Z^2 = 0$, but maps this entire null cone to infinity. In effect, (1.20) "exchanges" the null cone of the origin with the null cone at infinity, and illustrates the fact that the elements at infinity must actually have the structure of a null cone.

As an analogue of the sphere S^2, we could consider the intersection of K with the hyperplane T = 1. This intersection has the structure of a de Sitter space. However, the de Sitter space does not represent the entire compact space G, there still being points at infinity (corresponding to the generators of K in T = 0). Similar remarks apply to any hyperplane section of K. In particular, the "anti-de Sitter space" , defined by the section of K with W = 1, also requires points at infinity to be added in order to become compact. Thus, none of these space-times is an adequate model for the compactified Minkowski space G. As an alternative, we can consider the intersection of K with the 5-sphere defined by

$$T^2 + V^2 + W^2 + X^2 + Y^2 + Z^2 = 2 .$$

This gives a compact space-time model $\mathcal{H}$ which (by (1.19)) is clearly the topological product of a 3-sphere in (W, X, Y, Z) - space

$$W^2 + X^2 + Y^2 + Z^2 = 1$$

with a circle (1-sphere) in (T, V) - space

$$T^2 + V^2 = 1 .$$

However, this is not quite a model of G, but a twofold covering of it. This is because each point of H is represented once as (T,V,W,X,Y,Z) and once as (-T,-V,-W,-X,-Y,-Z) on $\mathcal{H}$. The space-time $\mathcal{H}$ is connected, so the two-fold nature of the covering is "essential". The topology of G is $S^1 \times S^3$, as is its twofold covering.

The pseudo-orthogonal group O(2,4) acts on the 6-space. Since K is invariant, the group also acts on K. Each O(2,4) transformation must induce a conformal map of G to itself, and a homomorphism O(2,4) → C(1,3) is thereby obtained. The group O(2,4) has 15 parameters; C(1,3) cannot have more parameters than 15 (this being the maximum for local conformal self-transformations of a 4-manifold [7]). The fact that the homomorphism is a local isomorphism is not hard to establish from this. The fact that it is (2-1) arises because O(2,4) transformations can reverse the directions of the generators of K, whereas G is the space of <u>unoriented</u> generators. The O(2,4) transformation which is a reflection in the origin represents the identity on G. This reflection is continuous with the identity because O(2) x O(4) is contained in O(2,4) and the reflection in the origin for each of O(2) and O(4) is continuous with the identity. The essentially (2-1) nature of the homomorphism O(2,4) → C(1,3) follows from this. A similar remark applies to each O(p + 1, q + 1) for which both p and q are odd. If both are even, the local isomorphism O(p + 1, q + 1) → C(p, q) is still (2-1) but inessentially so; if p + q is odd the local isomorphism is (1-1) and is therefore a global isomorphism.

A smooth map from one space-time to another may be characterized as conformal by the fact that it takes null cones into null cones, or that it takes null geodesics into null geodesics. It is of some interest, therefore, to observe that the null line and null cone structure of (compactified) Minkowski space is reflected in the linear structure of K. There are, in fact ∞^5 2-planes through the origin, in the 6-space, which are completely contained in K. These planes are <u>completely null</u> in the sense that the distance (as defined by (1.18)) between <u>any</u> two points on each of the planes is zero. The intersection of such a plane with M^4 is therefore a null straight line in M^4. The generators of the null cone at infinity for M^4 also arise from such 2-planes, in this case lying in V = W. The 5-plane V = W is a tangent 5-plane to *K*, which touches *K* all along the particular generator V - W = T = X = Y = Z = 0 corresponding to the vertex of the null cone at infinity for M^4. The intersection of V = W with *K* defines this null cone. All "finite" null cones on M^4 also arise as intersections with tangent 5-planes to *K* (i.e. null 5-planes through the origin of 6-space). A different approach to the geometry of compactified Minkowski space will be discussed in the next section.

2. ASYMPTOTICALLY SIMPLE SPACE-TIMES

The construction of a "conformal infinity" for Minkowski space can be approached from another point of view which lends itself more readily to a generalization to curved space-times. According to this alternative approach [6] we simply rescale the metric of the space-time *M*, replacing the original physical metric ds by a new "unphysical" metric $d\hat{s}$, which is conformally related to it

$$d\hat{s} = \Omega ds \, , \qquad (2.1)$$

Ω being a suitably smooth, everywhere positive function defined on *M*. The metric tensor* g_{ab} and its inverse g^{ab} are accordingly rescaled by

$$g_{ab} \to \hat{g}_{ab} = \Omega^2 g_{ab} \, , \; g^{ab} \to \hat{g}^{ab} = \Omega^{-2} g^{ab} \; . \qquad (2.2)$$

Provided that the asymptotic structure of *M* is suitable, and that Ω is chosen appropriately, it is possible to adjoin more points to the manifold in such a way that the "unphysical" metric $\hat{g}_{ab}$ extends smoothly to these new points. The function Ω can also be extended smoothly, but becomes zero at these extra points. This implies that the physical metric would have to be infinite at the new points, so it cannot be so extended. Thus the new points are, from the point of view of the physical metric, infinitely distant from their neighbours. Physically, they represent points "at infinity".

It should be emphasized that the conformal rescaling (2.1), (2.2) is not a conformal transformation (conformal mapping) of the kind considered in Section 1. Here the points of the manifold are not transformed. It is only that one metric on a fixed manifold is

---

* From now on I shall be adopting the "abstract index" conventions according to which the symbol "g_{ab}" actually stands for the metric tensor itself rather than its components in some coordinate system. This makes little difference at this stage, but it has the effect of simplifying some of the formulae arising when the spinor formalism is employed towards the end of this section and in Section 4. See ref. [8] for details.

replaced by another. Conformal rescalings on a given space-time form an (uninteresting) infinite parameter Abelian group, namely the multiplicative group of smooth positive functions on the manifold. (This group, as applied to Minkowski space, has no element in common with the 15-parameter conformal group - except, perhaps, for the identity element - and should not be confused with that conformal group.) It is sometimes convenient to accompany a conformal rescaling with a coordinate change. This is merely in order that the newly adjoined points be assigned finite coordinates. No transformation of the points of the space-time is involved. Coordinates are, in any case, just a matter of convenience. If desired, the new coordinates may be introduced first, before any conformal rescaling takes place. This serves to emphasize that the transformation involved (namely the conformal rescaling) has nothing to do with a change in coordinates. The calculations that one performs using this technique are, in any event, often of the invariant, basically coordinate-free type.

The utility of this technique derives, to a large extent, from the fact that many important physical concepts are actually invariant under conformal rescalings [6]. In particular, the concepts of causality and of a null geodesic (or light ray) are invariant; so also are the zero rest-mass free-field equations for each spin, and electromagnetic interactions. On the other hand, timelike (or spacelike) geodesics, fields of non-zero rest-mass, and gravitational interactions are not conformally invariant. I shall discuss some of these matters a little more fully in due course. The conformal technique is valuable particularly when radiation properties of zero rest-mass fields are to be discussed. The incoming and outgoing radiation fields may be defined precisely, for an asymptotically flat, curved space-time, in terms of the values of the fields at the adjoined "points at infinity". The conformal technique also affords a very convenient and "coordinate-free" definition of asymptotic flatness in general relativity. It becomes unnecessary to make statements about asymptotic properties of the space in terms of complicated limiting procedures. Invariance properties are readily discerned; complicated coordinate transformation properties are avoided. Finally, as we shall see in Section 3, the Bondi-Metzner-Sachs asymptotic symmetry group can be given a clear geometrical interpretation, and its relation to gravitational radiation readily understood.

Let us first see how the technique may be applied in the case of Minkowski space. The physical metric, in spherical polar coordinates, is

$$ds^2 = dt^2 - dr^2 - r^2(d\theta^2 + \sin^2\theta \, d\phi^2) . \tag{2.3}$$

For convenience, introduce a retarded time parameter $u = t - r$ and an advanced time parameter $v = t + r$ to obtain

$$ds^2 = dudv - \tfrac{1}{4} (v - u)^2 (d\theta^2 + \sin^2\theta \, d\phi^2). \tag{2.4}$$

There is much freedom in the choice of conformal factor Ω. However, for the type of space-time that we shall consider here ("asymptotically simple" space-times), our choice of Ω must be such that along any null geodesic it approaches zero, both in the past and in the future, like the reciprocal of an affine parameter λ on the geodesic (i.e. $\Omega\lambda \to$ const. as $\lambda \to \infty$, and $\Omega\lambda \to$ const. as $\lambda \to -\infty$, along the geodesic). Each u = const. hypersurface is a future light cone, generated by null geodesics (straight lines) given when θ and ϕ are also constant. The coordinate v serves as an affine parameter into the future on each of these null geodesics. Similarly, the coordinate u serves as an affine parameter into the past on these radial null geodesics. Thus, we shall require $\Omega v \to$ const. as $v \to \infty$ on u, θ, ϕ = const. and $\Omega u \to$ const. as $u \to -\infty$ on v, θ, ϕ = const. If we wish also to keep Ω smooth over the finite parts of the space-time, then the choice

$$\Omega = (1 + u^2)^{-1/2} \ (1 + v^2)^{-1/2}$$

suggests itself, so

$$d\hat{s}^2 = \Omega^2 ds^2 = \frac{dudv}{(1+u^2)(1+v^2)} - \frac{(v-u)^2}{4(1+u^2)(1+v^2)} (d\theta^2 + \sin^2\theta d\phi^2).$$

Many other choices of Ω are equally possible, but this one is especially convenient, as we shall see shortly.

In order that our "points at infinity" may be assigned finite coordinates, we can replace u and v by p and q, where

$$u = \tan p, \quad v = \tan q .$$

Then

$$d\hat{s}^2 = dpdq - \tfrac{1}{4} \sin^2(q-p) \ \{d\theta^2 + \sin^2\theta d\phi^2\} . \qquad (2.5)$$

The range of the variables p, q is as indicated in Figure 4. The vertical line $q - p = 0$ represents the spatial origin ($r = 0$) and is just a <u>coordinate</u> singularity. The space-time is, of course,

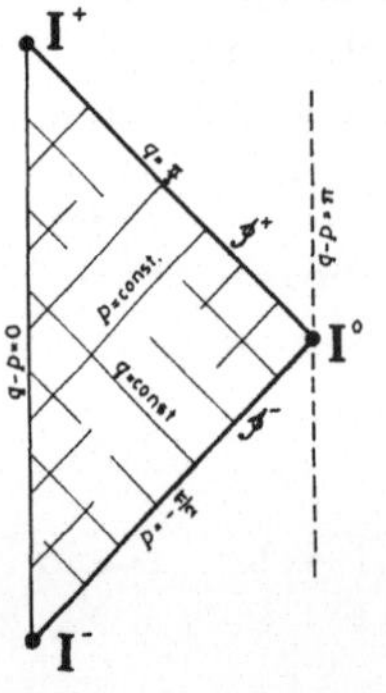

Fig. 4. The range of the coordinates p, q for Minkowski space.

non-singular on this line. The sloping lines $p = -\pi/2$ $(-\pi/2 < q < \pi/2)$ and $1 = \pi/2$ $(-\pi/2 < p < \pi/2)$ represent (null) infinity (denoted $\mathcal{I}^-$ and $\mathcal{I}^+$, respectively) for Minkowski space (since they correspond to $u = -\infty$ and to $v = \infty$). However, the metric (2.5) is evidently perfectly regular on these regions. Indeed, the space-time and its metric $d\hat{s}$ can clearly be extended beyond these regions in a non-singular fashion. The vertical line $q - p = \pi$ is again a coordinate singularity - of precisely the same type as that at $q - p = 0$. The entire vertical strip $0 \leq q - p \leq \pi$ may be used to define a space-time $\mathcal{E}$ whose global structure is that of the product of a space-like 3-sphere with an infinite timelike line (an "Einstein static universe"). To see this, we choose new coordinates

$$T = \tfrac{1}{2}(p + q), \; \rho = q - p$$

and we have

$$d\hat{s} = dT^2 - \tfrac{1}{4}\{d\rho^2 + \sin^2\rho(d\theta^2 + \sin^2\theta\, d\phi^2)\} .$$

The part in curly brackets represents the metric of a unit 3-sphere.

The portion of $\mathcal{E}$ which is conformal to the original Minkowski space may be described as that lying between the light cones of two points I^- and I^+. The point I^- is given by $p = q = -\pi/2$, and I^+ by $p = q = \pi/2$. This portion "wraps around" $\mathcal{E}$ to meet at the "back" in the single point I° (given by $q = -p = \pi/2$). (Note that $\sin^2(q-p) = 0$ at I°, indicating that I° should, in fact, be regarded as a single point, rather than a 2-sphere.) The situation is illustrated in Figure 5 in the two-dimensional case. Minkowski 2-space is conformal to the interior of a square (represented as

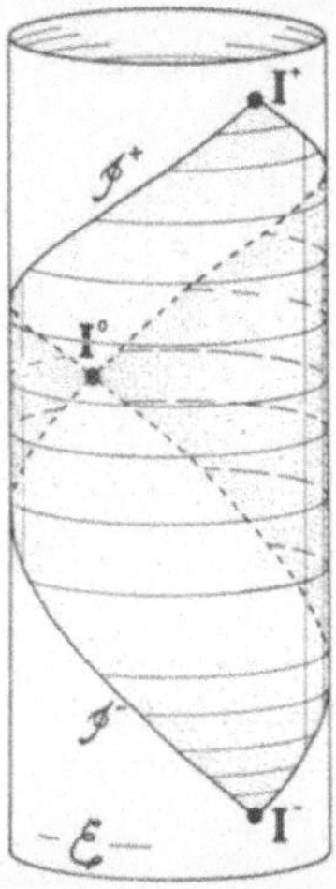

Fig. 5. Minkowski space is conformally identical to a portion (shown shaded) of an "Einstein static universe". The boundary of this portion represents the conformal infinity of Minkowski space.

tipped at 45°). This square wraps around the cylinder which is the two-dimensional version of the Einstein static universe. In higher dimensions the situation is similar. Near the point I^-, the relevant region lies in the interior of the future light cone of I^-. This light cone (i.e. the set of null geodesics directed into the future from I^-) is focussed around the back of the Einstein universe to a single point I° (which is spatially the antipode of I^-). Near I° the relevant ("Minkowski") region lies in spacelike directions from I°. The future light cone of I° is focussed back again to a single point I^+, whose spatial location corresponds to that of I^-. Near I^+, the relevant region lies in the interior of the past light cone of I^+ (see Figure 6).

The null geodesic segments which connect I^- to I° sweep out the portion of the boundary of the Minkowski space region that has been denoted* $\mathcal{I}^-$. Similarly the null geodesic segments from I° to I^+ sweep out $\mathcal{I}^+$. The points I^-, I°, I^+ themselves are considered not to belong to $\mathcal{I}^-$ or to $\mathcal{I}^+$. Physically, we interpret I^- as representing past temporal infinity, $\mathcal{I}^-$ as past null infinity, I° as spatial infinity, $\mathcal{I}^+$ as future null infinity, and I^+ as future temporal infinity. The reason for this terminology is made clear if we examine the behaviour of straight lines in Minkowski space (straight, that is, according to the Minkowski metric ds). A timelike straight line acquires a past end-point I^- and a future end-point I^+. A null straight line acquires a past end-point on $\mathcal{I}^-$ and a future end-point on $\mathcal{I}^+$. A spacelike straight line becomes a closed curve through I° when the point I° is adjoined. (The detailed verification of these facts is left as an exercise.) Since null geodesics remain null geodesics after conformal rescaling, we have

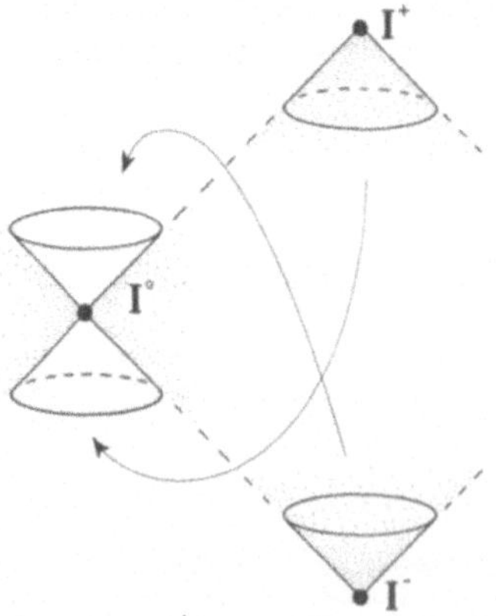

Fig. 6. The regions shaded, near each of I^-, I°, I^+, represent parts of the Minkowski space. If I^-, I°, and I^+ are identified as a single point, these three regions fit together.

* In order to distinguish verbally between $I^\pm$ and $\mathcal{I}^\pm$ in a concise fashion, it is convenient to pronounce $\mathcal{I}$ as "scri" - a contraction of "script I". See figures for the usual notation for $\mathcal{I}$.

the fact that the <u>null</u> straight lines become null geodesics according to the $d\hat{s}$ metric (but the timelike or spacelike straight lines are not, in general, geodesics with respect to $d\hat{s}$).

When we come to consider asymptotically flat space-times shortly we shall see that much of the above discussion will also apply to asymptotically flat space-times. However, one property which is very specific to the Minkowski space model is the fact that <u>every</u> null geodesic which originates at some point a^- on $\mathcal{I}^-$ will pass through the <u>same</u> point a^+ on $\mathcal{I}^+$ (see Figure 7). This property may seem surprising at first, but it becomes clear when we realize that the future light cone of a point of $\mathcal{I}^-$ is simply a null hyperplane in the Minkowski space. (It is the limit of a light cone when the vertex recedes into the past along a null straight line.) Similarly the past light cone of any point on $\mathcal{I}^+$ is also a null hyperplane. So a null hyperplane will acquire a past "vertex" on $\mathcal{I}^-$ (say a^-) and a corresponding future "vertex" (say a^+) on $\mathcal{I}^+$. We can also see this in terms of Einstein universe model $\mathcal{E}$. The future light cone of a^- will be focussed at a point which is spatially antipodal to a^-. This will be the point a^+.

Having this natural association between points of $\mathcal{I}^-$ and points of $\mathcal{I}^+$, for Minkowski space, it is reasonable to perform an identification between $\mathcal{I}^-$ and $\mathcal{I}^+$, the point a^- being identified with a^+. If we do this, then for the sake of continuity we should also identify I^- with I°, and I° with I^+. The three points $I^\pm$, I° thus become one - which we label I. We see from Figure 6 that the Minkowski regions fit neatly together at the point I, so that I becomes simply a normal interior point of the identified manifold. In fact, the manifold that we have constructed by performing these identifications is simply the "compactified Minkowski space" that we obtained in another way in Section 1. Since a^- is identified with a^+, each null geodesic becomes closed, with the topology of a circle S^1. Regarding only the conformal structure, every point of this space is on an equal footing with every other point, whether it be I itself, on the light cone $\mathcal{I}^- = \mathcal{I}^+$ of I, or in the finite portion of the Minkowski space.

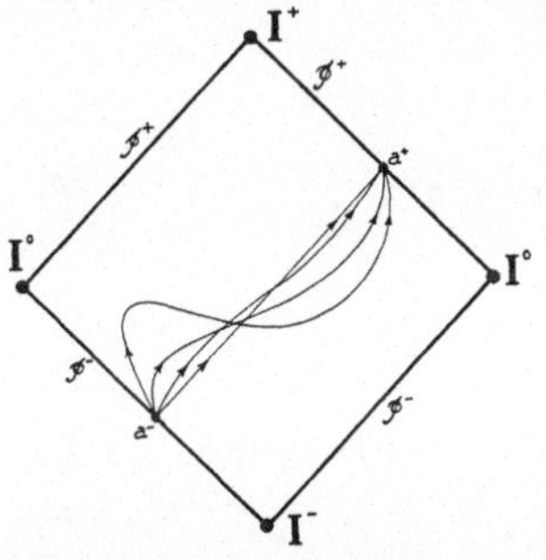

Fig. 7. Conformal infinity for Minkowski space has the special property that every null geodesic in the space which originates at $a^- \in \mathcal{I}^-$ must terminate at the same point $a^+ \in \mathcal{I}^+$.

When discussing Minkowski space it is sometimes convenient to perform the above identifications; sometimes it is more convenient not to and to leave $\mathscr{I}^-$ and $\mathscr{I}^+$ as distinct boundary hypersurfaces ("conformal infinity"). In the case of curved asymptotically flat space-times, however, only the second course of action is reasonable. There are two reasons for this. In the first place there appears to be, in general, no natural association of a point a^- on $\mathscr{I}^-$ with some unique point a^+ on $\mathscr{I}^+$. For example, the null geodesics from a^- will not focus cleanly at a point of a^+, but will tend to cross over one another (and to encounter "caustics") before $\mathscr{I}^+$ is reached. However, the situation is worse than this. Suppose some suitable "canonical" scheme for making identifications is found. It turns out that even in the simplest cases (e.g. the Schwarzschild solution) the performing of any identification of $\mathscr{I}^-$ with $\mathscr{I}^+$ will result in singularities in the metric $d\hat{s}$ along $\mathscr{I}$, whatever the choice of Ω (although $d\hat{s}$ can, in some cases, be made C^2 - but not C^3).

Let us examine conformal infinity of the Schwarzschild solution. The familiar form of the metric is

$$ds^2 = (1 - \frac{2m}{r})\, dt^2 - \frac{dr^2}{1 - \frac{2m}{r}} - r^2(d\theta^2 + \sin^2\theta d\phi^2) \; .$$

Rather than attempt to obtain $\mathscr{I}^+$ and $\mathscr{I}^-$ simultaneously, as was done for Minkowski space, it is simpler to introduce a retarded time coordinate

$$u = t - r - 2m \log (r - 2m)$$

and an advanced time coordinate

$$v = t + r + 2\, m \log (r - 2m)$$

separately. In the first case the metric becomes

$$ds^2 = (1 - \frac{2m}{r}) du^2 + 2dudr - r^2(d\theta^2 + \sin^2\theta d\phi^2) \tag{2.6}$$

and in the second

$$ds^2 = (1 - \frac{2m}{r}) dv^2 - 2dvdr - r^2(d\theta^2 + \sin^2\theta d\phi^2) \; . \tag{2.7}$$

In each case we can choose $\Omega = r^{-1} = w$, say.
The "unphysical metric" is

$$d\hat{s}^2 = \Omega^2 ds^2 = (w^2 - 2mw^3) du^2 - 2dudw - d\theta^2 - \sin^2\theta d\phi^2 \tag{2.8}$$

in the first case and

$$d\hat{s}^2 = (w^2 - 2mw^3) dv^2 + 2dvdw - d\theta^2 - \sin^2\theta d\phi^2 \tag{2.9}$$

in the second. The metrics (2.8) and (2.9) are manifestly regular (and analytic) on their respective hypersurfaces $w = 0$. (Clearly the determinants are non-zero at $w = 0$.) The physical space-time is given when $w > 0$ in (2.8) and we can extend the manifold to include the boundary hypersurface $\mathcal{I}^+$, given when $w = 0$. Similarly, in (2.9), the physical space-time has $w > 0$ and can be extended to include $\mathcal{I}^-$, given when $w = 0$. In fact, we could if desired extend the space-time across $w = 0$ to negative values of w, but this will not be done here. Only the boundary $\mathcal{I} = \mathcal{I}^- \cup \mathcal{I}^+$ will be adjoined to the space-time.

It is worth noting, at this point, a difficulty that is encountered if we try to identify $\mathcal{I}^-$ and $\mathcal{I}^+$. If we do extend the region of definition of (2.8) to include negative values of w, then making the replacement $w \to -w$ we see that the metric has just the form (2.9) (with u in place of v) but where the mass m has been replaced by -m. Thus, the extension across $\mathcal{I}$ involves a reversal of the sign of the mass. In fact the derivative, at $\mathcal{I}$, of the (conformal) curvature contains the information of the mass. (I shall elaborate on this point later, cf. (3.44).) It follows, therefore, that if we attempt to identify $\mathcal{I}^+$ with $\mathcal{I}^-$, with the same sign of the (non-zero) mass occurring on the two sides, then there must be a discontinuity in the derivative of the curvature across $\mathcal{I}$ (so that the metric $d\hat{s}$ must fail to be C^3 at $\mathcal{I}$).

Accepting, then, that it is not reasonable to identify $\mathcal{I}^-$ with $\mathcal{I}^+$, we are led to a picture closely resembling the one we obtained earlier in this section for Minkowski space. The only essential difference occurs with the points I^-, I°, I^+. It turns out that with mass present, the point I°, and normally $I^\pm$ also, must, if adjoined to the manifold, be singular for the conformal geometry. (I shall not go into the argument here.) It is therefore reasonable not to attempt to include these points, in the general case, as part of the conformal infinity. The picture, then, is as indicated in Figure 8. We have two disjoint boundary hypersurfaces $\mathcal{I}^-$ and $\mathcal{I}^+$ each of which is a "cylinder" with topology $S^2 \times \mathbb{R}$. It is clear from (2.8) and (2.9) that each of $\mathcal{I}^\pm$ is a null hypersurface (the induced metric, at $w = 0$, being degenerate). These null hypersurfaces are generated by null geodesics (given by θ, ϕ = const., $w = 0$) the tangents to which are the normals to the hypersurfaces. (Being null, the normals are also tangential.) These null geodesics may be taken to be the " $\mathbb{R}$'s" of the topological product $S^2 \times \mathbb{R}$ (i.e. the inverse images of the points of S^2 in the natural projection $S^2 \times \mathbb{R} \to S^2$).

We have obtained this structure explicitly in the case of the Schwarzschild metric. But it is clear that many other suitably asymptotically flat space-times will also give rise to such a structure. Let us start from a metric of the form

$$ds^2 = r^{-2}A dr^2 + 2B_i dx^i dr + r^2 C_{ij} dx^i dx^j , \qquad (2.10)$$

the coordinates being r, x^1, x^2, x^3. Each of A, B_i, C_{ij} is a suit-

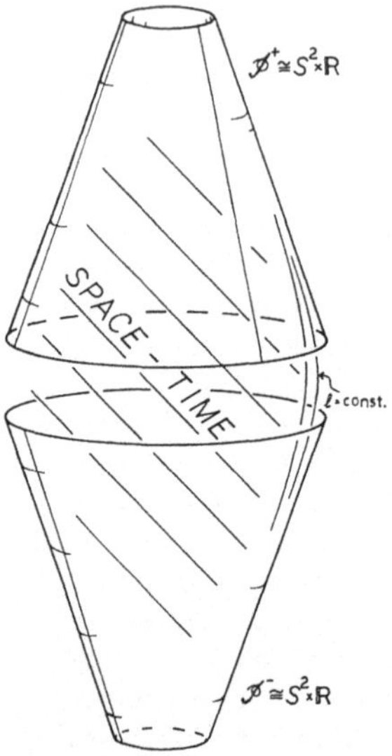

Fig. 8. Conformal infinity for an asymptotically flat space-time (e.g. the Schwarzschild solution). The points I^-, I°, I^+ have become singular and so are deleted; $\mathscr{I}^-$ and $\mathscr{I}^+$ are null cylinders $\cong S^2 \times \mathbb{R}$.

ably smooth function of x^0, x^1, x^2, x^3, with $x^0 = r^{-1}$ - these functions being smooth also at $x^0 = 0$. Then setting $\Omega = r^{-1}$ we have

$$d\hat{s}^2 = \Omega^2 ds^2 = A(dx^0)^2 - 2B_i dx^i dx^0 + C_{ij} dx^i dx^j \ . \tag{2.11}$$

Provided the relevant determinant formed from A, B_i and C_{ij} does not vanish, the metric (2.11) will be perfectly regular at $x^0 = 0$. Thus, a "conformal infinity" will exist for the space-time whose metric is given by (2.10).

Many metrics used in the study of gravitational radiation do in fact have the form (2.10). In particular, this applies to the metrics of Bondi (and his coworkers) [11], and Sachs [12], Robinson-Trautman and Newman-Unti [13]. These metrics describe a situation where there is an isolated source (with asymptotic flatness) and outgoing gravitational radiation. Incoming gravitational radiation (of a suitably curtailed duration) may also be present. So may non-gravitational (i.e. electromagnetic or neutrino) zero rest-mass radiation. For such a situation, therefore, we expect a future-null conformal infinity $\mathscr{I}^+$ to exist. For the time-reversed situation we would expect $\mathscr{I}^-$ to exist. There will also be a wide class of "physically reasonable" situations in which both $\mathscr{I}^+$ and $\mathscr{I}^-$ exist. I feel there is likely to be no real loss of generality involved in the assumption that both $\mathscr{I}^+$ and $\mathscr{I}^-$ should exist for an asymptotically flat space-time. It has occasionally been argued [9] that the assumption of the existence of I^- may impose unnecessarily severe restrictions on the behaviour of the outgoing radiation in the infinite past. However, I do not think that this is really so. One of the major difficulties in gravitational radiation theory which one encounters, if an approach such as the one I am setting forth here is not adopted, is to find a reasonably clear-cut coordinate independent definition of what one should mean by incoming (or outgoing) gravitational radiation. If I^- exists, then the incoming field can be recognized in terms of the curvature at I^- (cf. later). If I^- does not exist, then the situation is much less clear. Also, it seems that both I^+ and I^- do in fact exist for "finite" scattering problems, for which a finite number of particles come in and go out along hyperbolic-type orbits. There seems no reason why I^- and I^+ should not exist in many other "reasonable" scattering problem situations - although examples can obviously be concocted involving infinite wave trains in which either or both of $I^{\pm}$ fail to exist. Whether such examples are regarded as "physically reasonable" is clearly a matter of taste. Asymptotic flatness is, in any case, a mathematical idealization. I feel this is a subject worthy of further study, however.

Asymptotically flat space-times of the type I am considering constitute the most important subclass of those space-times which are termed "asymptotically simple" (or, more generally, "weakly asymptotically simple"). A space-time M is asymptotically simple [6] if a conformal factor Ω exists for which the metric $d\hat{s} = \Omega ds$ remains smooth (say C^4) on some extension $\bar{M}$ of the manifold M which includes a boundary I, Ω being smooth (say C^4) throughout $\bar{M} = M \cup I$, becoming zero on I and having non-zero gradient at I, the global assumption being also made that every maximal null geodesic in M acquires both a past and a future end-point on I. It is sometimes convenient to weaken the global assumption that all maximal null geodesics should reach conformal infinity both in the future and in the past. This is in order to admit the possibility that some null geodesics might not escape to conformal infinity (as would be the case with a black hole or some other similar situations). The space-time M is said to be weakly asymptotically simple [8] if, roughly speaking, it possesses the conformal infinity of an asymptotically simple space-time, but it may possess other "infinities" as well. More precisely, the condition is that some asymptotically simple M' should exist such that for some neighbourhood K' of I' in $\bar{M}'$, the portion $K' \cap M'$ should be isometric with a subset of M.

The boundary I must in all cases be a smooth hypersurface. If it happens to be null, then it can be shown [6] to have the same topological structure that we have obtained for the Minkowski and Schwarzschild cases (cf. Figure 8). Furthermore, if Einstein's vacuum equations without cosmological term (for the physical metric ds) are assumed to hold near I (i.e. throughout $K-I$, where K is some neighbourhood of I in $\bar{M}$), then it follows that I is everywhere null. The same conclusion also follows under much weaker assumptions. For example, near $\mathcal{I}$, we might impose Einstein's equations without cosmological term, with zero rest-mass fields as source. If, on the other hand, a cosmological term is present, then $\mathcal{I}$ is spacelike or timelike according as the cosmological constant is positive or negative.

These properties are simple consequences of the transformation formula for the Ricci tensor. Let us write this formula as

$$P_{ab} = \hat{P}_{ab} + \Omega^{-1}\hat{\nabla}_a\hat{\nabla}_b\Omega - \tfrac{1}{2}\,\Omega^{-2}\hat{g}_{ab}\hat{g}^{cd}\hat{\nabla}_c\Omega\hat{\nabla}_d\Omega \tag{2.12}$$

where

$$P_{ab} = \frac{1}{12}\,R\,g_{ab} - \frac{1}{2}\,R_{ab} \tag{2.13}$$

is a tensor (often occurring in conformal transformation expressions) containing the same information as the Ricci tensor R_{ab} for the ds metric, where $\hat{P}_{ab}$ is the corresponding tensor for $d\hat{s}$, and where $\hat{\nabla}_a$ denotes covariant derivative for the $d\hat{s}$ metric. Since $d\hat{s}$ and Ω are regular on $\mathcal{I}$, the quantities $\hat{P}_{ab}$, $\hat{g}_{ab}$, $\hat{g}^{ab}$, $\hat{\nabla}_a\Omega$ and $\hat{\nabla}_a\hat{\nabla}_b\Omega$ all remain finite and continuous at $\mathcal{I}$. Now if Einstein's vacuum equations hold without cosmological term (near $\mathcal{I}$) we have $R_{ab} = 0$. This is equivalent to $P_{ab} = 0$. Thus, multiplying (2.12) by Ω^2 and using the condition $\Omega = 0$ on $\mathcal{I}$, we obtain the fact that the vector

$$\hat{n}_a = \mp\hat{\nabla}_a\Omega \tag{2.14}$$

is <u>null</u> at $\mathcal{I}^{\pm}$ (sign chosen so that $\hat{n}_a$ is future-pointing). But $\hat{n}_a$ is normal to Ω = const.; in particular, it is normal to $\mathcal{I}$ (and does not vanish at $\mathcal{I}$, by one of the assumptions of asymptotic simplicity). Hence $\mathcal{I}$ is a <u>null</u> hypersurface. This same conclusion will also follow if we assume merely that the <u>trace</u> R of the physical Ricci tensor vanishes near $\mathcal{I}$ (the normal situation if all fields present are massless and there is no cosmological term). For we need only apply the above argument to the transvection of (2.12) with $\hat{g}^{ab}$ (i.e. to the "trace" of (2.12)) and the result is the same. Furthermore, if a cosmological constant <u>is</u> present, essentially the same argument will yield the other results mentioned above.

As a final consequence of (2.12), let us derive one further geometrical property of the hypersurface $\mathcal{I}$. For simplicity we assume that there is vacuum near $\mathcal{I}$, and no cosmological constant. The same property would actually follow under weaker assumptions (e.g. Einstein - Maxwell equations, without cosmological term, near $\mathcal{I}$), but this is more difficult to derive. So take $P_{ab} = 0$ near $\mathcal{I}$, and consider the "trace-free" part of (2.12) (i.e. transvect with $\delta_u^a\delta_v^b - \frac{1}{4}\,\hat{g}^{ab}\hat{g}_{uv}$). Multiplying by Ω and setting $\Omega = 0$, we obtain the equation

$$\hat{\nabla}_a\hat{n}_b = \{\tfrac{1}{4}\,\hat{g}^{cd}\hat{\nabla}_c\hat{n}_d\}\hat{g}_{ab} \quad \text{at } \mathcal{I} \tag{2.15}$$

(cf. (2.14)). Equation (2.15) tells us that the generators of I are non-rotating and shear-free. (These generators are the integral curves of normal vectors $\hat{n}^a$ to I - which, being null, are also tangent vectors to I.) The conclusion that they are non-rotating is trivial since I is a null hypersurface (symmetry of $\hat{\nabla}_a\hat{n}_b$). On the other hand, the shear-free nature of the generators ($\hat{\nabla}_a\hat{n}_b$ trace-free) tells us that small shapes are preserved as we follow these generators along I. That is to say, if we take any two cross-sections S_1, S_2 of I^+ or of I^-, then the correspondence between S_1 and S_2 established by the generators is a <u>conformal</u> one. This is the same situation as we encountered earlier, in Section 1, with the light cone of a point (cf. Figure 2). It implies that the space of generators of I^+ or of I^- has a conformal structure.

The fact that I^+ and I^- have this type of structure is essential for the definition of the B.M.S. group, as we shall see in the next section. We can take as the appropriate definition of asymptotic flatness, therefore, that a conformal infinity I (or at least a I^+) should exist with the structure as defined above. Weak asymptotic simplicity, together with Einstein-Maxwell equations (without cosmological term) holding near I will be sufficient to ensure this.

3. THE B.M.S. GROUP

In special relativity, the Poincaré Group arises in a natural way, as the group of symmetries of (Minkowski) space-time preserving its metric structure. In general relativity, however, no interesting group arises as the group of symmetries of general space-time. (I do not consider the "general coordinate group" or equivalently, the "group of diffeomorphisms of space-time" to be an interesting group in the context of general relativity theory. Since it preserves only <u>smoothness</u>, its relevance would be as much to any other space-time theory involving differential equations as to general relativity. And the group consisting of the identity element alone is even less interesting!) If we restrict attention to space-times which are asymptotically flat, however, then an interesting new concept does arise, namely that of an <u>asymptotic symmetry group</u>. In their original approach to the problem, Bondi and Metzner [11], and later Sachs [12,14], defined their asymptotic symmetry group (the B.M.S. group) as a group of transformations between asymptotically flat coordinate systems of a certain type. However, the geometrical approach to asymptotic flatness described in the last section affords us a much more vivid picture of the significance of this group as a transformation group and of its relevance in the presence of gravitational radiation [10].

Consider Minkowski space first. Since the transformations belonging to the Poincaré group are metric preserving, they are certainly conformal. They therefore induce transformations of I to itself which are also, in the appropriate sense, conformal (I being invariant). The Poincaré group may thus be thought of as a certain group of conformal symmetries of the Minkowski conformal infinity. In the same way, we might expect to be able to identify the Poincaré group as a group of conformal symmetries of the conformal infinity $\mathcal{I}$ of an asymptotically flat space-time; or we might expect that there would be some obstruction to making such an identification.

We must, however, be somewhat careful when it comes to defining exactly what structure on $\mathcal{I}$ is to be preserved by the transformations of this group. The most obvious structure that we can assign to $\mathcal{I}$ is its "inner conformal metric". But, $\mathcal{I}$ being a null hypersurface, this turns out to be rather too weak a structure. Distances along generators of $\mathcal{I}$ are just zero, so their ratios cannot be assigned. The group of self-transformations of $\mathcal{I}$ preserving this weak inner conformal metric does, however, have some significance as an asymptotic symmetry group. Like the B.M.S. group, it is the same group for every asymptotically flat space-time. It is considerably larger than the B.M.S. group and is referred to as the Newman-Unti group [13]. Before considering the strengthening of the structure of $\mathcal{I}$ that is required for the B.M.S. group, it is worthwhile to examine the nature of the inner conformal metric of $\mathcal{I}$ and these self-transformations which preserve it.

In Section 2 the structure of $\mathcal{I}$ for an asymptotically flat space-time M was described as the disjoint union of two smooth null hypersurfaces $\mathcal{I}^-$ and $\mathcal{I}^+$ (where $\mathcal{I}^-$ consists of past end-points of null geodesics in M and $\mathcal{I}^+$ of future end-points), the topology of each of $\mathcal{I}^{\pm}$ being $S^2 \times \mathbb{R}$, where the "$\mathbb{R}$'s" may be taken as the null-geodesic generators of $\mathcal{I}^{\pm}$. These generators are shear-free and so establish a conformal mapping between any two S^2 cross-sections of $\mathcal{I}^+$ or of $\mathcal{I}^-$. Now it is a theorem that any conformal 2-surface with the topology of a sphere S^2 is in fact conformal to the unit 2-sphere in Euclidean 3-space. Thus, without loss of generality we can assume that the conformal factor Ω has been chosen so that some cross-section S of $\mathcal{I}^+$, say, has the (unphysical) metric $-d\hat{s}^2$ of a unit 2-sphere. For, given one choice of Ω, we can always make a new choice $\Omega' = \Theta\,\Omega$ which again has the required properties of vanishing at $\mathcal{I}$ with non-zero gradient there, the factor Θ being an arbitrary smooth positive function on $\mathcal{I}$ which may be taken to rescale the metric on $\mathcal{I}$ as we please. We can in fact use this freedom in Ω to scale the metric $d\hat{s}$ along every generator of $\mathcal{I}^+$ so that the divergence of these generators vanishes. Put another way, we arrange that a continuous succession of cross-sections of $\mathcal{I}^+$ have metrics agreeing with that of S, as mapped along the generators. (These cross-sections are already conformal to S by this map.) Thus, without loss of generality, we can assume the (unphysical) metric of $\mathcal{I}^+$ to be given by

$$d\ell^2 = -\,d\hat{s}^2 = d\theta^2 + \sin^2\theta d\phi^2 + 0.du^2 \qquad (3.1)$$

where θ and ϕ are spherical polar coordinates for S, chosen to be constant along the generators of $\mathcal{I}^+$, and where u is a parameter defined along each generator of $\mathcal{I}^+$ (increasing monotonically with time from $-\infty$ to ∞) the surfaces u = const. being cross-sections of $\mathcal{I}^+$. It is clear from (3.1) that every cross-section (given by u = a function of θ, ϕ) of $\mathcal{I}^+$ now has the metric of a unit 2-sphere.

We can also choose Ω so that $\mathcal{I}^-$ has a metric of the form (3.1) as well, but where we use a coordinate v in place of u, with v also increasing into the future. Then u is a retarded time coordinate on $\mathcal{I}^+$ and v an advanced time coordinate on $\mathcal{I}^-$. These coordinates, and θ and ϕ also can, if desired, be extended into the finite space-time M. But this serves no real purpose here - from the point of view solely of asymptotic symmetry groups.

Let us consider the group of self-transformation of $\mathcal{I}^+$ which preserves its inner conformal metric. (I shall ignore time-reversing transformations. These should, strictly speaking, interchange $\mathcal{I}^+$ with $\mathcal{I}^-$.) Note first that any smooth transformation of $\mathcal{I}^+$ to itself which sends each generator into itself (and preserves the orientation on each generator) will be allowable. In terms of θ, ϕ, u these are given by

$$\begin{aligned} &\theta \to \theta \\ &\phi \to \phi \\ &u \to F(u, \theta, \phi) \end{aligned} \tag{3.2}$$

where F is smooth with $\partial F/\partial u > 0$. Clearly (3.1) is unchanged. In addition, we can allow conformal transformations of the (θ, ϕ)-sphere to itself. These were discussed in Section 1. For convenience we introduce

$$\zeta = e^{i\phi}\cot\frac{\theta}{2}, \tag{3.3}$$

as in (1.10), and the metric (3.1) becomes

$$d\ell^2 = \frac{4d\zeta d\bar{\zeta}}{(1+\zeta\bar{\zeta})^2} + 0.du^2 \tag{3.4}$$

(cf. (1.11)). The conformal maps of the sphere are now given by (1.12)(or this composed with $\zeta \to \bar{\zeta}$), so the general Newman-Unti transformation takes the form

$$\begin{aligned} &\zeta \to \frac{\alpha\zeta+\beta}{\gamma\zeta+\delta} \\ &u \to F(u, \theta, \phi) \end{aligned} \tag{3.5}$$

$(\partial F/\partial u > 0)$ or, in the case of spatially-reflective transformations, (3.5) composed with $\zeta \to \bar{\zeta}$.

There is a great deal of freedom in the function F and we

want to be able to reduce this. In fact it is possible to assign rather more geometric structure to $\mathcal{I}^+$ than we have given hitherto. The preservation of this additional structure will reduce the freedom in F to that of a function of θ and ϕ only. This will give us the B.M.S. group. To obtain this additional structure, let us consider the geometrical interpretation of a point of $\mathcal{I}^+$. Assume, first, that M is Minkowski space. We saw that in this case each point p of $\mathcal{I}^+$ is associated with a null hyperplane π in M. Every generator of π is a null geodesic and these all attain the same future end-point p on $\mathcal{I}^+$. In physical terms, we may think of π as a constant phase hypersurface in a plane wave. The other constant phase hypersurfaces belonging to the same plane wave will be null hyperplanes parallel to π. These other null hyperplanes will terminate at other points on $\mathcal{I}^+$, but the totality of all these points, for the given plane wave (i.e. for a given null direction in Minkowski space) will constitute the generator γ of $\mathcal{I}^+$ through p. (Parallel null geodesics terminate on $\mathcal{I}^+$ at points of one generator of $\mathcal{I}^+$.) The different null geodesics through p (apart from γ) are the different generators of π. Now for fixed π the space of these generators has the structure of a Euclidean plane E^2. This is because any cross-section (in particular, any cross-section by a space-like plane) of π has the intrinsic metric of a Euclidean plane, and the projection along the generators of π maps these planes isometrically. (As an exercise verify these facts.) In the neighbourhood of p, these null geodesics generate the past light cone of p. We can see the Euclidean E^2 structure arising if we take a "parabolic" section of the past null cone of p by a null 3-plane near p which is parallel to γ (see Figure 9). The situation is similar to that depicted in Figure 3 (except that here we are choosing a past rather than a future null cone). The $-d\hat{s}^2$ metric on this section is that of a Euclidean plane - conformal to E^2. This correspondence will be made more explicit shortly.

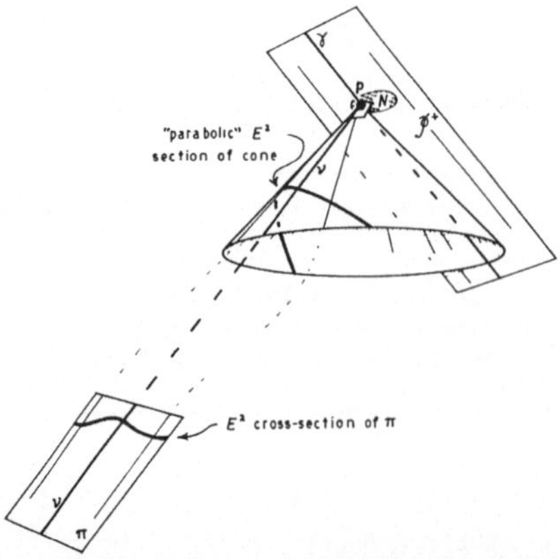

Fig. 9. The null hyperplane π becomes the past light cone of p. The generator ν is represented by an orthogonal 2-plane N tangent to $\mathcal{I}^+$ at p.

Now, by taking orthogonal compliments, we can refer this E^2 structure of the generator space of π to the tangent space to $\mathcal{I}^+$ at p. We associate any generator ν of π (i.e. null geodesic through p) with the 2-plane element at p which is spanned by the directions of γ and of ν at p. Since γ is normal to $\mathcal{I}^+$ at p, the orthogonal compliment of this 2-plane element will be another 2-plane element N, which is now tangent to $\mathcal{I}^+$ (and which does not contain the direction of γ). The 2-plane element N at p is characterized uniquely by the fact that it is tangent to $\mathcal{I}^+$ and orthogonal to ν. Thus, we may use tangent 2-plane elements to $\mathcal{I}^+$, to represent null geodesics in M. (This will also work if M is curved and asymptotically flat.) If we consider a 3-plane element, at p, which contains the direction of γ, we see that the intersection with the null cone at p gives us a one-dimensional system of null geodesics in π. It is not hard to see, M being Minkowski space, that this system corresponds to a straight line in E^2. Furthermore, the orthogonal compliment of this 3-plane element at p is a line element at p which is tangent to $\mathcal{I}^+$ (and not in the direction of γ). Thus line elements at p represent straight lines in E^2. Directions tangent to $\mathcal{I}^+$ at p represented oriented straight lines in E^2.

With very little change, this discussion can in fact be carried over to an asymptotically flat M. The past light cone of a point p on $\mathcal{I}^+$ (i.e. the locus of null geodesics in M which terminate at p) is a null hypersurface π in M which is asymptotically plane in the future. (Physically, this is a constant phase hypersurface of an outgoing asymptotically plane wave. The different constant phase hypersurfaces for one wave will be the past light cones with vertices on one generator of $\mathcal{I}^+$.) The cross-sections of π will not now be Euclidean planes exactly. Nor will the different cross-sections be isometric with one another. On the other hand, by taking the limit of such cross-sections as they recede into the future, we can recover an exact Euclidean plane E^2 to represent the space of generators of π. This may be seen by considering Figure 9 again. Our "parabolic" section of the past cone of p may be taken as close to $\mathcal{I}^+$ as we please - indeed we may consider it to be actually in the tangent space at p. Then we get an exact E^2 structure. This parabolic section can be given, in the tangent space, by the equation

$$\hat{x}^a \hat{n}_a = -1 \tag{3.6}$$

$\hat{x}^a$ being the position vector of a point on the section (and subject also to $\hat{x}^a \hat{x}^b \hat{g}_{ab} = 0$, so that it lies on the null cone), the null vector $\hat{n}_a$ being given, as in (2.14), by

$$\hat{n}_a = -\hat{\nabla}_a \Omega \tag{3.7}$$

at p. Now consider a point p' on π which lies in the remote future along π. In terms of $\hat{M}$, p' will be a point lying just to the past of p on the light cone of p. We can (using "classical" notation) label p' by the position vector $d\hat{x}^a$, relative to p. The conformal factor at p' is thus

$$d\Omega = d\hat{x}^a\hat{\nabla}_a\Omega \ , \qquad (3.8)$$

since $\Omega = 0$ at p. To pass from "unphysical" to "physical" distances at p' we must therefore divide by the factor (3.8). This is equivalent to expanding by the factor $\hat{x}^a : d\hat{x}^a$, where $\hat{x}^a$ is chosen in the same null direction from p as p', but subject to (3.6) (with (3.7)). It follows that to measure distances between points on π which lie in the remote future (i.e. "near" to p) we can simply refer to the "parabolic" section E^2, given by (3.6), it being ultimately relevant only on which generator of π the point lies, not how "far" it is up this generator. Hence, this E^2 does describe the geometry of cross-sections of π, but taken in the limit as these cross-sections recede into the remote future.

As remarked above, null geodesics in *M* can be represented by 2-plane elements tangent to I^+, similarly to the case for Minkowski space. Furthermore, a line element tangent to I^+ at p (which is not in the direction of γ) represents a one-dimensional system of generators of π - of the special type which corresponds to a straight line in E^2 , so asymptotically they generate a null 2-plane in π. A tangent direction to I^+ at p corresponds to an oriented straight line in E^2.

Let us now consider the relevant structure of I^+. The conformal geometry of I^+ affords us a definition of the angle between two line elements tangent to I^+ at p. Representing each line element by a straight line in E^2 in accordance with the above, we can interpret this angle simply as the Euclidean angle between the two corresponding straight lines in E^2. Provided the two line elements at p do not span a 2-plane containing the direction of γ, this will be an ordinary finite Euclidean angle between two intersecting straight lines in E^2. In special cases, however, the two straight lines will become parallel. Then the angle between the corresponding line elements becomes zero. This is the situation occurring when the 2-plane spanned by the line elements does contain the direction of γ (neither of the line elements being actually in the direction of γ). Although the angle is zero, there is still an invariant concept of separation between such line elements. I shall call this concept a null angle [10]. A null angle has units of a distance (or equivalently a time) since it describes the distance in E^2 between the associated parallel straight lines (Figure 10). Thus, we may attribute to I^+ a physically meaningful strong conformal geometry. In addition to the concept of angle between (non-null) tangent directions at a point of I^+ (which concept is the content of the inner conformal metric of I^+) the strong conformal geometry assigns a measure of separation whenever this angle becomes zero, termed the null angle between these directions, and having the dimensions of a distance (or time).

There is another (equivalent) way of specifying the strong

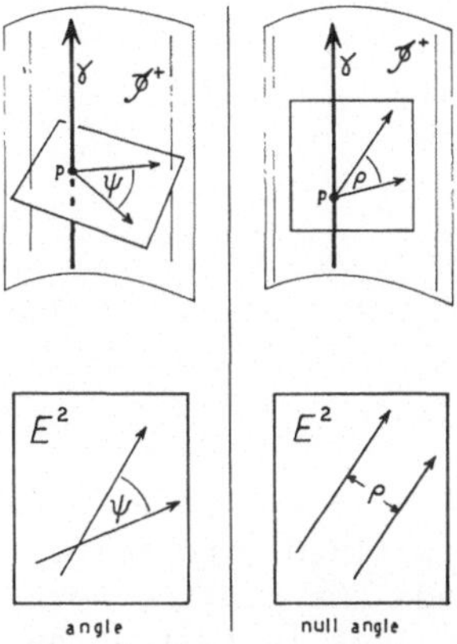

Fig. 10. The inner conformal metric of $\mathscr{I}^+$ assigns a measure of angle between any two (non-null) tangent directions to $\mathscr{I}^+$. When this angle is zero, the strong conformal geometry of $\mathscr{I}^+$ is needed to define the null angle between the directions. Representing directions on $\mathscr{I}^+$ by straight lines in E^2 we obtain finite angles as the angles between the lines, and null angles as the distance between them when these angles reduce to zero.

conformal geometry of $\mathscr{I}^+$. Observe that if we replace Ω by $\Omega' = \Theta\Omega$ (Θ being smooth and positive on $\bar{M}$), the normal vector $\hat{n}^a = \hat{g}^{ab}\hat{\nabla}_b\Omega$ is replaced by

$$\hat{n}'^a = \hat{g}'^{ab}\hat{\nabla}'_b\Omega' = \Theta^{-2}\hat{g}^{ab}\hat{\nabla}_b(\Theta\Omega) = \Theta^{-1}\hat{g}^{ab}\hat{\nabla}_b\Omega = \Theta^{-1}\hat{n}^a \tag{3.9}$$

on $\mathscr{I}^+$ (since $\Omega = 0$). Hence (setting $d\ell^2 = -d\hat{s}^2$ for the metric on $\mathscr{I}^+$),

$$\hat{n}^a d\ell = \hat{n}'^a d\ell' \tag{3.10}$$

so the quantity $\hat{n}^a d\ell$ (or $\hat{n}^a\hat{n}^b d\hat{s}^2$ if preferred) defines an invariant structure on $\mathscr{I}^+$. The vectors $\hat{n}^a$ are tangent to the generators of $\mathscr{I}^+$ and define a scaling for special parameters u along these generators according to

$$\frac{\partial}{\partial u} = \hat{n}^a\hat{\nabla}_a\,, \quad \text{i.e.} \quad \hat{n}^a\hat{\nabla}_a u = 1 \tag{3.11}$$

(the operators acting on scalars). The invariance of $\hat{n}^a d\ell$ may be restated as the invariance of the ratio $du:d\ell$. This ratio is, in effect, a null angle (see Figure 11), the invariance of the ratios of the $d\ell$'s in different directions giving the inner conformal metric also.

If we make some specific choice of metric $d\ell$ for cross-sections of $\mathscr{I}^+$, then this singles out a specific scaling for the parameter u. It is usual to choose the metric $d\ell$ to be that of a unit

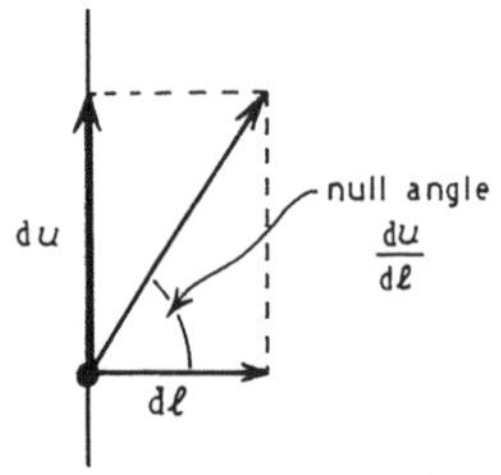

Fig. 11. The ratio du:dℓ defines a null angle.

sphere, so that the entire metric for $\mathcal{I}^+$ takes the form (3.1). The scalings for the associated parameters u are now fixed. The value of the null angle du/dℓ (in seconds, say) defines the scaling for u (in seconds). The only arbitrariness in u now lies in fixing the origin of the u-coordinate on each generator of $\mathcal{I}^+$.

A transformation of $\mathcal{I}^+$ to itself which preserves $\hat{n}^a d\ell$ - or, equivalently, which preserves angles and null angles - must have the effect that any expansion (or contraction) of the spatial distances dℓ is accompanied by an equal expansion (or contraction) of the scaling of the special u-parameters. The allowed transformations must have the form (3.5) (since these are the ones preserving the inner conformal metric), but the function F must now have the special form that allows the ratio du:dℓ to remain invariant. The sphere of cross-section of $\mathcal{I}^+$ undergoes a conformal mapping, so

$$d\ell \to K d\ell$$

where K is a positive function of the angular coordinates θ, φ or ζ, $\bar{\zeta}$. We must therefore also have

$$du \to K\, du.$$

Since K is independent of u, this means u transforms according to

$$u \to K(u + a(\zeta,\bar{\zeta})) \tag{3.12}$$

where a is some real function defined on the ζ-sphere. Using (3.4) and the explicit form (3.5) for the conformal transformations of the sphere, we obtain

$$K = \frac{1 + \zeta\bar{\zeta}}{(|\alpha\zeta+\beta|^2 + |\gamma\zeta+\delta|^2)} ,$$

α, β, γ, δ being complex parameters subject to

$$\alpha\delta - \gamma\beta = 1 \tag{3.13}$$

The general form of the transformation is then

$$\zeta \to \frac{\alpha\zeta+\beta}{\gamma\zeta+\delta}$$
$$u \to \frac{(1+\zeta\bar{\zeta})\{u+a(\zeta,\bar{\zeta})\}}{(|\alpha\zeta+\beta|^2+|\gamma\zeta+\delta|^2)} \tag{3.14}$$

where the arbitrary function $a(\zeta,\bar{\zeta})$ may be taken to be suitably smooth on the sphere*. The transformations (3.14) (or these composed with $\zeta \to \bar{\zeta}$) define the Bondi-Metzner-Sachs (or B.M.S.) group, this being the group of self-transformations of $\mathcal{I}^+$ which preserves strong conformal geometry.

The particular transformations of the form

$$\zeta \to \zeta$$
$$u \to u+a(\zeta,\bar{\zeta}) \tag{3.15}$$

are called supertranslations. Here each generator of $\mathcal{I}^+$ is sent into itself, being simply shunted along itself by an amount $a(\zeta,\bar{\zeta})$. Among these transformations are the ones called translations for which a is composed solely of spherical harmonics of orders 0 and 1. In terms of $\zeta,\bar{\zeta}$, these are

$$a(\zeta,\bar{\zeta}) = \frac{A+B\zeta+\bar{B}\bar{\zeta}+C\zeta\bar{\zeta}}{1+\zeta\bar{\zeta}} \tag{3.16}$$

where A, C are real and B complex. These transformations are in fact the ones which translations in Minkowski space induce on $\mathcal{I}^+$ (exercise). The supertranslations form an infinite parameter subgroup of the B.M.S. group and the translations form a four-parameter subgroup. In each case it is a normal subgroup. In fact, Sachs has shown [14] that the translation subgroup is the only four-parameter normal subgroup of the B.M.S. group; the supertranslation group is the largest proper normal subgroup of the B.M.S. group. Thus, each of these subgroups is singled out by its group-theoretic properties alone.

The factor group of the B.M.S. group by the supertranslation group is the conformal group on the sphere S^2. This sphere may be taken to be the space of generators of $\mathcal{I}^+$. As we saw in Section 1, this factor group is the orthochronous Lorentz group. Note that whereas the translation group occurs canonically as a subgroup of the B.M.S. group, we are obtaining the Lorentz group canonically only as a factor group. We cannot fit these two concepts together to obtain the Poincaré group in a canonical way from the B.M.S.

---

* There is a certain amount of choice as to the degree of smoothness or continuity taken for $a(\zeta,\bar{\zeta})$. This can make a considerable difference when it comes to the representation theory of the B.M.S. group. Representation theory for the B.M.S. group has been studied by Sachs [14], Cantoni [15,16] and McCarthy [17,18,19, 20].

group.

The Lorentz group occurs here as a factor group of the B.M.S. group by the infinite-parameter Abelian group of supertranslations. In the case of the Poincaré group, the Lorentz group appears as a factor group of it by the four-parameter Abelian group of translations. Each of the two groups, the B.M.S. group and the (orthochronous) Poincaré group, is what is known as a semidirect product [17,21] of the (orthochronous) Lorentz group with the appropriate (super) translation group. However, the fact that the translation group sits in the supertranslation group does not, in itself, allow us to find the (orthochronous) Poincaré group as a subgroup of the B.M.S. group in a canonical way. The difficulty occurs with defining what would be meant by a "supertranslation-free" Lorentz transformation.

Let us revert to Minkowski space M and see how the Poincare group arises in that case as a subgroup of the B.M.S. group. It should be clear that the orthochronous Poincaré group on this M does in fact arise as a subgroup of the B.M.S. group on I^+. For the strong conformal geometry of I^+ was defined entirely in terms of the metric structure of the space-time. Thus, any self-transformation of M preserving this metric structure (and time-sense) must also preserve the strong conformal geometry of I^+. However, the B.M.S. group, being much larger than the Poincaré group, preserves less structure on I^+ than does the Poincaré group. Let us examine the nature of this additional structure of I^+ whose preservation, in the case of Minkowski space M, allows us to restrict the B.M.S. transformations to Poincaré transformations.

The significance of the Poincaré transformations in this context is, of course, that they can be applied actually to the points of M, not just to I^+. So the question arises as to how to recognize points of M in terms of I^+. Clearly, the future light cone Q of any point $q \in M$ can be used to represent the point q. The intersection $\bar{Q} \cap I^+$ is a cross-section of I^+ which is itself sufficient to characterize Q and therefore the point q. (For $\bar{Q} \cap I^+$ is a spacelike 2-surface. Any spacelike 2-surface is locally the intersection of two uniquely defined null hypersurfaces. These null hypersurfaces are generated by the two systems of null geodesics which meet the 2-surface orthogonally. In the present situation these two null hypersurfaces are simply I^+ and $\bar{Q}$.) Let us call a cross-section of I^+ which arises in this way a good cross-section. A bad cross-section of I^+ is, on the other hand, the intersection of I^+ with some null hypersurface which does not come together cleanly at a single vertex [22]. Instead, the null geodesic generators of this hypersurface will encounter caustics and crossover regions somewhere in the interior of M (Figure 12). The B.M.S. transformations of I^+ which send good cross-sections into good cross-sections can be regarded as acting on the points of M. In fact they will be (orthochronous) Poincaré transformations, as we shall see.

We can use standard coordinates (r, u, θ, ϕ) for Minkowski

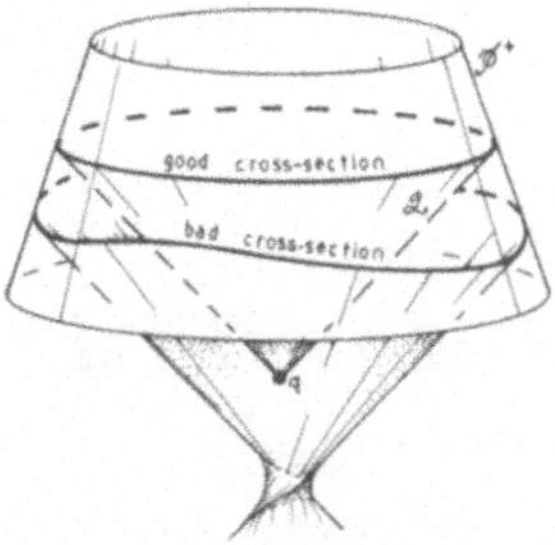

Fig. 12. A good cross-section of $\mathcal{I}^+$ for Minkowski space M is one arising as the intersection of $\mathcal{I}^+$ with the light cone of a point in M. The B.M.S. transformations sending good cross-sections to good cross-sections constitute the (orthochronous) Poincaré group on M.

space, as in (2.3), (2.4), (2.6), where r is radial, u a retarded time, and θ, ϕ spherical polar coordinates, the physical metric taking the form

$$ds^2 = du^2 + 2du dr - r^2(d\theta^2 + \sin^2\theta d\phi^2). \tag{3.17}$$

The parameter u is of the special type required for (3.11) as is readily verified. In fact the metric (3.17) is an example of the Bondi-Sachs form, which is (2.10) with $x^1 = u$, $x^2 = \theta$. $x^3 = \phi$ and

$$(C_{ij}) = \begin{pmatrix} 0 & 0 & 0 \\ 0 & -1 & 0 \\ 0 & 0 & -\sin^2\theta \end{pmatrix} + O(r^{-1}) \ , \tag{3.18}$$

so that the $d\ell^2$ metric of $\mathcal{I}^+$ takes the form (3.1) (with $\Omega = r^{-1}$); and where

$$B_1 = 1 + O(r^{-1}), \tag{3.19}$$

from which it follows directly that u is of the special type required for (3.11). (For, by (2.14), we have $\hat{n}_a = -\delta_a^0$; also $\hat{g}^{0b} = -\delta_1^b$ by (3.18), (3.19), (2.11), so $\hat{n}^a = \delta_1^a$ giving (3.11).) A u-parameter of this kind, for which u-const. are null hypersurfaces in M, is sometimes referred to as a Bondi-type retarded time. It was by consideration of the coordinate group preserving (3.18) and (3.19) that Bondi and Metzner [11] (in the axi-symmetric case) and Sachs [12] (in the general case) were first led to the B.M.S. group.

Returning to the special case of Minkowski space and the special form (3.17), we see that each cross-section of $\mathcal{I}^+$ given by u = const. is a good cross-section, since it arises from the light cone of a point on the origin-axis r = 0. The remaining good cross-sections are obtained from u = 0 by means of the translations (3.15), (3.16), i.e.

$$u = \frac{A+B\zeta+\bar{B}\bar{\zeta}+C\zeta\bar{\zeta}}{1+\zeta\bar{\zeta}} \tag{3.19a}$$

Setting $\zeta = e^{i\phi}\cot\frac{\theta}{2}$, we can re-express (3.19a) in terms of spherical polar coordinates

$$u = \left(\frac{A+C}{2}\right) + \left(\frac{C-A}{2}\right)\cos\theta + \left(\frac{B+\bar{B}}{2}\right)\sin\theta\cos\phi + \\ + i\left(\frac{B-\bar{B}}{2}\right)\sin\theta\sin\phi\ . \tag{3.20}$$

If we represent $\mathcal{I}^+$ as the spherical cylinder in $\mathbb{R}^4$ given by (u, sinθcosϕ, sinθsinϕ, cosθ) then the good cross-sections (3.20) are seen to be simply the sections of this cylinder by 3-planes in $\mathbb{R}^4$ (not parallel to the axis).

A (restricted) Lorentz transformation of *M*, about the origin r = 0, u = 0, leaves the particular good cross-section of $\mathcal{I}^+$ given by u = 0 invariant. Such a transformation - in fact the general (restricted) B.M.S. transformation leaving u = 0 invariant - is given by

$$\zeta \rightarrow \frac{\alpha\zeta+\beta}{\gamma\zeta+\delta}, \\ u \rightarrow \frac{(1+\zeta\bar{\zeta})u}{|\alpha\zeta+\beta|^2+|\gamma\zeta+\delta|^2} \tag{3.21}$$

where $\alpha\delta-\beta\gamma = 1$. These transformations must send good cross-sections into good cross-sections, so they must preserve the form (3.19a) (as, indeed, is easy to verify). Now, the <u>general</u> (restricted) B.M.S. transformation which sends good cross-sections to good cross-sections must obtain the particular good cross-section u = 0 from some other good cross-section. We can therefore express the B.M.S. transformation as the composition of the translation which sends this other good cross-section into u = 0, with a Lorentz transformation (i.e. (3.21)) which leaves u = 0 invariant. Thus the B.M.S. transformation can be expressed as:

$$\zeta \rightarrow \frac{\alpha\zeta+\beta}{\gamma\zeta+\delta} \\ u \rightarrow \left(\frac{1+\zeta\bar{\zeta}}{|\alpha\zeta+\beta|^2+|\gamma\zeta+\delta|^2}\right)\left(u + \frac{A+B\zeta+\overline{B\zeta}+C}{1+\zeta\bar{\zeta}}\right) \tag{3.22}$$

the parameters $\alpha,\beta,\gamma,\delta$,B being complex, with $\alpha\delta-\beta\gamma = 1$, and A,C being real. The B.M.S. transformations sending good cross-sections into good cross-sections thus form a ten-real-parameter (connected) group. Since any restricted Poincaré transformation on *M* belongs to this group, the group (3.22) must actually <u>be</u> the restricted

Poincaré group on M.

There are, however, many other subgroups of the B.M.S. group which can be expressed in the form (3.22) and which are therefore isomorphic with the (restricted) Poincaré group. This arises from the fact that the group (3.22) is not a normal subgroup of the B.M.S. group; the transformations (3.22) do not commute with general supertranslations, so conjugate Poincaré subgroups with respect to supertranslations may be formed. Put another way, we may choose any other allowable u-coordinate, say u', where u' = 0 is a bad cross-section of $\mathcal{I}^+$. (Thus, u' is obtained from u by a supertranslation which is not a translation.) Using u' in place of u in (3.22) we get another subgroup isomorphic with the restricted Poincaré group. To see that this subgroup is indeed different from the previous one, we may examine the "rotation group" subgroup for which each cross-section u' = const. is invariant (i.e., for which $A = B = C = \alpha\bar{\beta}+\gamma\bar{\delta} = 0$, $\alpha\bar{\alpha} = \delta\bar{\delta} = 1-\beta\bar{\beta} = 1-\gamma\bar{\gamma}$). The only (continuous) subgroups of the Poincaré group which are isomorphic with the rotation group are, in fact, the groups of rotation about some timelike line of origins in Minkowski space. Thus, if the above "rotation group" were to be contained in the Poincaré group on M, the invariant cross-sections of $\mathcal{I}^+$ would all have to be good cross-sections (corresponding to the points of the timelike line of origins). Since all the u' = const. cross-sections are bad cross-sections and invariant, it follows that the copy of the Poincaré group that we have obtained is not the same as the Poincaré group on M.

In the general case the two groups will have only their translation subgroups in common. In particular cases, however, the bad cross-section u' = 0 may have axial symmetry. Then there will be rotations about the symmetry axis which are also common to the two groups. These particular rotations happen to be "supertranslation-free" in the sense that they send good cross-sections into good cross-sections, but even then most rotations will not have this property.

Let us turn, now, to the case when M is a general asymptotically flat space-time. The problem of identifying a particular subgroup of the B.M.S. group as the (restricted) Poincaré group may be phrased in terms of deciding which elements of the B.M.S. group are to be regarded as "supertranslation-free" rotations (or Lorentz rotations). As we have seen, the translations may be distinguished from the remaining supertranslations in a B.M.S.-invariant way; it is the "supertranslation-free" rotations which cannot be so distinguished from the general B.M.S. rotations. Since "supertranslation-free" rotations are to be ones sending "good" cross-sections into one another, it would seem that a suitable definition of "goodness" of cross-sections of $\mathcal{I}^+$ is required.

The most obvious generalization, for an asymptotically flat space-time M, of the Minkowskian definition of a "good" cross-section, namely as the intersection of $\mathcal{I}^+$ with the light cone of a point in M, is totally inappropriate. In the first place, there are many perfectly reasonable asymptotically flat space-times in which no cross-sections of $\mathcal{I}^+$ whatever would arise in this way. A model describing two stars a large distance from each other would be an example. If the stars are far enough apart, then every point q of the space-time would have the property that it is "beyond the focal length" of one or other of the stars. That is to say, the null geodesics from q would be focussed by that star to the extent that the light cone of q would encounter caustics and crossover regions behind the star. This would have the effect that when these null geodesics reach $\mathcal{I}^+$, they would not give a proper cross-section of $\mathcal{I}^+$, but a certain surface which wraps backwards and forwards on $\mathcal{I}^+$ intersecting some generators more than once.

However, the situation is considerably worse than this. Even if we restrict attention only to asymptotically flat space-times which do contain a reasonable number of "good" cross-sections of this kind, we are not likely to obtain any group of B.M.S. transformations (apart from the identity) which sends this system of cross-sections into itself, let alone a group isomorphic with the Poincaré group. The difficulty is that the detailed irregularities of the structure of the interior of the space-time would be reflected in the definition of "goodness" of cross-sections of $\mathcal{I}'$. It would seem that this whole approach is quite wrong. When discussing an asymptotic symmetry group, we should use asymptotic properties only (i.e. properties defined in the neighbourhood of $\mathcal{I}^+$ alone) in order to specify the structure we require.

There is a more satisfactory way, in the case of Minkowski space M, of recognizing whether or not a cross-section of $\mathcal{I}^+$ is "good". This is to examine the asymptotic shear of the null hypersurface $\bar{Q}$ which intersects $\mathcal{I}^+$ in the given cross-section. The concept of "shear", for null geodesics, was mentioned briefly at the end of Section 2. I shall be a little more explicit shortly, but for the moment let me just say that in any asymptotically flat space-time M containing a congruence (i.e. a three-dimensional system) Γ of null geodesics there is a complex quantity σ°, defined at the intersection with $\mathcal{I}^+$ of each null geodesic of Γ, whose vanishing is necessary and sufficient for the shear of Γ to vanish asymptotically (or, strictly speaking, to fall off faster than the inverse square of an affine parameter) [24], so that small shapes are preserved asymptotically as we follow the lines of Γ. The quantity σ° can also be defined for a single null hypersurface; $\sigma^\circ = 0$ states that the null hypersurface is asymptotically shear-free. Since the null hypersurface is determined by its intersection with $\mathcal{I}^+$, we can refer the definition of σ° simply to this cross-section on $\mathcal{I}^+$. Now, it can be shown that, in the case of Minkowski space, those cross-sections of $\mathcal{I}^+$ which are shear-free (i.e. $\sigma^\circ = 0$) are precisely the good cross-sections that we defined earlier. Thus a definition of "goodness" is provided, for Minkowski space, which refers only to quantities defined asymptotically.

However, this definition encounters difficulties also, when applied to asymptotically flat space-times. In the first place, suppose we have a cross-section S of $\mathcal{I}^+$ which is shear-free in the above sense. If this is to be considered as a "good" cross-section then every cross-section obtainable from S by means of a translation ought also to be considered as a "good" cross-section (since the translation subgroup ought to belong to the "Poincaré group"). But when gravitational radiation is present, most translations of S will not be shear-free. It turns out, in fact [11,12], that if we define $\sigma°$ suitably for each u = const. cross-section of $\mathcal{I}^+$ (u being a Bondi-type retarded time) then $\partial^2\sigma°/\partial u^2$ is a measure of the outgoing gravitational radiation field expressed in terms of the asymptotic curvature. The quantity $\partial\sigma°/\partial u$ is the Bondi-Sachs "news function" whose squared modulus represents the flux of energy-momentum of gravitational radiation [10,11,12,13] across $\mathcal{I}^+$ (i.e. the energy-momentum which flows out of the system in the form of gravitational radiation). Thus, if S is given by u = 0 and is shear-free, the cross-sections u = const., which are translations of $\mathcal{I}^+$, will not be shear-free in the presence of gravitational radiation. I shall consider these formulae in a little more detail shortly.

There is a more serious difficulty even than this, however, namely that for a general asymptotically flat space-time, $\mathcal{I}^+$ will contain no shear-free cross-sections at all. This is to be expected because the freedom in choosing a cross-section of $\mathcal{I}^+$ is one real number per point of the section (i.e. the u-value on each generator of $\mathcal{I}^+$) whereas the quantity $\sigma°$ is complex, its vanishing, therefore, representing two real numbers per point of the section. Only for particular space-times can shear-free cross-sections be found. One class of space-times for which such sections do exist is that consisting of all stationary asymptotically flat space-times. It is of interest that in this case it is also true that every translation of a shear-free cross-section is again shear-free, from which it follows that a particular Poincaré subgroup of the B.M.S. group can be singled out canonically as the group of B.M.S. transformations which send shear-free cross-sections of $\mathcal{I}^+$ into shear-free cross-sections. But in the general case, when the space-time is not stationary, this will not work. It may be remarked, for example, that in a Bondi axi-symmetric reflection-symmetric radiating space-time, a 2-parameter system of good cross-sections exists but it is not invariant under translations.

We can attempt to weaken the condition $\sigma° = 0$ to something like "the real part of $\sigma°$ vanishes on the cross-section". For technical reasons ($\sigma°$ is a spin-weighted quantity [22]), we would have to put this a little differently and specify instead that, say, what is known as the "electric part" of $\sigma°$ should vanish [22]. Such cross-sections do then exist, but the other difficulty remains. Such sections are not sent into one another by translations.

In this connection it is worth mentioning an interesting new approach to the whole question of asymptotic symmetry groups for asymptotically flat space-times, which has been inititated by Newman, and his collaborators [23]. In this approach, instead of asking for shear-free cross-sections of $\mathcal{I}^+$ and for transformations between them, we ask instead for congruences of null geodesics for which $\sigma^\circ = 0$ and for transformations between them. In general, the rotation of such congruences will not vanish, so we do not get cross-sections of $\mathcal{I}^+$ defined, or retarded time hypersurfaces. (We may, however, view the question as that of finding "complex cross-sections" of $\mathcal{I}^+$. By allowing u to be complex, on a cross-section, we have the freedom, in effect, to make σ° zero). It seems possible, in this approach, that the Poincaré group (or a complex version of it) can be extracted as a kind of asymptotic symmetry group. The approach is, in a sense, complimentary to that of Bondi, Metzner and Sachs. It is closely related to the theory of twistors that will be described briefly in Section 4 - or, at least, to that version of twistor theory appropriate to asymptotically flat space-times [25]. The condition $\sigma^\circ = 0$ also seems to have some direct physical relevance to the definition of in- and out-states for zero rest-mass fields.

There is a method whereby it does appear to be possible to extract the Poincaré group in a reasonably canonical way, as a subgroup of the B.M.S. group, provided the space-time satisfies some very plausible additional requirements (near I°). Although no actual cross-section of $\mathcal{I}^+$ may be shear-free, it is reasonable to expect that in the limit $u \to -\infty$ on $\mathcal{I}^+$, such cross-sections (or at least, ones for which the "electric part" of σ° vanishes) will exist. Requiring that these limiting "shear-free cross-sections" be sent into one another, we can actually restrict the B.M.S. transformations to obtain a "canonically defined" subgroup of the B.M.S. group, which is isomorphic with the orthchronous Poincaré group [22].

The Poincaré group which emerges in this way may be thought of as that which has relevance to the outgoing field in the remote past, i.e. before any gravitational radiation has been emitted. In a similar way, we may extract another Poincaré subgroup of the B.M.S. group which has relevance to the remote future - after all the gravitational radiation has been emitted. (We simply replace the limit $u \to -\infty$ by $u \to +\infty$ in the above discussion.) There seems no reason to believe that these two Poincaré subgroups of the B.M.S. group will be the same, in general. The outgoing radiation which emerges between these limits could serve to "twist" one of these Poincaré subgroups in relation to the other. The asymptotic shear-free cross-sections of $\mathcal{I}^+$ in the remote future would in general be "supertranslations" and not translations) of those in the remote past. There might, likewise, be a period of quiescence between bursts of outgoing radiation in which the system is near enough to being stationary, say, that yet another Poincaré subgroup of the B.M.S. group may be extracted. We would expect that the Poincaré group extracted in the quiescent period would, in general, be different yet again from that extracted in the remote past or the remote future. (Furthermore, the entire discussion could be repeated for $\mathcal{I}^-$ if desired.) Only the translation group would be common as a subgroup to all these copies of the Poincaré group, in the general case, (and the Lorentz group, common as a factor group). The role played by the B.M.S. group from the physical point of view would seem to lie in its relation to matters such as these.

To end this section, I shall be a little more explicit about how the gravitational radiation field is to be expressed, in terms of conformal infinity, and its relation to the asymptotic shear σ°. It will facilitate matters greatly to be able to employ the 2-component spinor formalism, so I shall assume the reader is familiar with this. If not, reference [28] should be consulted.

To begin with, consider the zero rest-mass free field equations for spin n/2. We can use a spinor $\phi_{AB\ldots L}$ with n indices to describe the field, this spinor being symmetric

$$\phi_{AB\ldots L} = \phi_{(AB\ldots L)} \tag{3.23}$$

and satisfying the free-field equation

$$\nabla^{AP'}\phi_{AB\ldots L} = 0 \tag{3.24}$$

where $\nabla_a = \nabla_{AA'}$ denotes* covariant derivative.
When n = 1, this is the Weyl neutrino equation $\nabla^{AP'}\phi_A = 0$.
When n = 2, it is Maxwell's free-field equations

$$\nabla_{[a}F_{bc]} = 0, \quad \nabla_b F^{ab} = 0$$

with

$$F_{ab} = \phi_{AB}\varepsilon_{A'B'} + \varepsilon_{AB}\bar{\phi}_{A'B'} \ . \tag{3.25}$$

When n = 4, setting

$$K_{abcd} = \phi_{ABCD}\varepsilon_{A'B'}\varepsilon_{C'D'} + \varepsilon_{AB}\varepsilon_{CD}\bar{\phi}_{A'B'C'D'} \tag{3.26}$$

we get a tensor with the symmetries of a Riemann tensor satisfying the Einstein vacuum equations

$$K_{abcd} = K_{[ab][cd]} = K_{cdab} \ , \quad K_{[abc]d} = 0, \quad K^a{}_{bad} = 0 \ ,$$

the differential equation (3.24) taking the form of Bianchi's identity

$$\nabla_{[a}K_{bc]de} = 0 \ . \tag{3.27}$$

---

* Recall that the abstract index notation is being employed. The index label "a" stands for "AA'", so no connection symbols are required to connect tensor indices with spinor indices. This applies also to (3.25), (3.26) and later equations.

In flat space we can interpret solutions of (3.24) (or, equivalently (3.27)) as giving solutions of Einstein's vacuum equations in the weak field limit [29]. In curved space-time, the quantity K_{abcd} can be defined as before, but consistency requirements for (3.24) imply that K_{abcd} has to be related to the curvature tensor - actually to the Weyl conformal tensor defined by

$$C_{ab}{}^{cd} = R_{ab}{}^{cd} + 4\, P_{[a}{}^{[c} g_{b]}{}^{d]}$$

where P_{ab} is given by (2.13), so

$$C_{abcd} = C_{[ab][cd]} = C_{cdab}\ ,\quad C_{[abc]d} = 0\ ,\quad C^{a}{}_{bad} = 0\ .$$

We can put

$$C_{abcd} = \Psi_{ABCD}\varepsilon_{A'B'}\varepsilon_{C'D'} + \varepsilon_{AB}\varepsilon_{CD}\bar{\Psi}_{A'B'C'D'} \tag{3.28}$$

where

$$\Psi_{ABCD} = \Psi_{(ABCD)}$$

the consistency requirement for (3.24) being

$$\Psi_{ABC(M}\phi^{ABC}{}_{D\ldots L)} = 0\ . \tag{3.29}$$

If the space-time satisfies Einstein's vacuum equations (without cosmological term) then

$$C_{abcd} = R_{abcd}$$

so Bianchi's identity gives

$$\nabla_{[a}C_{bc]de} = 0$$

which is equivalent to

$$\nabla^{AP'}\Psi_{ABCD} = 0\ . \tag{3.30}$$

A particular solution of (3.24) is, in this case

$$\phi_{ABCD} = \Psi_{ABCD} \tag{3.31}$$

and (3.29) is then automatically satisfied. (For many vacuum space-times, the <u>only</u> solutions of (3.24) for $n = 4$ are those for which (3.31) is satisfied up to proportionality [30].) We may think of Ψ_{ABCD} as defining a spin 2 zero rest-mass field.

Let us now consider a conformal rescaling of the space-time metric according to

$$g_{ab} \to \hat{g}_{ab} = \Omega^2 g_{ab}\ ,\quad g^{ab} \to \hat{g}^{ab} = \Omega^{-2} g^{ab}\ . \tag{3.32}$$

We have

$$g_{ab} = \varepsilon_{AB}\varepsilon_{A'B'}\ , \quad g^{ab} = \varepsilon^{AB}\varepsilon^{A'B'}$$

so we can take

$$\hat{\varepsilon}_{AB} = \Omega\varepsilon_{AB}\ , \quad \hat{\varepsilon}_{A'B'} = \Omega\varepsilon_{A'B'}\ , \quad \hat{\varepsilon}^{AB} = \Omega^{-1}\varepsilon^{AB}\ ,$$
$$\hat{\varepsilon}^{A'B'} = \Omega^{-1}\varepsilon^{A'B'}\ . \qquad (3.33)$$

The covariant derivative $\hat{\nabla}_a$ for the rescaled metric differs from the original one ∇_a, when acting on some spinor, by one term corresponding to each index of the spinor according to the scheme

$$\hat{\nabla}_{AA'}\chi_{H\ldots M'}^{B\ldots G'} = \nabla_{AA'}\chi_{H\ldots M'}^{B\ldots G'} + \varepsilon_A{}^B T_{XA'}\chi_{H\ldots M'}^{X\ldots G'} + \ldots +$$
$$+ \varepsilon_{A'}{}^{G'} T_{AX'}\chi_{H\ldots M'}^{B\ldots X'} - T_{HA'}\chi_{A\ldots M'}^{B\ldots G'} - \ldots -$$
$$- T_{AM'}\chi_{H\ldots A'}^{B\ldots G'}\ , \qquad (3.34)$$

where

$$T_a = \Omega^{-1}\nabla_a\Omega = \nabla_a \log\Omega\ . \qquad (3.35)$$

From this follows at once the invariance of the zero rest-mass equations (3.24) under conformal rescaling [6,8], where we set

$$\hat{\phi}_{AB\ldots L} = \Omega^{-1}\phi_{AB\ldots L}\ . \qquad (3.36)$$

By (3.33), (3.25) and (3.26) we then have

$$\hat{F}_{ab} = F_{ab}\ , \quad \hat{K}_{abcd} = \Omega K_{abcd}\ .$$

If, for spin zero, we use the equation $(\nabla_a\nabla^a + \frac{1}{6}R)\phi = 0$ then again we have invariance with $\hat{\phi} = \Omega^{-1}\phi$.

Since zero rest-mass fields have this conformal invariance, we can refer them to the unphysical metric equally well as to the physical one, when discussing their properties. This allows us to consider the possibility of zero rest-mass fields defined on the whole of $\bar{M}$, where M is (weakly) asymptotically simple. If the field $\hat{\phi}_{AB\ldots L}$ is thus finite on I, we may re-interpret what asymptotic behaviour this entails in terms of the physical metric. It turns out [6] that such a physical field $\phi_{AB\ldots L}$ satisfies Sach's peeling-off property. Let ν be a null geodesic in M and choose a spinor dyad (or spin-frame) o^A, ι^A, normalized according to

$$o_A\iota^A = 1 \qquad (3.37)$$

and parallelly propagated along ν, the spinor o^A having its null ("flagpole") direction pointing along ν. The affine parameter r along ν is scaled according to

$$o^A \bar{o}^{A'} \nabla_{AA'} r = 1 \ .$$

We define the components of $\phi_{AB\ldots L}$ with respect to o^A, ι^A as follows

$$\phi_0 = \phi_{00\ldots0} = \phi_{AB\ldots L}\, o^A o^B \ldots o^L$$

$$\phi_1 = \phi_{10\ldots0} = \phi_{AB\ldots L}\, \iota^A o^B \ldots o^L$$

$$\vdots$$

$$\phi_n = \phi_{11\ldots1} = \phi_{AB\ldots L}\, \iota^A \iota^B \ldots \iota^L$$

then the future peeling-off property states

$$\phi_0 = \overset{0}{\phi}_0 r^{-n-1} + o(r^{-n-1})$$

$$\phi_1 = \overset{0}{\phi}_1 r^{-n} + o(r^{-n})$$

$$\vdots$$

$$\phi_n = \overset{0}{\phi}_n r^{-1} + o(r^{-1}) \ , \tag{3.38}$$

for large positive r where $\overset{0}{\phi}_0, \overset{0}{\phi}_1, \ldots, \overset{0}{\phi}_n$ are constant on ν. The past peeling-off property is similar, but stated for large negative r.

The behaviour (3.38) is actually a direct consequence of the assumption that $\hat{\phi}_{AB\ldots L}$ is continuous on $\mathcal{I}^+$ (or $\mathcal{I}^-$), where we assume from now on that M is asymptotically flat, and may be derived immediately once it is established [6] that the rescaled dyad $\hat{o}^A$, $\hat{\iota}^A$, given by

$$\hat{o}_A = o_A \ , \quad \hat{\iota}_A = \Omega \iota_A \ , \quad \hat{o}^A = \Omega^{-1} o^A, \quad \hat{\iota}^A = \iota^A \ ,$$

is continuous at $\mathcal{I}$. We have, in fact

$$\overset{0}{\phi}_0 = \hat{\phi}_{AB\ldots L} \hat{o}^A \hat{o}^B \ldots \hat{o}^L \ , \quad \overset{0}{\phi}_1 = \hat{\phi}_{AB\ldots L} \hat{\iota}^A \hat{o}^B \ldots \hat{o}^L, \ \ldots \ ,$$

$$\overset{0}{\phi}_n = \hat{\phi}_{AB\ldots L} \hat{\iota}^A \hat{\iota}^B \ldots \hat{\iota}^C$$

at $\mathcal{I}^+$, where Ω is chosen so that

$$\Omega = r^{-1} + o(r^{-1}) \ . \tag{3.39}$$

In fact, (3.39) is equivalent to

$$\hat{\iota}_A \bar{\hat{\iota}}_{A'} = \hat{n}_a = -\hat{\nabla}_a \Omega$$

at $p = \nu \cap \mathcal{I}^+$, as in (2.14). In any case, the "flagpole" of $\hat{\iota}^A$ necessarily points in the null direction in $\mathcal{I}^+$.

The quantity ϕ_n^0, being the "1/r part of the field" along ν, measures the (outgoing) radiation field along ν. In fact, since ϕ_n^0 does not involve $\hat{o}^A$, the radiation field is independent of the particular choice of null geodesic through p. Thus we may think of the outgoing radiation field as a function defined on $\mathcal{I}^+$, essentially dependent only on three parameters (u, θ, ϕ). In a similar way the incoming radiation field may be defined as the corresponding component defined on $\mathcal{I}^-$. For a retarded field, this incoming radiation field must vanish.

Suppose, now, that ν belongs to some congruence Γ of null geodesics. These could be the generators of the u = const. null hypersurfaces of a Bondi-type retarded time u, or they could constitute a more general system. We choose the spinor field o^A to have "flagpole" directions pointing along the null geodesics of the congruence. Then the complex shear σ and complex divergence ρ are defined by [24,8]

$$o^A o^B \nabla_{AA'} o_B = \sigma \bar{o}_{A'} \quad , \quad \bar{o}^{A'} o^B \nabla_{AA'} o_B = \rho o_A \ .$$

(Under $o^A \to \lambda o^A$, these rescale according to $\sigma \to \lambda^3 \bar{\lambda}^{-1} \sigma$, $\rho \to \lambda \bar{\lambda} \rho$.) The argument of σ defines the plane of maximum shear and the modulus of σ the magnitude of shear; the imaginary part of ρ measures the rotation of the congruence Γ and the real part its divergence. If the null geodesics generate null hypersurfaces, then $\rho = \bar{\rho}$.

Provided the null geodesics diverge into the future suitably (as they will if they generate Bondi-type hypersurfaces) we have

$$\sigma = \sigma^0 r^{-2} + O(r^{-3}) \quad , \quad \rho = -r^{-1} + O(r^{-2}) \tag{3.40}$$

for large positive r. (The situation in past directions is similar.) The quantity σ^0 in (3.40) is independent of r and defines the asymptotic shear of Γ, as was referred to earlier. In fact σ is a conformal density, with $\hat{\sigma} = \Omega^{-2}\sigma$. It follows that $\sigma^0 = \hat{\sigma}$ evaluated at $\mathcal{I}^+$.

Let us now assume that Einstein's vacuum equations hold near $\mathcal{I}^+$. We can adopt (3.31), and then the quantities

$$\Psi_0 = \phi_0 \ , \quad \Psi_1 = \phi_1 \ , \ \ldots \ , \ \Psi_4 = \phi_4 \tag{3.41}$$

measure the curvature. The (outgoing) gravitational radiation field is given by

$$\Psi_4^0 = \phi_4^0 \ .$$

It should be remarked that the spinor Ψ_{ABCD}, of which (3.41) are the components, describes the conformal curvature (cf. (3.28)) whether or not Einstein's vacuum equations hold. It is conformally invariant:

$$\hat{\Psi}_{ABCD} = \Psi_{ABCD} .$$

This should be contrasted with (3.36). With respect to the unphysical metric we obtain

$$\hat{\phi}_{ABCD} = \Omega^{-1}\hat{\Psi}_{ABCD} .$$

At first sight this would suggest that the assumption that $\hat{\phi}_{ABCD}$ is continuous at $\mathcal{I}^+$ is a very strong one. However, it can be shown [6] that the assumptions of Einstein's vacuum equations and asymptotic simplicity imply that $\hat{\Psi}_{ABCD} = 0$ on $\mathcal{I}^+$, and hence that $\hat{\phi}_{ABCD}$ is actually continuous at $\mathcal{I}^+$ as required. Incidentally, the B.M.S. group owes its existence to the (comparative) "flatness" of $\mathcal{I}^+$. This, in turn, is related to the vanishing of the conformal curvature at $\mathcal{I}^+$.

Taking Γ to be the system of generators of the u const. hypersurfaces in a Bondi-Sachs type coordinate system, we have the following formula [11,12,13] relating the asymptotic shear to the gravitational radiation field:

$$\frac{\partial^2\sigma^0}{\partial u^2} = -\bar{\Psi}_4^0 . \tag{3.42}$$

The time-integral of Ψ_4^0

$$N = -\frac{\partial\bar{\sigma}^0}{\partial u} \tag{3.43}$$

is essentially the Bondi-Sachs news function [11,12]. It can also be obtained [8,10] from the unphysical Ricci tensor R_{ab} on $\mathcal{I}^+$ by means of

$$\hat{R}_{AA'BB'}\hat{\iota}^A\hat{\iota}^B = 2N\,\bar{\hat{\iota}}_{A'}\bar{\hat{\iota}}_{B'} .$$

The quantity N enters into the Bondi-Sachs definition [11,12,10] of mass-momentum, evaluated at the retarded time u = const.

$$P^{\underline{a}} = \frac{1}{4\pi}\int W^{\underline{a}}(\sigma^0 N - \Psi_2^0)\,d\theta\,\sin\theta d\phi \tag{3.44}$$

where

$$W^0 = 1,\ W^1 = \sin\theta\cos\phi,\ W^2 = \sin\theta\sin\phi,\ W^3 = \cos\theta.$$

The rate of energy-momentum loss due to gravitational radiation is

$$\frac{\partial P^{\underline{a}}}{\partial u} = -\frac{1}{4\pi}\int W^{\underline{a}}N\bar{N}d\theta\,\sin\theta d\phi . \tag{3.45}$$

The intimate relation between the asymptotic shear and gravitational radiation should be clear from all this.

4. TWISTOR THEORY

The theory of twistors [25,26,27] is a formalism for relativistic physics which affords a new approach to the description of quantized fields and to the treatment of space-time itself. In the twistor formalism, space-time points need not be employed as the primary objects in terms of which all else is to be expressed. Instead, the primary objects can be the twistors themselves.

The basic twistors form a complex four-dimensional vector space (considering the case of flat-space twistors only). This may be regarded as the vector space on which SU(2,2) matrices act, these matrices corresponding to conformal transformations of Minkowski space (compactified) according to the local isomorphism $SU(2,2) \rightarrow C(1,3)$. Such a twistor may be pictured in a classical physical way as a zero rest-mass particle with intrinsic spin. (There are three real parameters for the momentum, three for the spatial location, one for the degree of helicity, and one for a phase - which may be pictured in terms of a polarization plane. Thus there are eight real parameters in all, and they can be re-expressed to give four complex components defining a complex vector space.) Perhaps rather more accurate is to regard the twistor as a kind of "square-root" of the energy-momentum-angular-momentum structure of a zero-rest-mass particle. This is in a sense similar to the way that a two-component spinor may be regarded as a "square root" of a null vector. Space-time points may be interpreted, in twistor terms, as linear subspaces (of complex dimension two) of twistor space. This amounts to specifying a point in space-time, in terms of the totality of zero rest-mass particles which pass through that point.

The twistor formalism may also be adapted to curved space-times - especially those which are asymptotically flat. The twistor complex structure can be retained, but twistor space becomes, in a sense, curved (as a Kähler manifold). The complex numbers which describe twistor space are closely related to those more familiar complex numbers which play such a basic role in quantum theory. This points to a unification between quantum theory and space-time structure. The picture is also presented that in a quantized general relativity, instead of the null cones becoming "smeared" by the uncertainty principle (which has been a usual viewpoint), the points of space-time should, as a logical alternative, be themselves "smeared". (This amounts to the above-mentioned linear subspaces of twistor space being "smeared", according to the uncertainty principle.) But even in the classical theory the complex structure of twistor space plays a basic geometric role. This is in relation to the (asymptotic) shear-free condition on congruences of null geodesics that was referred to in Section 3. The twistor space complex structure is related also (via a contour integration) to the zero rest-mass free-field equations (3.24). This leads to a twistor formalism, involving complex contour integration, for the calculation of quantum scattering amplitudes and results in a form of massless quantum electrodynamics which appears to be free of the usual divergences.

I shall not discuss many of these matters in detail here. The interested reader is referred especially to ref. [25]. Only flat-space twistors will be discussed, and these primarily in relation to their role in the construction of different representations (finite or infinite) for the group SU(2,2), or its Lie algebra. Twistors seem to provide a very convenient means for expressing such representations generally. They also give an explicit realization of the local isomorphism SU(2,2) $\rightarrow$ C(1,3).

Let us consider the definition of a basic flat-space twistor. Suppose we are given a classical (finite) system in special relativity. Let P^a be the total 4-momentum of the system, and M^{ab} ($= -M^{ba}$) be its angular momentum with respect to some given space-time origin O. If we pass to a new origin $\tilde{O}$, whose position vector relative to O is X^a, then we have

$$P^a \rightarrow \tilde{P}^a = P^a\ , \quad M^{ab} \rightarrow \tilde{M}^{ab} = M^{ab} - 2X^{[a}P^{b]}\ . \tag{4.1}$$

The spin vector

$$S_a = \tfrac{1}{2}\, e_{abcd}\, P^b M^{cd} \tag{4.2}$$

is, like the momentum, displacement invariant:

$$S_a \rightarrow \tilde{S}_a = S_a \tag{4.3}$$

(The tensor e_{abcd} is the alternating tensor: $e_{abcd} = e_{[abcd]}$, with $e_{0123} = (-g)^{1/2}$ in a right-handed frame.) If the system is to be equivalent (in respect of its energy-momentum-angular-momentum structure) to a zero rest-mass particle, then we require

$$P_aP^a = 0, \quad \underline{\text{with } P^a \text{ future-pointing, and}}\ S_a = sP_a \tag{4.4}$$

where s is the helicity, $|s|$ being the spin. (If we choose units with $\hbar = 1$, then for quantum systems s takes half integer values.) The conditions (4.4) imply [25] that

$$P_{AA'} = \bar{\pi}_A \pi_{A'}, \quad M^{AA'BB'} = i\bar{\pi}^{(A}\omega^{B)}\varepsilon^{A'B'} - i\varepsilon^{AB}\pi^{(A'}\bar{\omega}^{B')} \tag{4.5}$$

for some $\pi_{A'}$, ω^A. The information contained in the two spinors ω^A, $\pi_{A'}$ is the same as that contained in P_a and M^{ab} except that the spinors contain the extra information of a "phase". That is to say, P_a and M^{ab} are invariant under

$$(\omega^A, \pi_{A'}) \rightarrow (e^{i\theta}\omega^A, e^{i\theta}\pi_{A'})$$

(θ real) but apart from this freedom, they uniquely define (and are uniquely defined by) ω^A and $\pi_{A'}$.

The two spinors ω^A, $\pi_{A'}$ serve to define a twistor Z^α

$$Z^{\alpha} \leftrightarrow (\omega^{A}, \pi_{A'}) \ . \tag{4.6}$$

The four complex components of Z^{α} are (with respect to the chosen origin 0 and given spin-frame)

$$Z^{0} = \omega^{0}, \quad Z^{1} = \omega^{1}, \ Z^{2} = \pi_{0'}, \ Z^{3} = \pi_{1'} \ . \tag{4.7}$$

Under change of origin $0 \to \tilde{0}$, we have

$$\omega^{A} \to \tilde{\omega}^{A} = \omega^{A} - ix^{AA'}\pi_{A'} \ , \quad \pi_{A'} \to \tilde{\pi}_{A'} = \pi_{A'} \ , \tag{4.8}$$

this being consistent with (4.1) and (4.5). The (abstract) twistor itself is unaffected by the change in origin, but the representation (4.7) of the twistor components undergoes the replacement (4.8) when 0 is replaced by $\tilde{0}$.

We can also view a twistor in a slightly different way. Let us regard $\tilde{0}$ as a variable point at which the spinor $\tilde{\omega}^{A}$ is defined. Thus $\tilde{\omega}^{A}$ gives us a spinor field. It is convenient to drop the tilde and write this field simply as ω^{A}, rewriting the first equation (4.8) as

$$\omega^{A} = \underset{\circ}{\omega}^{A} - ix^{AA'}\underset{\circ}{\pi}_{A'} \tag{4.9}$$

where the "$\circ$" placed under a spinor symbol means that this is a quantity defined at the origin, and where x^{a} is the position vector relative to this origin. The field (4.9) satisfies the equation

$$\nabla_{B'}^{(B}\omega^{A)} = 0 \tag{4.10}$$

and it is not hard to see that, conversely, the general solution of (4.10) is (4.9), for some (unique) $\underset{\circ}{\omega}^{A}$, $\underset{\circ}{\pi}_{A'}$. Thus, we can think of the twistor Z^{α} as being, in effect, a spinor field satisfying (4.10). This viewpoint has the virtue that the twistor concept is made conformally invariant. For it follows from (3.34) that (4.10) is conformally invariant (with $\hat{\omega}^{A} = \omega^{A}$) under the conformal rescaling (3.32). We can thus think of the field ω^{A} as being defined over the whole compactified Minkowski space. There is, however, a difficulty. It turns out that the field ω^{A} must jump by a factor i (or -i) as we cross the light cone at infinity. Thus we must, strictly speaking, regard the field ω^{A} as defined on the fourfold covering space of compactified Minkowski space (or, what amounts to the same thing, as a four-valued field on compactified Minkowski space). The reason for this four-valuedness is the (4-1) nature of the homomorphism $SU(2,2) \to C(1,3)$.

Suppose we choose a conformal rescaling $g_{ab} \to \hat{g}_{ab} = \Omega^{2} g_{ab}$ which is of the type that renders the $\hat{g}_{ab}$ to be (like g_{ab}) flat. This means that Ω has a form proportional to

$$\Omega = \frac{Q_a Q^a}{(Q_b - x_b)(Q^b - x^b)} \tag{4.11}$$

(with Q^a constant). Since we can extract $\pi_{A'}(= \mathfrak{J}_{A'})$ from the field ω^A by

$$\nabla_{BB'}\omega^A = -i\varepsilon_B{}^A \pi_{B'} \tag{4.12}$$

[see (4.9)] either using the g_{ab} metric or the $\hat{g}_{ab}$ metric, we get (by (3.34))

$$\omega^A \to \hat{\omega}^A = \omega^A, \quad \pi_{A'} \to \hat{\pi}_{A'} = \pi_{A'} + iT_{AA'}\omega^A \tag{4.13}$$

under the conform rescaling (3.32), where

$$T_a = \nabla_a \log \Omega = \frac{2(Q_a - x_a)}{(Q_b - x_b)(Q^b - x^b)} . \tag{4.14}$$

Note that the transformations (4.8) and (4.13) are similar in form. They are each linear and unimodular in the components of the twistor Z^α (with (4.13) evaluated at the origin, so that it can be taken as referring to the components (4.7); also $\Omega = 1$ at the origin in (4.11), so the spin-frame normalization is unaffected, cf. (3.37)). Furthermore, they each leave invariant the expression

$$\begin{aligned} Z^\alpha \bar{Z}_\alpha &= \omega^A \bar{\pi}_A + \pi_{A'} \bar{\omega}^{A'} \\ &= 2\mathrm{Re}[\omega^0 \bar{\pi}_0 + \omega^1 \bar{\pi}_1] \end{aligned} \tag{4.15}$$

where the complex conjugate twistor to Z^α is defined by

$$\bar{Z}_\alpha \leftrightarrow (\bar{\pi}_A, \bar{\omega}^{A'}) . \tag{4.16}$$

The signature of the Hermitian form (4.15) is (++--), so that the transformations of components of Z^α given by (4.8) and (4.13) all belong to the group SU(2,2) whose invariant Hermitian form is $Z^\alpha \bar{Z}_\alpha$.

The same remark applies to the transformations of Z^α components induced by a Lorentz transformation (leaving the origin 0 invariant). For these are given by SL(2,C) transformations applied to ω^A, coupled with the corresponding dual conjugate SL(2,C) transformations applied to $\pi_{A'}$:

$$\omega^A \to \omega^B t_B{}^A, \quad \pi_{A'} \to \pi_{B'}(-\bar{t}^{B'}{}_{A'}), \tag{4.17}$$

the matrix of components of $t_B{}^A$ being unimodular (i.e., being a spin-matrix) so that $-t^B{}_A$ is inverse to $t_B{}^A$:

$$\varepsilon_{AB} t_C{}^A t_D{}^B = \varepsilon_{CD}, \quad \text{i.e.} \quad t_B{}^A(-t^C{}_A) = \varepsilon_B{}^C = (-t^A{}_B)t_A{}^C . \tag{4.18}$$

Clearly the transformation (4.17) is linear and unimodular on Z^{α} and preserves (4.15) (since (4.15) is a spin-scalar) as required. Furthermore, the dilations also belong to this SU(2,2). For expanding the Minkowski space by a factor k^2 $(k > 0)$ we have $x^a \rightarrow k^2 x^a$ and $\Omega = k^2$ (= const.). This corresponds to

$$\omega^A \rightarrow k\omega^A \quad \text{and} \quad \pi_{A'} \rightarrow k^{-1}\pi_{A'} \tag{4.19}$$

at the origin in order that the relation between abstract spinors and vectors be preserved under $x^a \rightarrow k^2 x^a$. (The conformal factor does not enter into (4.19) because of (4.13).) Choosing the spinor dyad o^A, ι^A as our spin-frame, with normalization

$$\varepsilon_{01} = \varepsilon_{AB}o^A\iota^B = o_A\iota^A = 1 \tag{4.20}$$

(cf. (3.37)), and using the fact that

$$\varepsilon_{AB} \rightarrow \Omega k^{-2}\varepsilon_{AB} = \varepsilon_{AB} \tag{4.21}$$

we get $o^A \rightarrow o^A$, $\iota^A \rightarrow \iota^A$, $o_A \rightarrow o_A$, $\iota_A \rightarrow \iota_A$ and the normalization (4.20) is preserved. Thus (4.19), when referred to this normalized spin-frame becomes

$$\omega^0 \rightarrow k\omega^0, \quad \omega^1 \rightarrow k\omega^1, \quad \pi_{0'} \rightarrow k^{-1}\pi_{0'}, \quad \pi_{1'} \rightarrow k^{-1}\pi_{1'} \tag{4.22}$$

(with $\pi_{0'} = \pi_{A'}\bar{o}^{A'}$, $\pi_{1'} = \pi_{A'}\bar{\iota}^{A'}$, $\omega^0 = \omega^A(-\iota_A)$, $\omega^1 = \omega^A o_A$). Clearly (4.22) is linear and homogeneous and preserves (4.15).

We have considered four types of transformation and seen that each of these induces a linear substitution of twistor components which belongs to the SU(2,2) for which (4.15) is the invariant Hermitian form. Three of these types of transformation (namely the translations, Lorenz transformations and dilations) clearly belong to C(1,3). In fact, the transformation (4.13), with (4.14), can also be regarded as belonging to C(1,3), but this requires a word of explanation. The transformation was given as simply a conformal rescaling from a flat space to a flat space. Notice that, by (4.11), Ω becomes infinite at $x^a = Q^a$. We can follow our conformal rescaling by a point mapping in which the light cone of $x^a = Q^a$ is sent to infinity and in which the metric regains the Minkowski form. This is given by

$$x^a \rightarrow \frac{x^a Q_c Q^c - Q^a x_c x^c}{(Q_b - x_b)(Q^b - x^b)} \tag{4.23}$$

These transformations form a 4-parameter subgroup of C(1,3) which is conjugate, with respect to inversion in the origin (cf. (1.20)) to the translation subgroup of C(1,3). The origin is transformed to itself under (4.23) and so is the tangent space at the origin.

Thus, the (active) C(1,3) transformation given by (4.23) is represented by an SU(2,2) transformation of twistor components, namely that obtained by taking the components of (4.13) at the origin. These four types of C(1,3) transformation will generate the whole of C(1,3). In fact, the infinitesimal transformations of these four types span the entire Lie algebra of C(1,3).

We can express the infinitesimal C(1,3) transformations as

$$x^a \to x^a + \epsilon\xi^a \tag{4.24}$$

where ξ^a satisfies the conformal Killing equation

$$\nabla_{(a}\xi_{b)} = \tfrac{1}{4}\, g_{ab}(\nabla_c\xi^c) \tag{4.25}$$

(i.e., the Lie derivative of the metric with respect to ξ^a is a scalar multiple of itself). The general solution of (4.25) turns out to be

$$\xi_a = S_{ab}x^b + T_a + Kx_a + L_a(x_bx^b) - 2x_a(x_bL^b) \tag{4.26}$$

where S_{ab} is skew and each of S_{ab}, T_a, K, L_a is constant. The infinitesimal Lorentz transformations are described by S_{ab} (6 parameters), the translations by T_a (4 parameters), dilations by K (1 parameter), and special conformal transformations of the type (4.23) by L_a (4 parameters), giving 15 parameters in all. The twistor representations of these infinitesimal transformations are readily obtained from the foregoing discussion of the finite ones (cf. (4.17), (4.8), (4.22) and (4.13),respectively).

The invariant Hermitian form (4.15) has an interesting physical interpretation. By referring back to (4.2), (4.4) and (4.5) we can obtain

$$Z^\alpha\bar{Z}_\alpha = 2s . \tag{4.27}$$

The helicity (or spin) is thus a conformal invariant. When $s = 0$ we have

$$Z^\alpha\bar{Z}_\alpha = 0 \tag{4.28}$$

and Z^α (or $\bar{Z}_\alpha$) is referred to as a <u>null</u> twistor.

Null twistors can be pictured in a particularly simple way. Suppose, first, that $\pi_{A'}$ is not a multiple of $\bar{\omega}_{A'}$ (at the origin O), but that (4.28) holds. Choose $\tilde{O}$ to have position vector x^a relative to O and set

$$x^{AA'} = i(\bar{\omega}^{B'}\pi_{B'})^{-1}\omega^A\bar{\omega}^{A'} + h\bar{\pi}^A\pi^{A'} . \tag{4.29}$$

The vector x^a is real (h being real), the Hermitian nature of (4.29) following because $\bar{\omega}^{B'}\pi_{B'}$ is pure imaginary by (4.28) (see (4.15)). Substituting (4.29) into (4.8) we see that $\tilde{\omega}^A = 0$. Since h can be any real number, the locus of such points $\tilde{O}$ is a null straight line Z in the direction of P^a. The spinor $\tilde{\omega}^A$ vanishes all along the null line Z and nowhere else (as one readily verifies). Thus, Z is the locus of points $\tilde{O}$, with respect to which the angular momentum $\tilde{M}^{ab}$ vanishes (see (4.5)). We can think of Z as the world line of the zero rest-mass particle to which the original classical system under consideration is equivalent. Knowledge of the momentum P^a and the location of Z (with Z parallel to P^a) will uniquely determine the energy-momentum-angular-momentum of the system. The twistor Z^α itself requires the one further piece of information given by the phase of $\pi_{A'}$. This null twistor may be pictured in terms of the world line Z together with the spinor $\pi_{A'}$, with the "flagpole" direction of $\pi_{A'}$ parallel to Z. The "flag plane" of $\pi_{A'}$ gives the remaining "phase" information, and can be pictured geometrically in terms of a kind of oriented "polarization plane" for the zero rest-mass particle.

We have considered only the case when $\pi_{A'}$ is not a multiple of $\bar{\omega}_{A'}$. If it is such a multiple, but $\pi_{A'} \neq 0$, the situation is no different (as is not hard to see) except that the null line Z lies in a null hyperplane through O and the specific form (4.29) cannot be used. However, if $\pi_{A'}$ vanishes (but still $\omega^A \neq 0$) then we do not get a finite line Z. Instead (as can be seen using a limiting argument or a conformal transformation), we obtain a null geodesic Z which generates the null cone at infinity (i.e. Z is a null geodesic through the point I at infinity). Again, the twistor may be pictured (actually up to a multiple ±1, ±i) as the null line Z, together with a spinor $\pi_{A'}$ defined at points of Z, the "flagpole" of $\pi_{A'}$ pointing along Z, where $\pi_{A'}$ is taken parallelly propagated along Z. This description is conformally invariant.

If X^α and Y^α are two null twistors corresponding, respectively to null lines X and Y, the necessary and sufficient condition for them to intersect (possibly at infinity - which would mean, when X and Y are finite lines, that they belong to the same null hyperplane, cf. Section 3) is twistor orthogonality:

$$X^\alpha \bar{Y}_\alpha = 0. \tag{4.30}$$

This is not hard to verify, using (4.8).

If the orthogonality condition (4.30) holds, then every twistor representing a null line through the intersection point R of X and Y will have the form

$$Z^\alpha = \lambda X^\alpha + \mu Y^\alpha . \tag{4.31}$$

(Clearly Z^α is null and orthogonal to each of X^α and Y^α.) Thus, the space-time point R may be represented in twistor space as the linear set (4.31) (a complex 2-plane through the origin $Z^\alpha = 0$ of twistor space).

We have considered the geometrical picture of a null twistor only, so far. If Z^α is non-null ($Z^\alpha \bar{Z}_\alpha \neq 0$) then there is no such simple picture. We cannot localize the zero rest-mass particle, representing our system, to a single world-line when the intrinsic spin is non-zero. A non-local description can be given, however, in terms of a certain twisting system of null straight lines known as a Robinson congruence [26]. The Robinson congruence associated with a given non-null twistor Z^α can be obtained either as the system of null lines represented by null twistors X^α satisfying

$$X^\alpha \bar{Z}_\alpha = 0$$

or, equivalently, as the set of integral curves of the vector field W^a, where $W^{AA'} = \omega^A \bar{\omega}^{A'}$, with ω^A being the solution of (4.10) which represents Z. (Incidentally, such vector fields are precisely the solutions of the conformal Killing equation (4.25) which are everywhere null.) The limiting case of a Robinson congruence which occurs when Z is null, is simply the system of null straight lines meeting the null line Z. If Z is also at infinity then this becomes a system of parallel null straight lines.

One property of a Robinson congruence is that it is everywhere shear-free. This property is a consequence of a theorem due to Kerr [26] which states that every congruence of null lines Z in Minkowski space which is defined by an equation

$$\Phi(Z^\alpha) = 0 \ ,$$

where Φ is holomorphic and homogeneous in Z^α, is shear-free. Conversely every shear-free null congruence in Minkowski space arises in this way - or else as a limiting case of such a construction. Thus the holomorphic nature of functions on twistor space is closely related to the shear-free condition which played such a role in Section 3. (We may think of the Cauchy-Riemann equations $\partial\Phi/\partial\bar{Z}_\alpha = 0$ as being essentially the shear-free condition.) There is also an analogue of this result for curved asymptotically flat space-times [23,25], the shear-free condition being taken asymptotically as $\sigma^0 = 0$.

Another role played by homogeneous holomorphic functions in twistor space is the generation of solutions of the zero rest-mass free-field equations (3.27) in flat space-time [25,26]. Let $f(W_\alpha)$ be holomorphic and homogeneous of degree -n-2 (n = 0,1,2,....) in W_α with suitable positioned singularities. Consider the contour integral

$$\phi_{AB\ldots L}(x^q) = \frac{1}{2\pi i} \oint \lambda_A \lambda_B \ldots \lambda_L \ f(\lambda_Q, -ix^{QQ'}\lambda_Q)\lambda^R d\lambda_R \qquad (4.32)$$

where $\lambda_A, \ldots, \lambda_L$ are n in number. The contour is a closed curve for each x^q, but can vary (continuously) with x^q. This defines a spinor field throughout Minkowski space (or part of Minkowski space - depending on the location of the singularities of f). The spinor field is symmetric, $\phi_{AB\ldots L} = \phi_{(AB\ldots L)}$, and (as follows at once upon differentiating (4.32) with respect to $x^{QQ'}$) satisfies the zero rest-mass free-field equations $\nabla^{AA'}\phi_{AB\ldots L} = 0$ (cf. (3.23) and (3.24)) automatically. The holomorphic (and homogeneous) nature of f is required in an essential way for this. In order to be able simply to differentiate under the integral sign, it is necessary that the value of the integral be unaltered if the path of integration is slightly deformed. We require, indeed, that (4.32) is a proper contour integral in this sense. If the contour is moved continuously over any region throughout which the integrand remains non-singular, then the value of the integral remains unaltered. It is only the way in which the contour "links" the singularity region (i.e., technically, the homology class to which the contour belongs, in the space over which f is non-singular) that is relevant in the position of the contour. And the condition that this be the case is that the exterior derivative [31] of the expression following the integral sign in (4.32) should vanish. This will be the case provided f is holomorphic (except at the specified regions of singularity) and homogeneous of degree -n-2.

There is another way of writing (4.32). Let us express the field $\phi_{AB\ldots L}(x^q)$ itself in twistor terms. The point defined by x^q may be specified by a two-dimensional linear subspace of twistor space. Equivalently, we may work in the dual twistor space and specify this point by the space spanned by two dual twistors U_α and V_α, given by

$$U_\alpha \leftrightarrow (\xi_A, -ix^{AA'}\xi_A), \quad V_\alpha \leftrightarrow (\eta_A, -ix^{AA'}\eta_A). \tag{4.33}$$

These twistors represent null lines U, V through the point defined by x^a (see Figure 13). For it follows from (4.8) that

$$Z^\alpha \leftrightarrow (ix^{AA'}\pi_{A'}, \pi_{A'})$$

if Z passes through the point with position vector x^a (i.e. $\tilde{\omega}^A = 0$ there). Taking the complex conjugate we get the form (4.33) (cf. (4.16)) provided the point is real. Without loss of generality we may assume ξ_A, η_A to be normalized thus:

$$\xi_A\eta^A = 1 \tag{4.34}$$

We can express λ_A as a linear combination of ξ_A and η_A (keeping

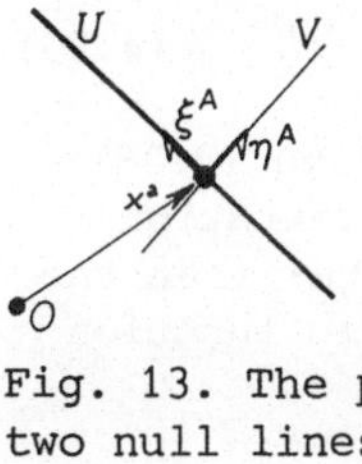

Fig. 13. The point with position vector x^a is the intersection of two null lines U, V. Twistors U^α and V^α, associated with these null lines, serve also to define a basis ξ^A, η^A for spin-vectors at that point.

ξ_A and η_A fixed for given x^a). Since the integrand in (4.32) is invariant under $\lambda_A \to \mu\lambda_A$ (where μ can be different at different points of the contour) - because of the homogeneity degree of f - we can put

$$\lambda_A = \lambda\xi_A + \eta_A \ .$$

Then $\xi_A\lambda^A = 1$ and $\lambda^R d\lambda_R = d\lambda$, whence

$$\xi_A\xi_B\ldots\xi_L\phi^{AB\ldots L}(x^q) = \frac{1}{2\pi i}\oint f(\lambda U_\alpha + V_\alpha)d\lambda = \phi(V_\alpha, U_\alpha) \tag{4.35}$$

say. We can use the integral (4.35) to define $\phi(V_\alpha, U_\alpha)$ whether or not the normalization (4.34) is maintained. The result is a function ϕ which is homogeneous of degree (-n-1) in V_α and homogeneous of degree -1 in U^α. This defines $\phi^{AB\ldots L}\xi_A\xi_B\ldots\xi_L$ at each point and for each ξ_A. Thus the field $\phi^{AB\ldots L}$, being symmetric, is defined by the twistor function ϕ. From (4.35) we derive the property

$$\phi(V_\alpha + kU_\alpha, U_\alpha) = \phi(V_\alpha, U_\alpha) \tag{4.36}$$

where k is arbitrary, with $|k|$ not too large. (This corresponds to the fact that (4.35) is independent of η_A.) Equation (4.36) can be restated as the property that ϕ can be expressed as a function of $V_{[\alpha}U_{\beta]}$ and U_α:

$$\phi(V_\alpha, U_\alpha) = F(V_{[\alpha}U_{\beta]}, U_\alpha) \tag{4.37}$$

(cf. ref. [35]). The significance of this is that $V_{[\alpha}U_{\beta]}$ effectively stands for the linear space spanned by V_α and U_α, so it represents the <u>point</u> with position vector x^a. The role of V is that it merely singles out this particular point on the line U. The direction of U has significance (provided $n \neq 0$) but not the direction of V. Another way of writing (4.36) (or (4.37)) may be derived:

$$U_\alpha \frac{\partial\phi}{\partial V_\alpha} = 0. \tag{4.38}$$

(Differentiate (4.36) with respect to k.) This equation may be compared with those (Euler's theorem) which state the homogeneity degrees of ϕ:

$$U_\alpha \frac{\partial\phi}{\partial U_\alpha} = -\phi \ , \quad V_\alpha \frac{\partial\phi}{\partial V_\alpha} = (-n-1)\phi. \tag{4.39}$$

Functions of twistors, which satisfy equations like (4.38) and (4.39) have an especial significance in the representation theory of U(2,2). Let us revert to twistor space, rather than the dual space, and consider, more generally, a holomorphic function Ψ of four twistors W^α, X^α, Y^α, Z^α which satisfies

$$X^\alpha \frac{\partial\Psi}{\partial W^\alpha} = Y^\alpha \frac{\partial\Psi}{\partial W^\alpha} = Z^\alpha \frac{\partial\Psi}{\partial W^\alpha} = 0, \ Y^\alpha \frac{\partial\Psi}{\partial X^\alpha} = Z^\alpha \frac{\partial\Psi}{\partial X^\alpha} = 0, \ Z^\alpha \frac{\partial\Psi}{\partial Y^\alpha} = 0 \tag{4.40}$$

and

$$W^{\alpha}\frac{\partial\Psi}{\partial W^{\alpha}} = w\Psi,\ X^{\alpha}\frac{\partial\Psi}{\partial X^{\alpha}} = x\Psi,\ Y^{\alpha}\frac{\partial\Psi}{\partial Y^{\alpha}} = y\Psi,\ Z^{\alpha}\frac{\partial\Psi}{\partial Z^{\alpha}} = z\Psi. \qquad (4.41)$$

By (4.41) Ψ is homogeneous in W^{α}, X^{α}, Y^{α}, Z^{α} of respective degrees w,x,y,z. By (4.40) we have

$$\Psi(W^{\alpha}+aX^{\alpha}+bY^{\alpha}+cZ^{\alpha},\ X^{\alpha}+pY^{\alpha}+qZ^{\alpha},\ Y^{\alpha}+tZ^{\alpha},\ Z^{\alpha}) = \Psi(W^{\alpha},\ X^{\alpha},\ Y^{\alpha},\ Z^{\alpha}) \qquad (4.42)$$

(for small enough $|a|,\ldots,|t|$) since the partial derivatives with respect to $a,b,\ldots,t$ all vanish and consequently Ψ can be re-expressed in terms of

$$W^{[\alpha}X^{\beta}Y^{\gamma}Z^{\delta]},\ X^{[\alpha}Y^{\beta}Z^{\gamma]},\ Y^{[\alpha}Z^{\beta]},\ Z^{\alpha}$$

(it follows that the dependence of Ψ on W^{α} is essentially trivial, Ψ being Δ^{-w} times a function of X^{α}, Y^{α}, Z^{α} only, Δ being as in (4.44).)

In fact, any such function can also be referred to the dual twistor space. Introduce

$$\tilde{W}_{\delta} = \Delta\varepsilon_{\alpha\beta\gamma\delta}\,W^{\alpha}X^{\beta}Y^{\gamma},\quad \tilde{X}_{\gamma} = \Delta\varepsilon_{\alpha\beta\gamma\delta}\,W^{\alpha}X^{\beta}Z^{\delta},$$
$$\tilde{Y}_{\beta} = \Delta\varepsilon_{\alpha\beta\gamma\delta}\,W^{\alpha}Y^{\gamma}Z^{\delta},\quad \tilde{Z}_{\alpha} = \Delta\varepsilon_{\alpha\beta\gamma\delta}\,X^{\beta}Y^{\gamma}Z^{\delta}, \qquad (4.43)$$

where

$$\Delta = (\varepsilon_{\alpha\beta\gamma\delta}\,W^{\alpha}X^{\beta}Y^{\gamma}Z^{\delta})^{-1} \qquad (4.44)$$

and where $\varepsilon_{\alpha\beta\gamma\delta}$ is the alternating twistor ($\varepsilon_{\alpha\beta\gamma\delta} = \varepsilon_{[\alpha\beta\gamma\delta]}$, $\varepsilon_{0123} = 1$). Setting $\tilde{\Delta} = \Delta^{-1}$ and $\tilde{\varepsilon}^{\alpha\beta\gamma\delta} = \varepsilon^{\alpha\beta\gamma\delta}$ ($=\varepsilon^{[\alpha\beta\gamma\delta]}$, with $\varepsilon^{0123} = 1$), we obtain expressions for W^{δ}, X^{γ}, Y^{β}, Z^{α} which are simply the "tilde versions" of (4.43). Thus we can re-express Ψ in terms of $\tilde{W}_{\alpha}$, etc. to obtain

$$\Psi(W^{\alpha},\ X^{\alpha},\ Y^{\alpha},\ Z^{\alpha}) = \chi(\tilde{W}_{\alpha},\ \tilde{X}_{\alpha},\ \tilde{Y}_{\alpha},\ \tilde{Z}_{\alpha}).$$

The function χ satisfies conditions analogous to (4.40), (4.41) and (4.42), the homogeneity degrees being $\tilde{w} = -z$, $\tilde{x} = -y$, $\tilde{y} = -x$, $\tilde{z} = -w$. For the case when $\tilde{W}_{\alpha} = \bar{W}_{\alpha}$, $\tilde{X}_{\alpha} = \bar{X}_{\alpha}$, $\tilde{Y}_{\alpha} = \bar{Y}_{\alpha}$, $\tilde{Z}_{\alpha} = \bar{Z}_{\alpha}$, such functions can be given rather direct space-time interpretations, but this matter will not be entered into here [34,35].

Functions of this type form a representation space for the Lie algebra of U(2,2) (and in some cases for the group itself) which, for fixed, w,x,y,z are eigenstates of the four Casimir operators. If w,x,y,z are non-negative integers, then the space of Ψ's contains as a subspace the space of polynomials in W^{α}, X^{α}, Y^{α}, Z^{α} homogeneous of respective degrees w,x,y,z. This subspace is finite-dimensional and in fact forms an irreducible representation space for the group U(2,2). We can write such Ψ's

$$\Psi(W^{\alpha}, \ldots Z^{\alpha}) = \Psi_{\alpha_1 \ldots \alpha_w \beta_1 \ldots \beta_x \gamma_1 \ldots \gamma_y \delta_1 \ldots \delta_z} W^{\alpha_1} \ldots W^{\alpha_w} \ldots Z^{\delta_1} \ldots Z^{\delta_z}$$

where $\Psi_{\ldots\ldots}$ has "Young tableau symmetry" corresponding to the partition $\{z,y,x,w\}$ (with $w \leqslant x \leqslant y \leqslant z$). That is,

$$\Psi_{\alpha_1 \ldots \beta_1 \ldots \gamma_1 \ldots \delta_1 \ldots} = \Psi_{(\alpha_1 \ldots)(\beta_1 \ldots)(\gamma_1 \ldots)(\delta_1 \ldots)}$$

and

$$\Psi_{(\alpha_1 \ldots \alpha_w \beta_1)\beta_2 \ldots} = 0, \ \ldots,$$

$$\Psi_{(\alpha_1 \ldots \alpha_w | \beta_1 \ldots \gamma_z | \delta_1)\delta_2 \ldots \delta_z} = 0, \ \ldots,$$

$$\Psi_{\alpha_1 \ldots\ldots (\gamma_1 \ldots \gamma_y \delta_1)\delta_2 \ldots \delta_z} = 0 \ .$$

(Indices between vertical bars are excluded from symmetrization.) The most general finite-dimensional irreducible representation of U(2,2) can in fact be put in this form. However, it is clear that non-polynomial Ψ's also exist for non-negative w,x,y,z, so the complete space of Ψ's is not irreducible. (A similar situation occurs for representations of SL(2,C) [32,33].) For other values of w,x,y,z the complete Ψ-space may be irreducible.

Let us see, in the general case, how the space of Ψ's may be regarded as a representation space of U(2,2) or SU(2,2) or of their Lie algebras. The group U(2,2) may be regarded as the group of linear transformations of twistor space to itself which preserves the Hermitian form $Z^{\alpha}\bar{Z}_{\alpha}$. These are given by

$$Z^{\alpha} \to t^{\alpha}{}_{\beta} Z^{\beta}$$

where

$$t^{\alpha}{}_{\beta}\bar{t}_{\gamma}{}^{\beta} = \delta^{\alpha}_{\gamma} \tag{4.45}$$

the twistor complex conjugate of $t^{\alpha}{}_{\beta}$ being $\bar{t}_{\alpha}{}^{\beta}$, so that in terms of components we have

$$\begin{pmatrix} \bar{t}_0{}^0 & \bar{t}_0{}^1 & \bar{t}_0{}^2 & \bar{t}_0{}^3 \\ \bar{t}_1{}^0 & \bar{t}_1{}^1 & \bar{t}_1{}^2 & \bar{t}_1{}^3 \\ \bar{t}_2{}^0 & \bar{t}_2{}^1 & \bar{t}_2{}^2 & \bar{t}_2{}^3 \\ \bar{t}_3{}^0 & \bar{t}_3{}^1 & \bar{t}_3{}^2 & \bar{t}_3{}^3 \end{pmatrix} = \begin{matrix}\text{complex}\\ \text{conjugate}\\ \text{of}\end{matrix} \begin{pmatrix} t^2{}_2 & t^2{}_3 & t^2{}_0 & t^2{}_1 \\ t^3{}_2 & t^3{}_3 & t^3{}_0 & t^3{}_1 \\ t^0{}_2 & t^0{}_3 & t^0{}_0 & t^0{}_1 \\ t^1{}_2 & t^1{}_3 & t^1{}_0 & t^1{}_1 \end{pmatrix} \tag{4.46}$$

To obtain SU(2,2) the matrices are taken to be unimodular; equivalently the twistor $\varepsilon^{\alpha\beta\gamma\delta}$ is invariant in the sense that

$$t^{\alpha}{}_{\kappa} t^{\beta}{}_{\lambda} t^{\gamma}{}_{\mu} t^{\delta}{}_{\nu} \varepsilon^{\kappa\lambda\mu\nu} = \varepsilon^{\alpha\beta\gamma\delta} \ .$$

Each $t^{\alpha}{}_{\beta}$ in U(2,2) is represented, in the space of functions ψ, by the linear transformation of such functions:

$$\psi(W^{\alpha}, X^{\alpha}, Y^{\alpha}, Z^{\alpha}) \rightarrow \psi(t^{\alpha}{}_{\beta}W^{\beta}, t^{\alpha}{}_{\beta}X^{\beta}, t^{\alpha}{}_{\beta}Y^{\beta}, t^{\alpha}{}_{\beta}Z^{\beta}). \qquad (4.47)$$

(This is clearly linear because $a\psi+b\phi \rightarrow a\psi'+b\phi'$, where $\psi \rightarrow \psi'$, $\phi \rightarrow \phi'$.) However ψ will in general possess singularities, the transform of ψ possessing singularities in different regions. There is thus no guarantee that a space of such functions can be constructed which are non-singular over some definite fixed region. If we consider infinitesimal transformations, on the other hand, this problem does not arise since the singularity regions remain fixed.

Putting

$$t^{\alpha}{}_{\beta} = \delta^{\alpha}{}_{\beta} + i\varepsilon h^{\alpha}{}_{\beta}$$

where ε is infinitesimal we have

$$h^{\alpha}{}_{\beta} = \bar{h}_{\beta}{}^{\alpha}$$

by (4.45), (4.46) where, in addition

$$h^{\alpha}{}_{\alpha} = 0$$

if $t^{\alpha}{}_{\beta}$ belongs to SU(2,2). The transformation (4.47) now becomes

$$\psi \rightarrow \psi + i\varepsilon h^{\alpha}{}_{\beta}L^{\beta}{}_{\alpha}\psi$$

where

$$L^{\beta}{}_{\alpha} = W^{\beta}\frac{\partial}{\partial W^{\alpha}} + X^{\beta}\frac{\partial}{\partial X^{\alpha}} + Y^{\beta}\frac{\partial}{\partial Y^{\alpha}} + Z^{\beta}\frac{\partial}{\partial Z^{\alpha}}\,.$$

A complete set of Casimir operators for U(2,2) is

$$L^{\alpha}{}_{\alpha},\ L^{\alpha}{}_{\beta}L^{\beta}{}_{\alpha},\ L^{\alpha}{}_{\beta}L^{\beta}{}_{\gamma}L^{\gamma}{}_{\alpha},\ L^{\alpha}{}_{\beta}L^{\beta}{}_{\gamma}L^{\gamma}{}_{\delta}L^{\delta}{}_{\alpha} \qquad (4.48)$$

or, equivalently,

$$K_1 = K_{\alpha}{}^{\alpha},\ K_2 = K_{\beta}{}^{\alpha}K_{\alpha}{}^{\beta},\ K_3 = K_{\gamma}{}^{\beta}K_{\beta}{}^{\alpha}K_{\alpha}{}^{\gamma},$$

$$K_4 = K_{\delta}{}^{\gamma}K_{\gamma}{}^{\beta}K_{\beta}{}^{\alpha}K_{\alpha}{}^{\delta} \qquad (4.49)$$

where

$$K_{\beta}{}^{\alpha} = L^{\alpha}{}_{\beta} + 4\,\delta^{\alpha}_{\beta}\,.$$

To obtain the Casimir operators for SU(2,2) we replace $K^{\alpha}{}_{\beta}$ or $L^{\alpha}{}_{\beta}$ by its trace-free part, in (4.48) or (4.49). The results are certain polynomials in $K_1,\ldots,K_4$, but I shall not bother with them here.

A. Qadir [34,35] has shown that when the operators act on a ψ satisfying (4.40) and (4.41), then

$$K_1 = 10 + S_1, \quad K_2 = 20 + 5S_1 + S_2,$$

$$K_3 = 35 + 15S_1 + \frac{1}{2}S_1{}^2 + \frac{11}{2}S_2 + S_3$$

$$K_4 = 36 + 35S_1 + 3S_1{}^2 + \frac{1}{3}S_1{}^3 + 18S_2 + \frac{20}{3}S_3 + S_4 \qquad (4.50)$$

where

$$S_1 = \omega + \xi + \eta + \zeta, \quad S_2 = \omega^2 + \xi^2 + \eta^2 + \zeta^2,$$

$$S_3 = \omega^3 + \xi^3 + \eta^3 + \zeta^3, \quad S_4 = \omega^4 + \xi^4 + \eta^4 + \zeta^4, \qquad (4.51)$$

the numbers of homogeneity ω,ξ,η,ζ being defined by

$$\omega = \omega, \ \xi = x+1. \ \eta = y+2, \ \zeta = z+3 \ .$$

This curious, but no doubt significant, fact that the eigenvalues of the Casimir operators are symmetric functions of the numbers of homogeneity [34,35], is related to the existence of certain contour integral expressions of which (4.35) is a special case. Consider (4.35) first and re-express it

$$\phi(Y^\alpha,Z^\alpha) = \oint \psi(\lambda Z^\alpha+Y^\alpha)d\lambda \ , \qquad (4.52)$$

$\psi(Z^\alpha)$ being homogeneous of degree z. Now ϕ satisfies (4.40) and (4.41) (or (4.42)) as does ψ (trivially), by (4.38), (4.39). The homogeneity degrees of ψ in Y^α and in Z^α are z+1 and -1, respectively. The numbers of homogeneity $\{\omega,\xi,\eta,\zeta\}$ are $\{0,1,2,z+3\}$ in the case of ψ, and are $\{0,1,z+3.2\}$ in the case of ϕ. Note that these numbers are merely permuted in the passage from ψ to ϕ. Thus the values of the Casimir operators are the same for ϕ as for ψ. In fact this can be seen more directly from (4.52) [35].

The expression (4.52) has been generalized by Qadir in the following way [35]. Let $\psi(W^\alpha,X^\alpha,Y^\alpha,Z^\alpha)$ be as in (4.40), (4.41) and put

$$\rho(W^\alpha,X^\alpha,Y^\alpha, Z^\alpha) = \oint \psi(Z^\alpha,W^\alpha+\lambda Z^\alpha,X^\alpha+\mu Z^\alpha, Y^\alpha+\nu Z^\alpha)d\lambda{}_\wedge d\mu{}_\wedge d\nu,$$

$$\sigma(W^\alpha,X^\alpha,Y^\alpha,Z^\alpha) = \oint \psi(W^\alpha,Z^\alpha,X^\alpha+\mu Z^\alpha,Y^\alpha+\nu Z^\alpha)d\mu{}_\wedge d\nu, \qquad (4.53)$$

$$\tau(W^\alpha,X^\alpha,Y^\alpha,Z^\alpha) = \oint \psi(W^\alpha,X^\alpha,Z^\alpha,Y^\alpha+\nu Z^\alpha)d\nu.$$

the contours and singularity structure of ψ being suitably arranged. It can be verified that ρ,σ,τ are all functions of the required type, the numbers of homogeneity being $\{\zeta,\omega,\xi,\eta\}$, $\{\omega,\zeta,\xi,\eta\}$, $\{\omega,\xi,\zeta,\eta\}$. Since an arbitrary permutation of $\{\omega,\xi,\eta,\zeta\}$ can be built up by successive applications of permutations of these types (or even of the first two) and since, as can be simply shown, ρ, σ and τ must have the same eigenvalues of any Casimir operator as ψ, the invariance of the Casimir operators under permutation of the numbers of homogeneity follows at once. Thus these eigenvalues must be expressible in terms of (4.51) as was done in (4.50). It would be interesting to know whether the integrals (4.53) can, like (4.52), find important application within twistor theory itself. For a different kind of application of contour integration in twistor theory which relates to quantum scattering theory see ref. [25].

I have not attempted here to connect this twistor method of constructing representations of U(2,2) or SU(2,2) with previously existing work - for which the reader is referred, especially, to ref. [36]. In particular, the question of unitarity of the representations has not been discussed here. I feel that there is much scope for further study of the interrelations between the twistor formalism and other approaches.

ACKNOWLEDGEMENTS

I am grateful to Boston University for their hospitality and for providing typing facilities while this article was in preparation. I would also like to thank Judith Daniels for suggesting several improvements to the manuscript.

REFERENCES

1. Cartan, M.E., Ann. Sci. Ecole Normale Superieure, **31** (1914) 263.
2. H. Weyl, The Classical Groups, Princeton University Press, Princeton, New Jersey, 1939.
3. N.H. Kuiper, Ann. Math. 50 916 (1949).
4. H. Rudberg, Disseration, University of Uppsala, Sweden, 1958.
5. R. Penrose, In Proc. 1962 Conf. on Relativistic Theories of Gravitation, Warsaw, Polish Acad. Sci., Warsaw, 1965.
6. R. Penrose, Proc. Roy. Soc. (Lond.) **A 284** 159 (1965).
7. L.P. Eisenhart, Continuous Groups of Transformations, Princeton University Press 1933; reprinted by Dover 1961.
8. R. Penrose, in Battelle Rencontres 1967, C.M. DeWitt & J.A. Wheeler (eds.), Benjamin, New York, 1968.
9. J.M. Bardeen and W.H. Press, J. Math. Phys. **14** 7 (1973).
10. R. Penrose, in Relativity, Groups and Topology, C.M. DeWitt & B.S. DeWitt (eds.), Les Houches Lectures 1963, Blackie & Son, London, 1964.
11. H. Bondi, A.W.K. Metzner and M.J.G. van der Burg. Proc. Roy. Soc.(London) **A 269** 21 (1962).
12. R.K. Sachs, Proc. Roy. Soc. (London) **A 270** 103 (1962).
13. E.T. Newman and T.W.J. Unti, J. Math. Phys. **3** 891 (1962).
14. R.K. Sachs, Phys. Rev. **128** 2851 (1962).

15. V.J. Cantoni, J. Math. Phys. 7 1361 (1966); 8 1700 (1967).
16. V.J. Cantoni, Accad. Naz. Linc. (13) 43 30 (1967).
17. P.J. McCarthy, Proc. Roy. Soc. (London) A 330 517 (1972).
18. P.J. McCarthy, J. Mat. Phys. 13 1837 (1972).
19. P.J. McCarthy, Proc. Roy. Soc. (London) A 333, 317 (1973).
20. M. Crampin and P.J. McCarthy, Representations of the B.M.S. Group III (to appear).
21. R. Geroch and E.T. Newman, J. Math. Phys. 12 314 (1971).
22. E.T. Newman and R. Penrose, J. Math. Phys. 7 863 (1966).
23. B. Aronson, R. Lind, J. Messmer and E.T. Newman, J. Math. Phys. 12 2462 (1971).
24. E.T. Newman and R. Penrose, J. Math. Phys. 3 566 (1962); 4 998 (1963).
25. R. Penrose and M.A.H. MacCallum, Physics Reports 6C 242 (1973)
26. R. Penrose, J. Math. Phys. 8 345 (1967); 10 38 (1969).
27. R. Penrose, Int. J. Theor. Phys. 1 61 (1968).
28. F.A.E. Pirani, in Lectures on General Relativity, Brandeis. Summer Institute in Theoretical Physics 1964, Vol. 1, Prentice Hall, Englewood Cliffs, New Jersey, 1965.
29. R.K. Sachs and P.G. Bergmann, Phys. Rev. 112 674 (1958).
30. P. Bell and P. Szekeres, Int. J. Theor. Phys. 6 111 (1972).
31. H. Flanders, Differential Forms, Academic Press, New York, 1963.
32. M.A. Naimark, Linear Representations of the Lorentz Group, Pergamon, Oxford, 1964.
33. I.M. Gel'fand, M.I. Graev and N.Ya. Vilenkin, Generalized Functions, Vol. 5, Academic Press, New York, 1966.
34. A. Qadir, Ph.D. Thesis, London University, 1971.
35. A. Qadir, Trieste preprint 1972.
36. T. Yao, J. Math. Phys. 8 1931 (1967).

SL(2, C) SYMMETRY OF THE GRAVITATIONAL FIELD*

M. Carmeli
Department of Physics, University of the Negev,
Beer-Sheva, Israel°
and
Aerospace Research Laboratories,
Wright-Patterson AFB, Ohio

ABSTRACT. The theory of general relativity is presented in the form of a gauge field theory by use of the group SL(2, C). The lectures include the following sections: 1. Spinor representation of the group SL(2, C); 2. Connection between spinors and tensors; 3. Maxwell, Weyl and Riemann spinors; 4. Classification of Maxwell spinor; 5. Classification of Weyl spinor; 6. Isotopic spin and gauge fields; 7. Lorentz invariance and the gravitational field; 8. SL(2, C) invariance and the gravitational field; 9. Gravitational field equations.

1. SPINOR REPRESENTATION OF THE GROUP SL(2, C)

Spinors can most naturally be understood within the theory of representations of the group SL(2, C), the group of all 2 x 2 complex unimodular matrices. They describe the finite-dimensional representation of the group SL(2,C), the representation being nonunitary. To find the representation one proceeds as follows [1-3].

1.1 Spinor representation

Let $R_{m,n}$ describe all polynomials $p(z,\bar{z})$ of degree equal or less than m in the variable z, and of degree equal or less than n in

---

* Lectures given at the International Advanced Study Institute on Mathematical Physics: Group Theory in Nonlinear Problems, 7-18 August 1972, Istanbul, Turkey.

° Present address.

the variable $\bar{z}$, the complex conjugate of z, where m and n are non-negative integers. $R_{m,n}$, accordingly, provides a linear vector space. A representation of the group SL(2,C) in the space $R_{m,n}$ is then obtained by the formula

$$T_g p(z,\bar{z}) = (bz+d)^m \overline{(bz+d)}^n \, p\left(\frac{az+c}{bz+d}, \frac{\bar{a}\bar{z}+\bar{c}}{\bar{b}\bar{z}+\bar{d}}\right) . \tag{1.1}$$

Here g is an element of the group SL(2,C),

$$g = \begin{pmatrix} a & b \\ c & d \end{pmatrix} ; \quad ad - bc = 1 . \tag{1.2}$$

One can easily check that T_g is a linear operator which satisfies the conditions $T_{g1} T_{g2} = T_{g1g2}$ and $T_I = 1$, where I is the 2 x 2 unit matrix. Equation (1.1) is called the spinor representation of the group SL(2,C). The dimension of the representation is (m+1)(n+1).

1.2 Alternative method

The spinor representation can also be realized in a somewhat different way.

One considers all possible systems of numbers $\phi^{i_1 \ldots i_m j_1 \ldots j_n}$ which are symmetrical in the indices $i_1, \ldots, i_m$ and in $j_1, \ldots, j_n$, where the indices take the value 0, 1. The aggregate of all such systems of numbers forms a space of dimension (m+1)(n+1). Let us denote this space by $\tilde{R}_{m,n}$.

We now make a correspondence between the spaces $R_{m,n}$ and $\tilde{R}_{m,n}$. To each system $\phi^{i_1 \ldots i_m j_1 \ldots j_n}$ of $\tilde{R}_{m,n}$ one assigns a polynomial

$$p(z,\bar{z}) = \sum_{\substack{i_1 \ldots i_m \\ j_1 \ldots j_n}} \phi^{i_1 \ldots i_m j_1 \ldots j_n} z^{i_1 + \ldots + i_m} \bar{z}^{j_1 + \ldots + j_n} \tag{1.3}$$

of the space $R_{m,n}$. Since, furthermore, each polynomial

$$p(z,\bar{z}) = \sum_{p,q} a_{pq} z^p \bar{z}^q \tag{1.4}$$

of the space $R_{m,n}$ can be written in the form (1.3) by putting

$$\phi^{i_1 \ldots i_m j_1 \ldots j_n} = \frac{1}{m!n!} a_{pq} \tag{1.5}$$

with $i_1 + \ldots + i_m = p$ and $j_1 + \ldots + j_n = q$, it follows that the mapping between the spaces $R_{m,n}$ and $\tilde{R}_{m,n}$ is one-to-one.

The operator T_g can now be regarded as an operator defined in the space $\tilde{R}_{m,n}$. By using the notation $a = g_{11}$, $b = g_{10}$, $c = g_{01}$ and $d = g_{00}$, the spinor representation (1.1) will have the form

$$T_g p(z,\bar{z}) = \sum_{\substack{i_1 \ldots i_m \\ j_1 \ldots j_n}} \phi'^{i_1 \ldots i_m j_1 \ldots j_n} z^{i_1 + \ldots + i_m} \bar{z}^{j_1 + \ldots + j_n} , \qquad (1.6)$$

where

$$\phi'^{i_1 \ldots i_m j_1 \ldots j_n} = \Sigma g_{i_1 i_1'} \ldots g_{i_m i_m'} \bar{g}_{j_1 j_1'} \ldots g_{j_n j_n'} \phi^{j_1' \ldots i_m' j_1' \ldots j_n'} \qquad (1.7)$$

Changing notation, one then obtains the familiar spinor transformation law under the group translation:

$$\phi'_{AB \ldots C'D' \ldots} = g_{AF} g_{BG} \ldots \bar{g}_{C'H'} \bar{g}_{D'K'} \ldots \phi_{FG \ldots H'K'} \quad . \qquad (1.8)$$

2. CONNECTION BETWEEN SPINORS AND TENSORS

In the last section it was shown how 2-component spinors are associated with the finite-dimensional representations of the group SL(2,C) when the representation is realized in the space of polynomials. In particular, it was shown that spinors appear (up to factorial terms) as the coefficients of the polynomials of the space in which the representation is realized. Furthermore, it was shown that their transformation law under the group translation provides another form for the representation.

We now use 2-component spinors in the description of the gravitational field. Accordingly, these quantities become functions of space-time when they are applied in physics. To this end, one associates to each tensor describing a physical quantity in general relativity, a spinor.

2.1 Spinors in Riemannian space

2-component spinors are introduced in a Riemannian space at each space-time point, in a tangent two-dimensional complex space. The correspondence between tensors and spinors is then obtained by means of mixed indices quantities [4]. They are four 2 x 2 Hermitian matrices, denoted by $\tilde{\sigma}^{\mu}_{AB'}$. Greek letters are used for tensor indices running over 0, 1, 2, 3, and Roman capitals for spinor indices taking the values 0, 1. Prime indices refer to the complex conjugate. The Hermiticity of the matrices $\tilde{\sigma}^{\mu}$ means $\tilde{\sigma}^{\mu}{}_{AB'} = \bar{\tilde{\sigma}}^{\mu}{}_{B'A}$. When a locally cartesian coordinate frame is used, the $\tilde{\sigma}^{\mu}$ matrices may be taken (apart from a factor) as the unit matrix (for $\tilde{\sigma}^0$) and the three Pauli matrices (for $\tilde{\sigma}^k$). Other sets may be obtained from these by coordinate transformations. We will not need an explicit knowledge of any one set of $\tilde{\sigma}^{\mu}$.

The four matrices $\tilde{\sigma}^{\mu}$ satisfies the relation

$$g_{\mu\nu} \tilde{\sigma}^{\mu}{}_{AB'} \tilde{\sigma}^{\nu}{}_{CD'} = \varepsilon_{AC} \varepsilon_{B'D'} \quad , \qquad (2.1)$$

where $g_{\mu\nu}$ is the <u>geometrical</u> metric tensor, and ε_{AC} and $\varepsilon_{B'D'}$, along with ε^{AC} and $\varepsilon^{B'D'}$, are the skew-symmetric Levi-Civita symbols, given by

$$\varepsilon_{AB} = \begin{pmatrix} 0 & 1 \\ -1 & 0 \end{pmatrix} \tag{2.2}$$

Raising or lowering a spinor index is made by means of the above symbols ε, with the following conventions:

$$\begin{aligned} \xi^A &= \varepsilon^{AB}\xi_B \quad , \quad \xi_A = \xi^B \varepsilon_{BA} \\ \eta^{A'} &= \varepsilon^{A'B'}\eta_{B'} \quad , \quad \eta_{A'} = \eta^{B'}\varepsilon_{B'A'} \end{aligned} \tag{2.3}$$

2.2 Equivalence of Spinors and Tensors

The spinor equivalent of a tensor is a quantity which has an unprimed and a primed spinor index for each tensor index. The spinor representing the tensor $T^{\alpha\beta}{}_{\gamma}$, for example, is

$$T^{AB'CD'}{}_{EF'} = \tilde{\sigma}_{\alpha}{}^{AB'}\tilde{\sigma}_{\beta}{}^{CD'}\tilde{\sigma}^{\gamma}{}_{EF'}T^{\alpha\beta}{}_{\gamma} \quad . \tag{2.4}$$

The tensor representing the spinor $S^{AB'}{}_{CD'}$, on the other hand, is given by

$$S^{\alpha}{}_{\beta} = \tilde{\sigma}^{\alpha}{}_{AB'}\tilde{\sigma}_{\beta}{}^{CD'}S^{AB'}{}_{CD'} \quad . \tag{2.5}$$

Greek indices are lowered and raised, as usual, by the metric tensor $g_{\alpha\beta}$ and its inverse, $g^{\alpha\beta}$. The spinor expressions for the metric tensor are given by

$$g_{AB'CD'} = \varepsilon_{AC}\varepsilon_{B'D'} \quad , \quad g^{AB'CD'} = \varepsilon^{AC}\varepsilon^{B'D'} \quad . \tag{2.6}$$

When taking the complex conjugate of a spinor, unprimed indices become primed, and vice versa. The complex conjugate of the spinor $S^{AB'}$, for example, is $\bar{S}^{A'B}$. Accordingly, the condition for the vector S^{α} to be real is that its spinor equivalent be Hermitian:

$$S^{AB'} = \bar{S}^{B'A} \quad . \tag{2.7}$$

2.3 Covariant derivative of spinors

The covariant derivative, $\nabla_{\mu}\xi_A$, of a spinor ξ_A is

$$\nabla_{\mu}\xi_A = \partial_{\mu}\xi_A - \Gamma^{B}{}_{A\mu}\xi_B \quad , \tag{2.8}$$

where $\Gamma^{B}{}_{A\mu}$ is the spinor affine connection. The corresponding quantity $\bar{\Gamma}^{B'}{}_{A'\mu}$ deals with the spinor $\bar{\xi}_{A'}$. The spinor affine connection is fixed by the requirement that the covariant derivatives of $\tilde{\sigma}^{\mu}{}_{AB'}$, ε_{AB}, and $\varepsilon_{A'B'}$ shall all vanish [5]:

$$\begin{aligned} \nabla_{\alpha}\tilde{\sigma}^{\mu}{}_{AB'} &= 0 \ , \\ \nabla_{\alpha}\varepsilon_{AB} &= 0 \ , \\ \nabla_{\alpha}\varepsilon_{A'B'} &= 0 \ . \end{aligned} \tag{2.9}$$

3. MAXWELL, WEYL, AND RIEMANN SPINORS

We now find the spinors describing the electromagnetic and gravitational fields. They are obtained, using the procedure outlined in the last section, by associating a spinor to the Maxwell, Weyl, and Riemann tensors. However, since these tensors have some special symmetry properties, their spinor equivalent are simplified.

3.1 The electromagnetic field

Let $F_{\mu\nu}$ describe the Maxwell tensor, i.e., a real skew-symmetric tensor with two indices. Let us denote the spinor equivalent of this tensor by $F_{AB'CD'}$. It obviously satisfies

$$F_{AB'CD'} = -F_{CD'AB'} \ , \tag{3.1}$$

and, as a result, one obtains the identity

$$F_{AB'CD'} = (1/2)(F_{AB'CD'} - F_{CB'AD'}) + (1/2)(F_{CB'AD'} - F_{CD'AB'}).$$

Accordingly, one obtains (see Problem 2)

$$F_{AB'CD'} = (1/2)(\varepsilon_{AC}F_{GB'}{}^{G}{}_{D'} + \varepsilon_{B'D'}F_{CG'A}{}^{G'}) \ . \tag{3.2}$$

The last equation can be simplified. If one denotes $(1/2)F_{CG'A}{}^{G'}$ by ϕ_{AC}, then

$$\phi_{AC} = (1/2)F_{CG'A}{}^{G'} = -(1/2)F_{A}{}^{G'}{}_{CG'} = (1/2)F_{AG'C}{}^{G'} = \phi_{CA} \ ,$$

by the antisymmetry property of F. Hence the spinor ϕ_{AC} is symmetric. By taking the complex conjugate of ϕ_{BD}, on the other hand, gives

$$\bar{\phi}_{B'D'} = (1/2)\overline{F_{BP'D}{}^{P'}} = (1/2)\bar{F}_{B'PD'}{}^{P} = (1/2)F_{PB'}{}^{P}{}_{D'} \ ,$$

where the last equality was a consequence of the Hermiticity property of F. Using the above two equations in Equation (3.2) we obtain the rather simple decomposion of the spinor $F_{AB'DC'}$:

$$F_{AB'CD'} = \varepsilon_{AC}\bar{\phi}_{B'D'} + \phi_{AC}\varepsilon_{B'D'} \ . \tag{3.3}$$

We thus see that the Maxwell tensor is equivalent to a symmetric spinor with two indices. In other words, the six real components of the skew-symmetric tensor $F_{\mu\nu}$ are equivalent to the three complex components of the symmetric spinor ϕ_{AB}.

In the following ϕ_{AB} will be referred to as the Maxwell spinor.

3.2 The gravitational field

Let now $R_{\mu\nu\rho\sigma}$ be the Riemann tensor, i.e., a real four-index tensor having the symmetry properties [6]:

$$R_{\mu\nu\rho\sigma} = -R_{\nu\mu\rho\sigma} = -R_{\mu\nu\sigma\rho} \ ,$$
$$R_{\mu\nu\rho\sigma} + R_{\mu\rho\sigma\nu} + R_{\mu\sigma\nu\rho} = 0 \ . \tag{3.4}$$

Following the procedure outlined above for the Maxwell tensor, one obtains for the Riemann spinor [7]:

$$R_{AB'CD'EF'GH'} = \chi_{ACEG}\varepsilon_{B'D'}\varepsilon_{F'H'} + \phi_{ACF'H'}\varepsilon_{B'D'}\varepsilon_{EG} + \varepsilon_{AC}\bar{\phi}_{B'D'EG}\varepsilon_{F'H'} + \varepsilon_{AC}\varepsilon_{EG}\bar{\chi}_{B'D'F'H'} \ , \tag{3.5}$$

where

$$\chi_{ACEG} = (1/4)R_{AB'C}{}^{B'}{}_{EF'G}{}^{F'} \ ,$$

and

$$\phi_{ACF'H'} = (1/4)R_{AB'C}{}^{B'}{}_{EF'}{}^{E}{}_{H'} \ .$$

The two spinors χ_{ABCD} and $\phi_{ABC'D'}$ uniquely define the curvature spinors. From the symmetries of the Riemann tensor, Equations (3.4), it follows that the spinors χ_{ABCD} and $\phi_{ABC'D'}$ have the following symmetries properties:

$$\chi_{ABCD} = \chi_{BACD} = \chi_{ABDC} = \chi_{CDAB} \ , \tag{3.6}$$

and

$$\phi_{ABC'D'} = \phi_{BAC'D'} = \phi_{ABD'C'} = \bar{\phi}_{C'D'AB} \tag{3.7}$$

3.3 The Weyl spinor

The spinor χ_{ABCD} may be further decomposed. To this end one writes it as

$$\chi_{ABCD} = (1/3)(\chi_{ABCD}+\chi_{ACBD}+\chi_{ADBC})+(1/3)(\chi_{ABCD}-\chi_{ACBD})+ +(1/3)(\chi_{ABCD}-\chi_{ADBC}) \ ,$$

and hence, using Equation (3.6), one can write

$$\chi_{ABCD} = \psi_{ABCD} + (1/3)\varepsilon_{BC}\chi_{AE}{}^{E}{}_{D} + (1/3)\varepsilon_{BD}\chi_{AEC}{}^{E} , \tag{3.8}$$

where we have used the notation

$$\psi_{ABCD} = (1/3)(\chi_{ABCD} + \chi_{ACBD} + \chi_{ADBC}) . \tag{3.9}$$

But the expression $\chi_{AE}{}^{E}{}_{D}$ is skew-symmetric in the indices A, D since

$$\chi_{AE}{}^{E}{}_{D} = \chi^{E}{}_{DAE} = \chi_{D}{}^{E}{}_{EA} = -\chi_{DE}{}^{E}{}_{A} .$$

Therefore

$$\chi_{AE}{}^{E}{}_{D} = (1/2)(\chi_{AE}{}^{E}{}_{D} - \chi_{DE}{}^{E}{}_{A}) = (1/2)\varepsilon_{AD}\chi_{CD}{}^{EC} .$$

Accordingly, we obtain for Equation (3.8)

$$\chi_{ABCD} = \psi_{ABCD} + (1/6)(\varepsilon_{BC}\varepsilon_{AD} + \varepsilon_{BD}\varepsilon_{AC})\lambda , \tag{3.10}$$

where we have denoted $\chi_{AB}{}^{AB}$ by λ. The spinor ψ_{ABCD} is, of course, completely symmetric in its four indices. It corresponds uniquely to the Weyl conformal tensor $C_{\mu\nu\rho\sigma}$. It will be referred to as the Weyl spinor [8]. Moreover, using the second of Equations (3.4), one can show that λ is real (see Problem 6).

One thus obtains for the Riemann spinor, Equation (3.5), the following decomposition:

$$\begin{aligned} R_{AB'CD'EF'GH'} = {} & \psi_{ACEG}\varepsilon_{B'D'}\varepsilon_{F'H'} + \varepsilon_{AC}\varepsilon_{BG}\bar{\psi}_{B'D'F'H'} + \\ & + \frac{\lambda}{6}\{(\varepsilon_{CE}\varepsilon_{AG} + \varepsilon_{CG}\varepsilon_{AE})\varepsilon_{B'D'}\varepsilon_{F'H'} + \\ & + \varepsilon_{AC}\varepsilon_{EG}(\varepsilon_{D'F'}\varepsilon_{B'H'} + \varepsilon_{D'H'}\varepsilon_{B'F'})\} + \\ & + \phi_{ACF'H'}\varepsilon_{B'D'}\varepsilon_{EG} + \varepsilon_{AC}\varepsilon_{F'H'}\bar{\phi}_{B'D'EG}. \end{aligned} \tag{3.11}$$

Counting components, one finds five complex components for ψ_{ABCD}, three real and three complex components for $\phi_{ABC'D'}$, and one real λ. Their sum, is thus, equivalent to the twenty real components of the Riemann tensor.

3.4 Ricci's and Einstein's spinors

To conclude this section, we find below the spinors correspond to the Ricci and Einstein tensors.

The Ricci tensor $R_{\mu\nu} = R^{\sigma}{}_{\mu\sigma\nu}$ has the spinor form

$$R_{AB'CD'} = R^{EF'}{}_{AB'EF'CD'} = \lambda\varepsilon_{AC}\varepsilon_{B'D'} - 2\phi_{ACB'D'} . \tag{3.12}$$

The Ricci scalar (scalar curvature) $R = R^{\alpha}{}_{\alpha}$ is given by

$$R = R^{AB'}{}_{AB'} = 4\lambda \quad . \tag{3.13}$$

Hence the spinor $\phi_{ABC'D'}$ represents the trace-free part of the Ricci tensor. The Einstein tensor $G_{\mu\nu} = R_{\mu\nu}-(1/2)g_{\mu\nu}R$, using Equations (2.6), (3.12) and (3.13), therefore, takes the spinorial form

$$G_{AB'CD'} = -\lambda\varepsilon_{AC}\varepsilon_{B'D'} - 2\phi_{ACB'D'} \quad . \tag{3.14}$$

4. CLASSIFICATION OF MAXWELL SPINOR

We are now in a position to classify both the electromagnetic and gravitational fields. The classification of the gravitational field has been of great interest in relation with the study of gravitational radiation.

In this section we will discuss the classification of the Maxwell tensor, whereas, the gravitational field will be discussed in the next section. The discussion of the electromagnetic field will be made in such a way to emphasize the analogy to the classification of the Weyl tensor for the gravitational field case [9].

4.1 Complex 3-space

Let $F_{\mu\nu}$ be the Maxwell tensor and let $*F_{\mu\nu}$ be its dual (see Problem 4). Let us also define the tensor $F^{+}{}_{\mu\nu}$ by

$$F^{+}{}_{\mu\nu} = F_{\mu\nu} + i{*F}_{\mu\nu} \quad , \tag{4.1}$$

which satisfies $*F^{+}{}_{\mu\nu} = -i\, F^{+}{}_{\mu\nu}$. The spinor equivalent of the tensor $F_{\mu\nu}$ was found in the last section and is given by Equation (3.3), whereas that of the tensor $F^{+}{}_{\mu\nu}$ is given by

$$F^{+}{}_{AB'CD'} = 2\phi_{AC}\varepsilon_{B'D'} \quad , \tag{4.2}$$

where ϕ_{AC} is the Maxwell spinor.

Classification of the electromagnetic field can now be made through classifying ϕ_{AB}. To this end, one studies the eigenspinors and eigenvalues of the spinorial equation

$$\phi_{AB}\alpha^{B} = \lambda\alpha_{A} \quad . \tag{4.3}$$

To study this equation, one introduces a basis in spin space. Let the two spinors of the basis be denoted by ℓ_A and η_A, satisfying the normalization condition $\ell_A n^A = 1$. This basis induces another basis, given by

$$\xi_{0AB} = n_A n_B \quad , \quad \xi_{1AB} = -2\ell_{(A}n_{B)} \quad , \quad \xi_{2AB} = \ell_A \ell_B \quad , \tag{4.4}$$

in the 3-dimensional space, E_3, of bispinors. This means a bispinor ϕ_{AB} can be written in terms of the basis (4.4) as

$$\phi_{AB} = \sum_{m=0}^{2} \phi_m \, \xi_{mAB} \quad , \tag{4.5}$$

where ϕ_0, ϕ_1, and ϕ_2 are called dyad components of the bispinor and correspond to the six real components of the tensor $F_{\mu\nu}$.

The spin frame ℓ_A, n_A induces other bases in E_3, such as the one given by

$$\begin{aligned} \eta_{0AB} &= 2^{(1/2)} i \ell_{(A}n_{B)} \quad , \\ \eta_{1AB} &= 2^{(-1/2)} (\ell_A \ell_B + n_A n_B) \quad , \\ \eta_{2AB} &= 2^{(-1/2)} i (\ell_A \ell_B - n_A n_B) \quad . \end{aligned} \tag{4.6}$$

This basis satisfies the orthogonality relation

$$\eta_{mAB}\eta_n{}^{AB} = \delta_{mn} \tag{4.7}$$

In terms of this last basis ϕ_{AB} can now be written as

$$\phi_{AB} = \sum_{m=0}^{2} \chi_m \eta_{mAB} \quad . \tag{4.8}$$

The two sets of three components $\underline{\chi}$ and $\underline{\phi}$ are then related by

$$\begin{aligned} \chi_0 &= 2^{(1/2)} i \phi_1 \quad , \\ \chi_1 &= 2^{(-1/2)} (\phi_0 + \phi_2) \quad , \\ \chi_2 &= 2^{(-1/2)} i (\phi_0 - \phi_2) \quad . \end{aligned} \tag{4.9}$$

4.2 Classification

In terms of the dyad components ϕ_m, the eigenvalue Equation (4.3) becomes

$$\Phi\alpha = \lambda\alpha \tag{4.3a}$$

where Φ is a 2 x 2 matrix, and α is a column matrix, given by

$$\Phi = \begin{pmatrix} \phi_1 & \phi_2 \\ -\phi_0 & -\phi_1 \end{pmatrix} \quad , \tag{4.10a}$$

and

$$\alpha = \begin{pmatrix} \alpha^0 \\ \alpha^1 \end{pmatrix}, \tag{4.10b}$$

The two eigenvalues of Equation (4.3a) are $\lambda = \pm(\phi_1{}^2-\phi_0\phi_2)^{(1/2)}$. One, therefore, has two cases: (1) $\phi_1{}^2-\phi_0\phi_2 \neq 0$, in which case there are different eigenspinors. The spinor ϕ_{AB} is called algebraically general or non-null; and (2) $\phi_1{}^2-\phi_0\phi_2 = 0$, in which case $\lambda = 0$ and there is only one eigenspinor. The spinor ϕ_{AB} is then called algebraically special or null [10].

4.3 Changes of spin frame

Let us introduce another basis ℓ'_A, n'_A in spin space that is related to the original basis ℓ_A, n_A by

$$\begin{pmatrix} \ell'_A \\ n'_A \end{pmatrix} = \begin{pmatrix} a & b \\ c & d \end{pmatrix} \begin{pmatrix} \ell_A \\ n_A \end{pmatrix}. \tag{4.11}$$

Here, a, b, c, and d are complex numbers satisfying ad-bc = 1. Thus, the matrix

$$g = \begin{pmatrix} a & b \\ c & d \end{pmatrix} \tag{4.12}$$

is an element of the group SL(2,C). We can now write ϕ_{AB} in terms of the new basis,

$$\phi_{AB} = \phi'_m \xi'_{mAB} , \tag{4.13}$$

where ξ'_{mAB} is the induced basis in E_3 and is given in terms of ℓ'_A and n'_A in accordance with Equation (4.4). The dyad components ϕ'_m can then be obtained in terms of ϕ_n by

$$\phi' = (g^t)^{-1} \Phi g^t , \tag{4.14}$$

or by

$$\begin{pmatrix} \phi'_0 \\ \phi'_1 \\ \phi'_2 \end{pmatrix} = \begin{pmatrix} a^2 & 2ab & b^2 \\ ac & bc+ad & bd \\ c^2 & 2cd & d^2 \end{pmatrix} \begin{pmatrix} \phi_0 \\ \phi_1 \\ \phi_2 \end{pmatrix}. \tag{4.15}$$

The corresponding transformation law for χ's are obtained, using Equations (4.9) and (4.15):

$$\begin{pmatrix} \chi'_0 \\ \chi'_1 \\ \chi'_2 \end{pmatrix} = \begin{pmatrix} ad+bc & i(ac+bd) & ac-bd \\ -i(ab+cd) & (a^2+b^2+c^2+d^2) & -\frac{i}{2}(a^2-b^2+c^2-d^2) \\ ab-cd & (a^2+b^2-c^2-d^2) & \frac{1}{2}(a^2-b^2-c^2+d^2) \end{pmatrix} \begin{pmatrix} \chi_0 \\ \chi_1 \\ \chi_2 \end{pmatrix} . \quad (4.16)$$

Let us denote the two square 3 x 3 complex matrices in Equations (4.15) and (4.16) by Q and P, respectively. They give <u>three-dimensional representations</u> for the proper, orthochroneous, homogenous Lorentz group. The matrix in Equation (4.16) is <u>orthogonal</u>, $P^{-1} = P^t$, with determinent unity [11], whereas that in Equation (4.15) satisfies the relation

$$\sum_{k=0}^{2} \frac{(-1)^k}{k!(2-k)!} Q_{kj} Q_{(2-k),\ell} \equiv \frac{(-1)^\ell}{\ell!(2-\ell)!} \delta_{\ell,(2-j)} .$$

The spin frame transformation (4.11) also induces a proper, orthochroneous, Lorentz transformation on the null tetrad in the curve space constructed from the two spinors ℓ_A and n_A.

The null tetrad induced by the two spinors ℓ^A and n^A is given by [12]:

$$\begin{aligned} \ell^\mu &= \tilde{\sigma}^\mu{}_{AB'} \ell^A \bar{\ell}^{B'} \\ m^\mu &= \tilde{\sigma}^\mu{}_{AB'} \ell^A \bar{n}^{B'} \\ \bar{m}^\mu &= \tilde{\sigma}^\mu{}_{AB'} n^A \bar{\ell}^{B'} \\ n^\mu &= \tilde{\sigma}^\mu{}_{AB'} n^A \bar{n}^{B'} . \end{aligned} \quad (4.17)$$

Accordingly, a change of a null tetrad in the curve space is represented in the space E_3 by a proper orthogonal matrix, provided the basis in E_3 is chosen to be orthogenal as in Equation (4.7).

As has been shown [1], the matrix $g \in SL(2,C)$, given in Equations (4.11) and (4.12), can be written as a product of three matrices of the form

$$g_1(z) = \begin{pmatrix} 1 & 0 \\ z & 1 \end{pmatrix}, \quad g_2(z) = \begin{pmatrix} z & 0 \\ 0 & z^{-1} \end{pmatrix}, \quad g_3(z) = \begin{pmatrix} 1 & z \\ 0 & 1 \end{pmatrix} \quad (4.18)$$

where z is a complex number. The transformation $g_1(z)$ leaves the spinor ℓ_A, and hence the null vector ℓ_μ, invariant. It is called a one-(complex) parameter <u>null rotation</u> about ℓ_μ. The transformation $g_3(z)$ is also a one-parameter null rotation, but about the vector n_μ. The transformation $g_2(z)$ corresponds to an ordinary Lorentz transformation (boost) in the ℓ_μ-n_μ plane, along with a spatial rotation in the m_μ- $\bar{m}_\mu$ plane [13].

The matrices $Q_1(z)$, $Q_2(z)$, and $Q_3(z)$ obtained from Equation (4.15), corresponding to the three matrices $g_1(z)$, $g_2(z)$, and $g_3(z)$ of the group SL(2,C), can be obtained by putting the appropriate values in Q. The transformations of $\phi = (\phi_0, \phi_1, \phi_2)$ under $Q_1(z)$, $Q_2(z)$, and $Q_3(z)$ are then given by:

$$\begin{aligned} \phi'_0 &= \phi_0 \\ \phi'_1 &= z\phi_0 + \phi_1 \\ \phi'_2 &= z^2\phi_0 + 2z\phi_1 + \phi_2 \quad , \end{aligned} \tag{4.19}$$

$$\begin{aligned} \phi'_0 &= z^2\phi_0 \\ \phi'_1 &= \phi_1 \\ \phi'_2 &= z^{-2}\phi_2 \quad , \end{aligned} \tag{4.20}$$

and

$$\begin{aligned} \phi'_0 &= \phi_0 + 2z\phi_1 + z^2\phi_2 \\ \phi'_1 &= \phi_1 + z\phi_2 \\ \phi'_2 &= \phi_2 \quad , \end{aligned} \tag{4.21}$$

respectively.

4.4 Invariants

The matrix Φ given by Equations (4.3a) and (4.10) can also be written, using Equation (4.9), in terms of the components of the 3-vector $\underline{\chi} = (\chi_0, \chi_1, \chi_2)$ as:

$$\Phi = (i/\sqrt{2})X \quad , \tag{4.22}$$

$$X = \begin{pmatrix} -\chi_0 & \chi_2 - i\chi_1 \\ \chi_2 + i\chi_1 & \chi_0 \end{pmatrix} \tag{4.23}$$

Under a change of basis, the trace of the matrix X must be invariant. But $\mathrm{Tr}X = 0$, and thus it does not yield an interesting invariant. However, $\mathrm{Tr}X^2 = 2\chi_m\chi_m = 2\underline{\chi}\cdot\underline{\chi} = 2\phi_{AB}\phi^{AB}$ is an "obvious" invariant. In fact, from $X^2 = \underline{\chi}\cdot\underline{\chi}I$, where I is the 2 x 2 unit matrix, it follows that

$$\begin{aligned} \mathrm{Tr}X^{2n-1} &= 0 \quad , \\ \mathrm{Tr}X^{2n} &= 2(\underline{\chi}\cdot\underline{\chi})^n \quad , \end{aligned} \tag{4.24}$$

for any natural number n.

We thus see that the invariant $\chi.\chi$ plays an important role in the classification of the bivector. If $\chi.\chi$ vanishes, the bivector is null, otherwise, it is non-null.

4.5 Canonical forms

There are two canonical forms which correspond to the two types of bivectors.

If the bivector is null one can always choose a spin frame ℓ'_A, n'_A in such a way that the direction of $n'_A n'_B$ in E_3 coincides with that of the given null bispinor. To see this we proceed as follows. Let $\underline{\phi} = (\phi_0, \phi_1, \phi_2)$, with $\underline{\chi}.\underline{\chi} = 2(\phi_0\phi_2 - \phi_1^2) = 0$, be the components of the bispinor in the basis (4.4). Without loss of generality one can assume that $\phi_0 \neq 0$. (A null rotation $g_3(z)$ about n_μ, of the form (4.21), could always make it so.) Under a null rotation around ℓ_μ, the components of ϕ transform according to Equation (4.19). ϕ'_2 is a quadratic polynomial in z, whereas ϕ'_1 is proportional to the derivative of ϕ'_2 with respect to z. The condition $\phi_0\phi_2 - \phi_1^2 = 0$ yields a double root for ϕ'_2 given by $z = -\phi_1/\phi_0$. Choosing this root for z makes both ϕ'_2 and ϕ'_1 vanish simultaneously. Accordingly, in the new frame $\phi_{AB} = \phi'_0 n'_A n'_B$, and $\phi' = (\phi'_0, 0, 0)$. The matrix Φ and the eigenspinor α of Equation (4.10) will have the forms:

$$\begin{pmatrix} 0 & 0 \\ -\phi'_0 & 0 \end{pmatrix}, \tag{4.25}$$

and

$$\begin{pmatrix} 0 \\ 1 \end{pmatrix}, \tag{4.26}$$

i.e., n'_A. It will be noted that it is equally possible to make the direction $\ell'_A \ell'_B$ coincide with the given bispinor by making a null rotation around n_α instead.

If the given bispinor is non-null, we can make ϕ'_2, but not ϕ'_1, to vanish by choosing z to be one of the roots of the quadratic form ϕ'_2. Applying a null rotation around n_α with appropriate value for z will leave ϕ''_1 as the only non-zero component of $\underline{\phi}''$. Hence $\phi_{AB} = -2\phi''_1 \ell''_{(A} n''_{B)}$. The matrix Φ, Equation (4.10a), is then given by

$$\begin{pmatrix} \phi''_1 & 0 \\ 0 & -\phi''_1 \end{pmatrix}, \tag{4.27}$$

whereas the eigenspinors, Equation (4.10b), will be given by

$$\begin{pmatrix}1\\0\end{pmatrix} \text{ and } \begin{pmatrix}0\\1\end{pmatrix}, \qquad (4.28)$$

i.e., by the new basis spinors ℓ''_A and n''_A.

4.6 Spinor method

The classification of the Maxwell spinor ϕ_{AB} could also be made by decomposing it into the symmetrized product of spinors with one index. This is done as follows.

Let ξ^A be an arbitrary spinor, and consider the expression $\phi_{AB}\xi^A\xi^B$. This is a homogeneous polynomial of second degree in ξ^0 and ξ^1. This polynomial may be factored into two linear factors, thus getting the identity

$$\phi_{AB}\xi^A\xi^B = (\alpha_A\xi^A)(\beta_B\xi^B) \quad ,$$

or

$$\{\phi_{AB} - \alpha_{(A}\beta_{B)}\}\xi^A\xi^B = 0 \quad .$$

Therefore, since ξ^A is arbitrary, one obtains a decomposition of ϕ_{AB},

$$\phi_{AB} = \alpha_{(A}\beta_{B)} \quad , \qquad (4.29)$$

which is called the canonical decomposition of ϕ_{AB}.

The spinors α_A and β_B are determined up to a (complex) scalar factor. They are called principal spinors, and each of them, in turn, determines a real null direction. They need not be distinct. As a result, the decomposition (4.29) determines at least one and at most two real null directions, called the principal null directions of ϕ_{AB}. Classification of ϕ_{AB} may be based on counting the multiplicities of principal null directions. If α_A and β_B coincide, the bispinor ϕ_{AB} is null, otherwise it is general. This classification coincides with our previous discussion.

4.7 Tensor method

Finally, we briefly mention the tensor method of classification. For every skew-symmetric tensor there exist two null directions $\xi_\mu \neq 0$, which may or may not coincide, satisfying the equation

$$F^+{}_{\mu[\nu}\xi_{\rho]}\xi^\mu = 0 \quad . \qquad (4.30)$$

If the directions coincide, $F\mu\nu$ is null, otherwise it is non-null. Equation (4.30) is equivalent to the spinor equation

$$\phi_{AB}\xi^A\xi^B = 0 \quad .$$

Our previous discussion shows the tensor method to be equivalent to previous methods.

5. CLASSIFICATION OF WEYL SPINOR

In the last section bivectors were discussed as a preliminary to discussing the Weyl spinor. Moreover, bivectors occur as eigenvectors of the Weyl tensor. In this section, we discuss the Weyl tensor.

The Weyl tensor $C_{\alpha\beta\gamma\delta}$ has the same symmetry properties of the Riemann tensor, Equations (3.4). In addition, it satisfies

$$C_{\alpha\rho\ \beta}^{\ \ \rho} = 0 \quad . \tag{5.1}$$

These identities reduce the number of independent compounts of $C_{\alpha\beta\gamma\delta}$ to ten.

In Section 3 the spinor equivalent of $C_{\alpha\beta\gamma\delta}$ was found to be a symmetric spinor of four indices, ψ_{ABCD},

$$C_{AB'CD'EF'GH'} = \varepsilon_{AC}\varepsilon_{EG}\bar{\psi}_{B'D'F'H'} + \psi_{ACEG}\varepsilon_{B'D'}\varepsilon_{F'H'} \quad . \tag{5.2}$$

Corresponding to the Weyl tensor $C_{\alpha\beta\gamma\delta}$ one can define the tensor $C^+{}_{\alpha\beta\gamma\delta}$ by

$$C^+_{\alpha\beta\gamma\delta} = C_{\alpha\beta\gamma\delta} + i\ {}^*C_{\alpha\beta\gamma\delta} \quad , \tag{5.3}$$

where * denotes the (left- or right-hand) dual. The spinor equivalent to (5.3) is given by

$$C^+_{AB'CD'EF'GH'} = 2\varepsilon_{B'D'}\varepsilon_{F'H'}\psi_{ACEG} \tag{5.4}$$

5.1 Complex 5-space

In order to classify the Weyl tensor, we classify the Weyl spinor ψ_{ABCD} in terms of its eigenvalues and eigenspinors. The characteristic equation is now:

$$\psi_{ABCD}\phi^{CD} = \lambda\phi_{AB} \tag{5.5}$$

The basis ℓ_A, n_A in spinorial space induces the basis

$$\xi_{0ABCD} = \frac{1}{2}n_A n_B n_C n_D$$

$$\xi_{1ABCD} = -2\ell_{(A}n_B n_C n_{D)} \qquad \xi_{3ABCD} = -2\ell_{(A}\ell_B\ell_C n_{D)} \quad , \tag{5.6}$$

$$\xi_{2ABCD} = 3\ell_{(A}\ell_B n_C n_{D)}\ , \qquad \xi_{4ABCD} = \frac{1}{2}\ell_A\ell_B\ell_C\ell_D \quad ,$$

in the 5-dimensional complex space, E_5, of completely symmetric four-spinors. The Weyl spinor can now be written in terms of the basis (5.6) as

$$\psi_{ABCD} = \sum_{n=0}^{4} \psi_n \xi_{nABCD} , \qquad (5.7)$$

where ψ_n, with n = 0, 1,..., 4, are the dyad components of the Weyl spinor, and correspond to the ten real components of the Weyl tensor.

Since E_5 is a subspace of $E_3 \times E_3$, one can expand ψ_{ABCD} in terms of the basis $\eta_{mAB}\eta_{nCD}$ of $E_3 \times E_3$:

$$\psi_{ABCD} = \sum_{m,n=0}^{2} \psi_{mn}\eta_{mAB}\eta_{nCD} . \qquad (5.8)$$

One can then write the coefficients ψ_{mn} in terms of the dyad components $\psi_0,\ldots,\psi_4$ by use of Equations (4.6) and (5.6), to obtain a symmetric and trace-free matrix:

$$\psi_{mn} = \begin{pmatrix} -\psi_2 & \frac{i}{2}(\psi_1+\psi_3) & \frac{1}{2}(\psi_3-\psi_1) \\ \frac{i}{2}(\psi_1+\psi_3) & \frac{1}{4}(2\psi_2+\psi_0+\psi_4) & \frac{i}{4}(\psi_0-\psi_4) \\ \frac{i}{2}(\psi_3-\psi_1) & \frac{i}{4}(\psi_0-\psi_4) & \frac{1}{4}(2\psi_2-\psi_0-\psi_4) \end{pmatrix} . \qquad (5.9)$$

We have seen that the Weyl tensor can be regarded as a vector in a five-dimensional space. The space E_5 has properties similar to E_3 discussed in the last section. It will be useful to introduce an orthonormal basis in E_5. Such a basis is provided by the following five, completely symmetric, four-spinors:

$$\begin{aligned} \eta_{0ABCD} &= (1/\sqrt{2})(\ell_A\ell_B\ell_C\ell_D + n_An_Bn_Cn_D) , \\ \eta_{1ABCD} &= i\sqrt{2}(\ell_{(A}\ell_B\ell_Cn_{D)} + \ell_{(A}n_Bn_Cn_{D)}) , \\ \eta_{2ABCD} &= \sqrt{6}\,\ell_{(A}\ell_Bn_Cn_{D)} , \\ \eta_{3ABCD} &= \sqrt{2}\,(\ell_{(A}\ell_B\ell_Cn_{D)} - \ell_{(A}n_Bn_Cn_{D)}) , \\ \eta_{4ABCD} &= (i/\sqrt{2})(\ell_A\ell_B\ell_C\ell_D - n_An_Bn_Cn_D) . \end{aligned} \qquad (5.10)$$

As can be easily verified, they satisfy

$$\eta_{mABCD}\eta_n{}^{ABCD} = \delta_{mn} \quad ; \quad m, n = 0,\ldots, 4 , \qquad (5.11)$$

and an arbitrary element of the space E_5 can be written as linear combination of them:

$$\psi_{ABCD} = \sum_{m=0}^{4} \chi_m\eta_{mABCD} , \qquad (5.12)$$

in analogous to Equation (4.8) for the Maxwell spinor. The components χ_m can then be expressed in terms of the components ψ_n of Equation (5.7) [14]. We find:

$$\chi_0 = 2^{-\frac{3}{2}}(\psi_0+\psi_4) \quad , \quad \chi_1 = 2^{-\frac{1}{2}}i(\psi_1+\psi_3) \quad ,$$
$$\chi_2 = (3/2)^{\frac{1}{2}}\psi_2 \quad , \quad \chi_3 = 2^{-\frac{1}{2}}(\psi_1-\psi_3) \quad , \qquad (5.13)$$
$$\chi_4 = 2^{-\frac{3}{2}}i(\psi_0-\psi_4) \quad .$$

5.2 Classification

In terms of the matrix Ψ of Equation (5.9), the eigenvalue Equation (5.5) can be written as

$$\Psi\chi = \lambda\chi \quad , \qquad (5.14)$$

where χ is the column matrix whose elements are χ_m, m = 0, 1, 2, and χ_m are the components of ϕ_{AB} in the orthonormal basis η_{mAB}.

The Weyl spinor can now be classified according to the possible numbers of eigenvalues and eigenvectors of the matrix Ψ, Equation (5.9). The maximum number of eigenvalues for the matrix Ψ is three. Corresponding to every eigenvalue there is at least one eigenvector. Accordingly, we obtain Table I.

TABLE I

Distinct eigenvectors		3		2		1
Distinct eigenvalues	3	2	1	2	1	1
Petrov type	I	D	O	II	N	III

Remarks: 1. In the following, it will be shown that if there is only one distinct eigenvalue, then that eigenvalue is necessarily zero. Therefore, if there were three linearly independent eigenvectors corresponding to it, every vector of E_3 would also be an eigenvector. This is possible if and only if the Weyl spinor is identically zero.

Remark 2. Type I is also known as algebraically general, the others are known as algebraically special [15].

5.3 Change of frame

A change of the basis according to the transformation (4.11) induces changes in the various field components. Comparing Equations (4.8) and (5.8) shows that if the law of transformation of the vector χ is given by (4.16), then the law of transformation of the matrix Ψ

should be given by

$$\Psi' = P\Psi P^t \quad . \tag{5.15}$$

One then can obtain the law of transformation for the dyad components $\psi_0,\ldots,\psi_4$ which is found to be

$$\begin{pmatrix}\psi'_0\\ \psi'_1\\ \psi'_2\\ \psi'_3\\ \psi'_4\end{pmatrix}=\begin{pmatrix} a^4 & 4a^3b & 6a^2b^2 & 4ab^3 & b^4\\ a^3c & a^2(3bc+ad) & 3ab(ad+bc) & b^2(3ad+bc) & b^3d\\ a^2c^2 & 2ac(ad+bc) & 1+6abcd & 2bd(ad+bc) & b^2d^2\\ ac^3 & c^2(3ad+bc) & 3cd(ad+bc) & d^2(3bc+ad) & bd^3\\ c^4 & 4c^3d & 6c^2d^2 & 4cd^3 & d^4\end{pmatrix}\begin{pmatrix}\psi_0\\ \psi_1\\ \psi_2\\ \psi_3\\ \psi_4\end{pmatrix} \tag{5.16}$$

Using Equation (5.13) one finds the transformation law for the components χ_m, $m = 0, 1, \ldots, 4$. The result can be written in the form

$$\chi' = R\chi \quad . \tag{5.17}$$

The 5 x 5 complex matrix R is a function of the complex variables a, b, c and d of the matrix $g \in SL(2,C)$. One can show that the matrix R is orthogonal and has a determinant unity (see Problem aa). The 5 x 5 matrices in Equations (5.16) and (5.17) give five-dimensional representations for the proper, orthochroneous, homogeneous Lorents group.

The transformation law (5.15) can be applied for specific cases when the matrix $g \in SL(2,C)$ of Equation (4.12) is taken as $g_1(z)$, $g_2(z)$, and $g_3(z)$. Under a null rotation $g_1(z)$ around ℓ_μ, the dyad components $\psi_0,\ldots,\psi_4$ of the Weyl spinor transform into

$$\begin{aligned}\psi'_0 &= \psi_0 \quad ,\\ \psi'_1 &= z\psi_0 + \psi_1 \quad ,\\ \psi'_2 &= z^2\psi_0 + 2z\psi_1 + \psi_2 \quad ,\\ \psi'_3 &= z^3\psi_0 + 3z^2\psi_1 + 3z\psi_2 + \psi_3 \quad ,\\ \psi'_4 &= z^4\psi_0 + 4z^3\psi_1 + 6z^2\psi_2 + 4z\psi_3 + \psi_4 \quad ,\end{aligned} \tag{5.18}$$

$g_2(z)$ induces the transformation

$$\begin{aligned}\psi'_0 &= z^4\psi_0 \quad , \qquad \psi'_1 = z^2\psi_1 \quad ,\\ \psi'_2 &= \psi_2 \quad , \qquad \psi'_3 = z^{-2}\psi_3 \quad ,\\ \psi'_4 &= z^{-4}\psi_4 \quad ,\end{aligned} \tag{5.19}$$

whereas, $g_3(z)$ induces the null rotation around n_α:

$$
\begin{aligned}
\psi'_0 &= \psi_0 + 4z\psi_1 + 6z^2\psi_2 + 4z^3\psi_3 + z^4\psi_4 \quad , \\
\psi'_1 &= \psi_1 + 3z\psi_3 + 3z^2\psi_3 + z^3\psi_4 \quad , \\
\psi'_2 &= \psi_2 + 2z\psi_3 + z^2\psi_4 \quad , \\
\psi'_3 &= \psi_3 + z\psi_4 \quad , \\
\psi'_4 &= \psi_4 \quad .
\end{aligned}
\tag{5.20}
$$

5.4 Invariants

By writing the components $\psi_0,\ldots,\psi_4$ in terms of $\chi_0,\ldots,\chi_4$ as expressed by Equation (5.13), the matrix Ψ of Eq. (5.9) may be written in terms of the components of the Weyl tensor in the orthonormal basis as:

$$
\Psi = 2^{-\frac{1}{2}} \begin{pmatrix} -(2\chi_2/\sqrt{3}) & \chi_1 & -\chi_3 \\ \chi_1 & (\chi_2/\sqrt{3})+\chi_0 & \chi_4 \\ -\chi_3 & \chi_4 & (\chi_2/\sqrt{3})-\chi_0 \end{pmatrix} . \tag{5.21}
$$

As can be verified from the transformation law (5.15), the characteristics of the symmetric and traceless matrix (5.21) are independent of the spin frame.

Since the matrix (5.21) has a zero trace, we consider the invariant $\mathrm{Tr}\Psi^2$ which is equal to

$$
\mathrm{Tr}\Psi^2 = \underline{\chi}\cdot\underline{\chi} = \sum_{m=0}^{4} \chi_m\chi_m = \psi_{ABCD}\psi^{ABCD} \quad . \tag{5.22}
$$

If the invariant $\mathrm{Tr}\Psi^2$ vanishes, the Weyl tensor is null. Otherwise, it is non-null. Since an arbitrary orthogonal transformation in E_5 does not necessarily represent a change of spin frame, there is another invariant. It is $\mathrm{Tr}\Psi^3$,

$$
\mathrm{Tr}\Psi^3 = \psi^{AB}{}_{CD}\psi^{CD}{}_{EF}\psi^{EF}{}_{AB} \quad . \tag{5.23}
$$

Now the eigenvalues of Ψ satisfy the equation $|\Psi-\lambda I| = 0$, where I is the 3 x 3 unit matrix. This equation gives the cubic equation in λ:

$$
f(\lambda) = \lambda^3 - \frac{1}{2}A\lambda - \frac{1}{3}B = 0 \quad , \tag{5.24}
$$

where $A = \underline{\chi}\cdot\underline{\chi}$ and $B = 3 \det \Psi$. By the Cayley-Hamilton theorem,

$$
\Psi^3 - \frac{1}{2}A\Psi - \frac{1}{3}BI = 0 \quad ,
$$

and hence $\mathrm{Tr}\Psi^3 = B$. One also easily verifies that $\mathrm{Tr}\Psi^n$, where $n = 4, 5, \ldots$, can be expressed in terms of A and B, and therefore there are no further independent invariants.

Let λ_1, λ_2, and λ_3 be the eigenvalues of Ψ (which may or may not be distinct). From Equation (5.24) one then obtains

$$\lambda_1 + \lambda_2 + \lambda_3 = 0 \quad ,$$
$$-2(\lambda_1\lambda_2 + \lambda_2\lambda_3 + \lambda_3\lambda_1) = A \quad , \tag{5.25}$$
$$3\lambda_1\lambda_2\lambda_3 = B \quad .$$

Accordingly, if $\lambda_1 = \lambda_2 = \lambda_3$, then $\lambda_1 = \lambda_2 = \lambda_3 = 0$, and hence the two invariants A and B vanish. This is the case of the gravitational fields of types N and III. A gravitational field is of type II or D if two of the eigenvalues, let us say λ_1 and λ_2, are equal, $\lambda_1 = \lambda_2$, and $\lambda_3 \neq \lambda_1$. Equations (5.25) then show that $\lambda_1 = (A/6)^{\frac{1}{2}} = \lambda_2$, and $\lambda_3 = -(2A/3)^{\frac{1}{2}}$, and that $A^3 = 6B^2 \neq 0$. A Weyl tensor is of type I if and only if $\lambda_1 \neq \lambda_2 \neq \lambda_3$, and hence $A^3 \neq 6B^2$.

The classification given above is invariant under a change of frame. This is so since if $\underline{\chi}$ is an eigenvector of Ψ with eigenvalue λ then, because of the transformation law (5.15), $P\underline{\chi}$ is an eigenvector of Ψ' with the same eigenvalue. Conversely, if χ' is an eigenvector of Ψ', then $P^t\underline{\chi}'$ is an eigenvector of Ψ with the same eigenvalue. One can also show that if two Weyl tensors can be transformed to each other by a change of a basis, then they must be of the same type and have the same eigenvalues. The converse is also correct: If Ψ and Ψ' are of the same type and have the same eigenvalues, they can be transformed one into the other. This result enables us to put a matrix Ψ, corresponding to a non-zero Weyl tensor, in one of five canonical forms by choosing the spin frame in an appropriate way. Thus, every element of the space E_5 can be put into one of five standard forms.

5.5 Canonical forms

We can assume without loss of generality that $\psi_0 \neq 0$, since otherwise a transformation of the type (5.20) will allow us to make ψ_0 non-zero. We now consider ψ'_4 as a quartic in z, ψ'_3 a cubic in z, etc., and notice that ψ'_3, ψ'_2, ψ'_1, and ψ'_0 are proportional to the first, second, third and fourth derivative, respectively of ψ'_4 with respect to z. If ψ'_4 has a double root we make ψ'_3 vanish simultaneously with ψ'_4 by choosing z to be this root. If ψ'_4 has a triple root, we make ψ'_4, ψ'_3, ψ'_2 vanish simultaneously by choosing z to be this triple root. If ψ'_4 has a quadruple root, choosing z as this root will make ψ'_4, ψ'_3, ψ'_2, ψ'_1 zero. One then finds that a necessary and sufficient condition for ψ'_4 to have a quadruple of a triple root is $A = B = 0$. For one or two double roots the condition is $A^3 = 6B^2 \neq 0$, for no multiple roots it is $A^3 \neq 6B^2$.

After the transformation (5.18) has been performed, let us drop the primes from the components of the Weyl spinor. Equation (5.18) can now be followed up by a transformation of the type (5.20) describing a null rotation about n_α. If the quartic in (5.18) has a quadruple root, allowing us to make ψ_4, ψ_3, ψ_2 and ψ_1 vanish, no further transformation is necessary. The Weyl spinor is in the standard form

$$(\psi_0, 0, 0, 0, 0) \text{ with } \psi_0 \neq 0 \quad . \tag{5.26}$$

If the quartic had a triple root, so that $\psi_4 = \psi_3 = \psi_2 = 0$, $\psi_1 \neq 0$, then $\psi'_4 = \psi'_3 = \psi'_2 = 0$, $\psi'_1 = \psi_1$ and ψ'_0 can be made to vanish by choosing $z = -(\psi_0/4\psi_1)$, yielding the standard form

$$(0, \psi_1, 0, 0, 0) \text{ with } \psi_1 \neq 0 \quad . \tag{5.27}$$

If the quartic had a double root, so that $\psi_4 = \psi_3 = 0$, $\psi_2 \neq 0$, then $\psi'_4 = \psi'_3 = 0$, $\psi'_2 = \psi_2$. ψ'_0 is a quadratic in z and can be made to vanish by choosing z to be one of its roots. If this root is a double root ψ'_1 will vanish also for this choice of z, otherwise it will not. The former case occurs if the quartic had two double roots, the latter if it had one double root and two single ones. To show this is easy, but tedious. Thus we get, dropping the primes, the standard forms

$$(0, 0, \psi_2, 0, 0) \text{ with } \psi_2 \neq 0 \quad . \tag{5.28}$$

and

$$(0, \psi_1, \psi_2, 0, 0) \text{ with } \psi_1 \neq 0,\ \psi_2 \neq 0 \quad . \tag{5.29}$$

If the quartic has only single roots, then $A^3 \neq 6B^2$. This is also the condition that the quartic in (5.20) have no repeated roots. Hence only one of ψ'_0, ψ'_1, ψ'_2, ψ'_3 can be made to vanish by an appropriate choice of z. Thus we see that we can find a spin-frame ℓ_A, n_A which induces a corresponding basis (5.6) in E_5 such that the components of the given type I Weyl spinor take on the standard form

$$(0, \psi_1, \psi_2, \psi_3, 0) \tag{5.30}$$

where ψ_1, ψ_2, ψ_3 are all non-zero and satisfy the condition

$$A^3 - 6B^2 = \tfrac{1}{2}\psi_1\psi_3\ (9\psi^2_2 - 16\psi_1\psi_3) \neq 0 \quad .$$

The Weyl spinors (5.26) - (5.29) are, respectively, of type N, III, D, II, as can be seen by finding the corresponding matrices from (5.9) and calculating the eigenvectors and eigenvalues. The results are shown in the following table. Note that the algebraically special types are characterized by the existence of a null eigenbispinor... All the eigenbispinors of a type I Weyl spinor are non-null, as is best shown from the canonical form of the corresponding matrix (given below).

TABLE II

Equation	Eigenvalues	Eigenbispinors	Type
(5.26)	0, 0, 0	$\ell_{(A}n_{B)}$, $n_A n_B$	N
(5.27)	0, 0, 0	$n_A n_B$	III
(5.28)	$\frac{1}{2}\psi_2, \frac{1}{2}\psi_2$	$\ell_A \ell_B$, $n_A n_B$	D
	$-\psi_2$	$\ell_{(A}n_{B)}$	
(5.29)	$\frac{1}{2}\psi_2, \frac{1}{2}\psi_2$	$n_A n_B$	II
	$-\psi_2$	$\ell_{(A}n_{B)} - \frac{1}{3}\frac{\psi_1}{\psi_2} n_A n_B$	

For each type of Weyl spinor, we have found a standard form. For example, a Weyl spinor of type D can be put into the form (5.28) by an appropriate choice of spin-frame. Our method is essentially the matrix method; the matrix, however, was obtained using spinors. Corresponding to each type there is a canonical form of the matrix which is obtained by choosing the spin-frame appropriately. These canonical forms are listed below with the components of the Weyl spinor in the basis (5.6) also given:

$$\begin{pmatrix} 0 & 0 & 0 \\ 0 & 1 & i \\ 0 & i & -1 \end{pmatrix} \quad \text{for type N,} \quad (4,\ 0,\ 0,\ 0,\ 0)$$

$$\begin{pmatrix} 0 & 1 & 0 \\ 1 & 0 & i \\ 0 & i & 0 \end{pmatrix} \quad \text{for type III,} \quad (2, -i,\ 0,\ -i,\ -2)$$

$$\begin{pmatrix} \lambda_1 & 0 & 0 \\ 0 & \lambda_2+1 & i \\ 0 & i & \lambda_2-1 \end{pmatrix} \quad \text{for type II,} \quad (4,\ 0,\ 2\lambda_2,\ 0,\ 0)$$

$$\begin{pmatrix} -2\lambda_1 & 0 & 0 \\ 0 & \lambda_1 & 0 \\ 0 & 0 & \lambda_1 \end{pmatrix} \quad \text{for type D, } (0,\ 0,\ 2\lambda_1,\ 0,\ 0)$$

$$\begin{pmatrix} -(\lambda_1+\lambda_2) & 0 & 0 \\ 0 & \lambda_1 & 0 \\ 0 & 0 & \lambda_2 \end{pmatrix} \quad \text{for type I, } (\lambda_1-\lambda_2,\ 0,\ \lambda_1+\lambda_2,\ 0, \lambda_1-\lambda_2).$$

λ_1 and λ_2 are eigenvalues.

The standard forms for each type can be obtained from the approriate canonical form by a spin-frame transformation and vice versa. For instance, the type III Weyl spinor (2, -i, 0, -i, -2) can be transformed into (0, i, 0, 0, 0) by performing transformation (5.18) with $z = i$ followed by (5.20) with $z = i/2$. Conversely, performing (5.20) with $z = -\psi_1/2$ followed by (5.18) with $z = 1/\psi_1$, followed by (5.19) with $z^2 = i/\psi_1$ transforms $(0, \psi_1, 0, 0, 0)$ into (2, -i, 0, -i, -2).

5.6 Spinor method

The spinor method of classifying a Weyl tensor is analogous to that of classifying a bivector. The expression $\psi_{ABCD}\xi^A\xi^B\xi^C\xi^D$ can be written as a quartic polynomial in $C = \xi^0/\xi^1$:

$$\psi_{ABCD}\xi^A\xi^B\xi^C\xi^D = (\xi^1)^4\ [\psi_{0000}C^4 + 4\psi_{1000}C^3 + 6\psi_{1100}C^2 + \\ + 4C\psi_{0111} + \psi_{1111}] \qquad (5.31)$$

$$= (\xi^1)^4\ [\psi_0 C^4 + 4\psi_1 c^3 + 6\psi_2 C^2 + 4\psi_3 C + \psi_4]$$

where the dyad components are taken with respect to some spin-frame. This quartic polynomial can be factored:

$$\psi_{ABCD}\xi^A\xi^B\xi^C\xi^D = (\xi^1)^4(\alpha_1 C+\alpha_2)(\beta_1 C+\beta_2)(\gamma_1 C+\gamma_2)(\delta_1 C+\delta_2)$$

$$= \alpha_A\xi^A\beta_B\xi^B\gamma_C\xi^C\delta_D\xi^D \quad ,$$

where the spinors α_A, β_A, γ_A, δ_A, are determined up to a constant of proportionality. Since ψ_{ABCD} is symmetric,

$$\psi_{ABCD} = \alpha_{(A}\beta_B\gamma_C\delta_{D)} \quad . \qquad (5.32)$$

The directions of the null vectors corresponding to these spinors are called principal null directions. If two or more coincide, the Weyl spinor is said to be algebraically special; if they are distinct it is algebraically general. The Weyl spinor is said to be of type N if all four spinors coincide; of type III if three of the spinors coincide; of type II if two coincide with the remaining two distinct; of type D if the spinors coincide in pairs; of type I if the spinors are all distinct. This is usually expressed in the Penrose diagram

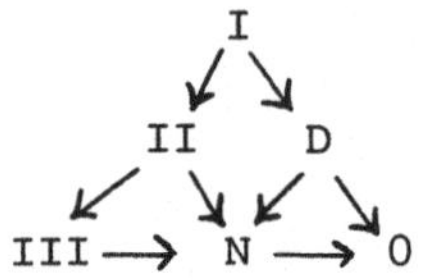

Fig. 1. Penrose diagram.

where A → B indicates that type B is obtained from type A by the confluence of two principal null directions. Type 0, a vanishing Weyl spinor, has been included here for completeness. That this way of classifying Weyl spinors is equivalent to the matrix method becomes clear when we note that the quartic polynomial appearing in (5.31) is precisely the one that was under discussion in the previous section.

5.7 Tensor method

The tensor method depends on the fact that for every Weyl tensor there exist four null directions $\xi_\mu \neq 0$, some of which may coincide, satisfying the equation

$$\xi_{[\alpha}C^{+}{}_{\beta]\gamma\delta[\rho}\xi_{\sigma]}\,\xi^{\gamma}\xi^{\delta} = 0 \quad . \tag{5.33}$$

A Weyl tensor is of Petrov type I if the four directions are dintinct; of type II if two coincide with the remaining pair distinct; of type D if they coincide in pairs; of type III if three directions coincide; of type N if all four directions coincide. Equation (5.33) is equivalent to the spinor equation

$$\psi_{ABCD}\xi^{A}\xi^{B}\xi^{C}\xi^{D} = 0 \quad .$$

Our discussion following (5.31) shows that the tensor method is equivalent to the matrix and spinor methods.

6. ISOTOPIC SPIN AND GAUGE FIELDS

In ordinary gauge invariance of a charged field which is described by a complex wave function ψ, a change of gauge [16] means a change of phase factor $\psi \to \psi'$, $\psi' = (\exp i\alpha)\psi$, a change that is devoid of any physical consequences. Since ψ depends on space-time points, the relative phase factor of ψ at two different points is completely arbitrary and α is, accordingly, a function of space-time. In other words, the arbitrariness in choosing the phase factor is local in character.

To preserve invariance, it is then necessary to counteract the variation of the phase α with space-time coordinates by introducing the electromagnetic potentials $A_\mu(x)$ which change under a gauge transformation as

$$A'_\mu = A_\mu + \frac{1}{e}\frac{\partial\alpha}{\partial x^\mu} ,$$

and to replace the derivative of ψ by a "covariant derivative" with the combination $(\partial_\mu + ieA_\mu)\psi$.

6.1 Isotopic spin

Historically, an <u>isotopic spin</u> parameter was first introduced by Heisenberg [17] in 1932 to describe the two charge states, namely neutron and proton, of a nucleon. The idea that the neutron and proton correspond to two states of the same particle was suggested at the time by the fact that their masses are nearly equal, and that the light stable even nuclei contain equal numbers of them. Later on it was pointed out that the p - p and n - p interactions are approximately equal in the ^1S state [18,19], and consequently it was assumed that the equality holds also in the other states available to both the n - p and p - p systems. Under such an assumption one arrives at the concept of a <u>total</u> isotopic spin [20] which is concerved in nucleon-nucleon interactions. Experiments on the energy levels of light nuclei strongly suggest that this assumption is indeed correct [21]. This implies that all strong interactions, such as the pion-nucleon interaction, should also satisfy the same conservation law. This, and the fact that there are three charge states for the pion, and that pions can be coupled to the nucleon field singly, lead to the conclusions that pions have isotopic spin unity. A verification of this conclusion was found in experiments which compare the differential cross section of the process $n + p \to \pi^\circ + d$ with that of the previously measured process $p+p \to \pi^+ + d$ [22].

6.2 Conservation of isotopic spin and invariance

The conservation of isotopic spin is identical with the requirement that all interactions be invariant under isotopic spin rotation, when electromagnetic interactions are neglected. This means that the orientation of the isotopic spin has no physical significance. Differentiation between a neutron and a proton is then an arbitrary process. Thus arbitrariness is subject to the limitation that once one chooses what to call a proton and what to call a neutron at one space-time point, one is then not free to make any other choices at other space-time points. It also seems not to be consistent with the localized field concept which underlies the usual physical theories.

6.3 Isotopic spin and gauge fields

The possibility of requiring that all interactions be invariant under independent rotations of the isotopic spin at all space-time points, so that the relative orientation of the isotopic spin at two space-time points becomes physically meaningless, was accordingly explored by Yang and Mills [23]. They introduced isotopic gauge as an arbitrary way of choosing the orientation of the isotopic spin axes at all space-time points, in analogy with the electromagnetic gauge with represents an arbitrary way of choosing the complex phase factor of a charged field at all space-time points. This suggests that all physical processes, which do not involve the electromagnetic field, be invariant under the isotopic gauge transformation $\psi \to \psi'$, $\psi' = S^{-1}\psi$, where S represents a space-time dependent isotopic spin rotation which is a 2 x 2 unitary matrix with determinant unity, i.e., an element of the group SU_2.

In an entirely similar manner to what is done in electrodynamics, Yang and Mills introduced a B-field in the case of the isotopic gauge transformation to counteract the dependence of the matrix S on the space-time coordinates. Accordingly, and in analogy with the electromagnetic case, all derivatives of the wave function ψ describing a field with isotopic spin $\frac{1}{2}$ should appear as "covariant derivatives" of the form $(\partial_\mu - iB_\mu)\psi$, where B_μ are four 2 x 2 Hermitian matrices. The field equations satisfied by the twelve independent components of the B-field, which is called the b field, and their interaction with any field having an isotopic spin, are fixed just as in the electromagnetic case.

6.4 Isotopic gauge transformation

Under an isotopic transformation, a two-components wave function ψ describing a field with isotopic spin $\frac{1}{2}$ transforms according to

$$\psi = S\psi' \quad . \tag{6.1}$$

Invariance then requires that the covariant derivative expression transforms as $S(\partial_\mu - iB'_\mu)\psi' = (\partial_\mu - iB_\mu)\psi$. When combined with Equation (6.1), we obtain the isotopic gauge transformation of the 2 x 2 potential matrix B_μ:

$$B'_\mu = S^{-1}B_\mu S + S^{-1}\partial_\mu S \quad . \tag{6.2}$$

In analogy to the procedure of obtaining gauge invariant

field strengths in the electromagnetic case, Yang and Mills define their field as

$$F_{\mu\nu} = \partial_\nu B_\mu - \partial_\mu B_\nu + [B_\mu, B_\nu] \quad , \tag{6.3}$$

where the commutator $[B_\mu, B_\nu] = B_\mu B_\nu - B_\nu B_\mu$. Under the transformation (6.1) the 2 x 2 field matrix (6.3) transforms as

$$F'_{\mu\nu} = S^{-1} F_{\mu\nu} S \quad . \tag{6.4}$$

Now Equation (6.2) is valid for any S and its corresponding B_μ. Furthermore, the matrix $S^{-1}\partial S/\partial x^\mu$ appearing in Equation (6.2) is a linear combination of the isotopic spin "angular momentum" matrices T^i, i = 1, 2, 3, corresponding to the isotopic spin of the field ψ under consideration. Accordingly, the matrix B_μ itself must also contain a linear combination of the matrices T^i; any part of B_μ in addition to this, denote it by B_μ, is a scalar or tensor combination of the T's, and must transform by the homogeneous part of (6.2), $B_\mu = S^{-1} B_\mu\, S$. Such a field is extraneous and was allowed by the very general form we took for the B-field but is irrelevant to the question of isotopic gauge. Therefore, the relevant part of the B-field can be written as a linear combination of the matrices T^i:

$$B_\mu = \underline{b}_\mu \cdot \underline{T} \quad , \tag{6.5}$$

where bold-face letters denote 3-component vectors in the isotopic space.

The isotopic-gauge covariant field matrices $F_{\mu\nu}$ can also be expressed as a linear combination of the T's. One obtains

$$F_{\mu\nu} = \underline{f}_{\mu\nu} \cdot \underline{T} \quad , \tag{6.6}$$

where

$$\underline{f}_{\mu\nu} = \frac{\partial \underline{b}_\mu}{\partial x^\nu} - \frac{\partial \underline{b}_\nu}{\partial x^\mu} - 2\, \underline{b}_\mu \times \underline{b}_\nu \quad . \tag{6.7}$$

One notices that $\underline{f}_{\mu\nu}$ transforms like a vector under an isotopic gauge transformation. The corresponding transformation of $\underline{b}_\mu$ is cumbersome. Under infinitesimal isotopic gauge transformation, $S = 1 - i\underline{T}\cdot\underline{\delta\omega}$. Then

$$\underline{b}'_\mu = \underline{b}_\mu + 2\underline{b}_\mu \times \delta\underline{\omega} + \partial\underline{\delta\omega}/\partial x^\mu \quad . \tag{6.8}$$

6.5 Field equations

In analogy to the electromagnetic case one can write down an isotopic gauge invariant Lagrangian density:

$$\mathcal{L} = -\frac{1}{4}\operatorname{Tr} F_{\mu\nu}F^{\mu\nu} \tag{6.9}$$

One can also include a field with isotopic spin ½ to obtain the following total Lagrangian density:

$$\mathcal{L} = -\frac{1}{4}\operatorname{Tr} F_{\mu\nu}F^{\mu\nu} - \bar{\psi}\gamma_\mu(\partial_\mu - iB_\mu)\psi - m\bar{\psi}\psi \quad . \tag{6.10}$$

The equations of motion obtained from the Lagrangian (6.10) are [24]:

$$\partial \underline{f}_{\mu\nu}/\partial\chi^\nu + 2(\underline{b}_\nu \times \underline{f}_{\mu\nu}) + \underline{J}_\mu = 0 \quad , \tag{6.11}$$

$$\gamma_\mu(\partial_\mu - i\underline{\sigma}\cdot\underline{b}_\mu)\psi + m\psi = 0 \quad , \tag{6.12}$$

where

$$\underline{J}_\mu = i\bar{\psi}\gamma_\mu\underline{\sigma}\psi \quad . \tag{6.13}$$

Since the divergence of J_μ does not vanish, one may define

$$\tilde{\underline{J}}_\mu = \underline{J}_\mu + 2\underline{b}_\nu \times \underline{f}_{\mu\nu} \quad , \tag{6.14}$$

which leads to the equation of continuity,

$$\partial\tilde{\underline{J}}_\mu/\partial x^\nu = 0 \quad . \tag{6.15}$$

Equation (6.15) guarantees that the total isotopic spin

$$T = \int \tilde{J}_0 \, d^3\chi \tag{6.16}$$

is independent of time and Lorentz transformation.

6.6 Nonlinearity of the field equations

Equation (6.14) shows that the isotopic spin arises from both the spin ½ field J_μ and from the $\underline{b}_\mu$ field itself. This fact makes the field equations for the B-field nonlinear, even in the absence of the spin ½ field. The situation here is different from that of the electromagnetic case whose field is chargeless, and hence satisfies linear equations.

6.7 Internal holonomy group of gauge fields

Group-theoretical considerations concerning the gauge theory discussed above involved so far the gauge group SU_2 only. In addition to the possibility of generalizing the gauge group to others [25], there exists another group which is defined by the potential B_μ [26]. One arrives at this group by the observation that the potential can be used to define "parallel displacement" of multiplets ψ at neighbouring points in the same way as the Christoffel symbols define parallelity of vectors in Riemann space. By making a parallel displacement of multiplets around a closed curve in space-time, one obtains a linear transformation of multiplets. Doing this for all closed curves passing through a given point x^μ results in a continuous set of linear multiplet transformations at x^μ. This set turns out to be a Lie group, called the internal holonomy group, in analogy to the ordinary holonomy group in the Riemann space [27]. To see this we proceed as follows.

We call the multiplets $\psi(x^\mu)$ and $\psi(x^\mu + dx^\mu)$ equivalent if

$$\psi(x^\mu + dx^\mu) - \psi(x^\mu) = B_\alpha(x^\mu)\psi(x^\mu)dx^\alpha \quad . \tag{6.17}$$

This equivalence relation which is invariant under the gauge transformation (6.1) because of Equation (6.2) can be used to execute an equivalence displacement of multiplets ψ along a curve in space-time. The question is whether such equivalence displacement is path dependent. If it is not, we have

$$\nabla_\kappa\psi = \partial_\kappa\psi - B_\kappa\psi = 0 \quad , \tag{6.18}$$

which leads to the integrability condition

$$\nabla_{[\lambda}\nabla_{\kappa]}\psi = -\tfrac{1}{2}F_{\lambda\kappa}\psi = 0 \quad . \tag{6.19}$$

Since this is true everywhere for linearly independent internal vectors ψ, it follows that $F_{\lambda\kappa} = 0$. Accordingly, in order to have nonvanishing gauge fields the equivalence transport of ψ must be path dependent. In the same way one can show that if the field $F_{\mu\nu}$ vanishes everywhere, then the potential B_μ can be transformed away by a gauge transformation (see Problem 17).

Now let C be a closed, piecewise, continuously differentiable, and has a sense of circumscription. Taking a multiplet ψ around C by equivalence displacement results in a linear transformation H(C):

$$\psi' = H(C)\psi \quad . \tag{6.20}$$

Doing this for all closed curves C through a point x one gets a set $\mathcal{H}$ of linear internal transformations. The inverse of H(C) is produced by equivalence displacement around C in the opposite direction. The composition $H(C_2)H(C_1)$ is the element $H(C_1 + C_2)$, where $C_1 + C_2$ describes the loop consisting of C_1 and C_2. Hence $\mathcal{H}$ is a group which is a subgroup of the full complex linear group GL(n,c). It is a connected Lie group. This is the internal holonomy group. It follows that the internal holonomy groups at different points are isometric (Problem 18).

Finally, an interesting result is obtained if we consider the relation between the internal holonomy group $\mathcal{H}$ and the gauge group

G. We will leave it to the reader (Problem 19) to show that compatibility of the groups $\mathcal{H}$ and G requires that either (1) $\mathcal{H}$ be a subgroup of G, or (2) G be an invariant subgroup of $\mathcal{H}$.

7. LORENTZ INVARIANCE AND THE GRAVITATIONAL FIELD

In the last section we saw that the existence of the electromagnetic field and the Yang-Mills field can be related to invariance properties. Thus, if the Lagrangian density is invariant under phase transformations $\psi \to (\exp i\alpha)\psi$, and if we wish to make it invariant under the general gauge transformations for which α is a function of x, then it is necessary to introduce a new field A_μ which transforms according to $A_\mu \to A_\mu + e^{-1}\partial_\mu\alpha$, and to replace the derivative of ψ by a "covariant derivative" $(\partial_\mu + ieA_\mu)\psi$. A similar argument was applied to isotopic spin rotations, by Yang and Mills, to yield a triplet of vector fields. It is thus an attractive idea to relate the existence of the gravitational field to Lorentz invariance.

7.1 Homogeneous Lorentz group and the gravitational field

Utiyama [28] has proposed a method which leads to the introduction of 24 field variables $A^{ij}{}_\mu$ by considering the homogeneous Lorentz transformations of the group O(1,3) specified by six parameters. One starts by assuming that the action integral

$$I = \int L(\psi^A, \psi^A{}_{,k})d^4x \quad , \tag{7.1}$$

where $\psi_{,k} = \partial_k\psi$, is invariant under Lorentz transformations. Besides the x-system one introduces an arbitrary system of curvilinear coordinates u^μ. We will use Latin and Greek indices to represent quantities defined with respect to the x-system (local Lorentz frame) and the u-system respectively. The square of the invariant length of the infinitesimal line element is given by

$$ds^2 = \eta_{ik}\, dx^i dx^k = g_{\mu\nu} du^\mu du^\nu \quad ,$$

where η_{ik} is the Minkowskian metric and $g_{\mu\nu}(u) = (\partial x^i/\partial u^\mu)(\partial x^k/\partial u^\nu)\eta_{ik}$. Defining the functions

$$h^k{}_\mu(u) = \partial x^k/\partial u^\mu \quad , \quad h_k{}^\mu(u) = \partial u^\mu/\partial x^k \quad , \tag{7.2}$$

then gives $\eta_{k\ell}h^k{}_\mu h^\ell{}_\nu = g_{\mu\nu}(u)$, $g_{\mu\nu}h_k{}^\mu h_\ell{}^\nu = \eta_{k\ell}$, $h_k{}^\mu h^\ell{}_\mu = \delta_k{}^\ell$, $h_k{}^\mu h^k{}_\nu = \delta^\mu{}_\nu$, and $\det g_{\mu\nu} = g = -h^2 = -(\det h^k{}_\mu)^2$. Raising or lowering both kind of indices is made by means of the matrices $g^{\mu\nu}$, $g_{\mu\nu}$ or $\eta_{k\ell}$ and $\eta^{k\ell}$. Under Lorentz transformation $x^k \to x^k + \varepsilon^k{}_\ell x^\ell$, where $\varepsilon^{k\ell} = -\varepsilon^{\ell k}$ are infinitesimal parameters, one has $h_k{}^\mu \to h_k{}^\mu + \delta h_k{}^\mu$, with $\delta h_k{}^\mu = -\varepsilon^\ell{}_k h_\ell{}^\mu$. Using the h function we can transform a world tensor into a corresponding local tensor defined with respect to the local frame, and vice versa. For example, $\psi^k(u) = h^k{}_\mu(u)\psi^\mu(u)$ and $\psi^\mu(u) = h_k{}^\mu(u)\psi^k(u)$, where $\psi^k(u)=\psi^k(x(u))$.

Accordingly, we can rewrite the action integral as

$$I = \int \mathcal{L}\ (\psi^A(u),\ \psi^A{}_{,\mu}(u),\ h^k{}_\mu(u))d^\mu u \quad . \qquad (.73)$$

where $\mathcal{L} = L(\psi^A(u),\ h_k{}^\mu(u)\psi^A{}_{,\mu}(u))h$, and $\psi^A{}_{,\mu} = \partial\psi^A(u)/\partial u^\mu$.

7.2 Invariance of the action integral

The action integral I is invariant under: (1) the Lorentz transformation which yields

$$\delta h^k{}_\mu = \varepsilon^k{}_\ell h^\ell{}_\mu \quad ,$$
$$\delta\psi^A = \tfrac{1}{2}T_{(k\ell)}{}^A{}_B\psi^B\varepsilon^{k\ell} \quad , \qquad (7.4)$$

where u^μ is unchanged, and $T_{(k\ell)}{}^A{}_B$ is the AB matrix element of the infinitesimal generator of the Lorentz group. The matrix $T_{(k\ell)}$ satisfies

$$[T_{(k\ell)},\ T_{(mn)}] = \tfrac{1}{2}f_{k\ell}{}^{ab}{}_{mn}T_{(ab)},\quad T_{(k\ell)} = -T_{(\ell k)} \quad ;$$

(2) the general point transformation

$$u^\mu \rightarrow u^\mu + \lambda^\mu(u) \quad ,$$

where $\lambda^\mu(u)$ is an arbitrary function of u, which yields

$$\delta h^k{}_\mu = -(\partial\lambda^\nu/\partial u^\mu)h^k{}_\nu \quad ,$$
$$\delta\psi^A(u) = \psi'^A(u') - \psi^A(u) = 0 \quad , \qquad (7.5)$$
$$\delta\psi^A{}_{,\mu} = -(\partial\lambda^\nu/\partial u^\mu)\psi^A{}_{,\nu} \quad .$$

In the following, the set $h^k{}_\mu$ shall be considered as 16 independent given functions.

7.3 Generalized Lorentz transformation

We now generalize the Lorents transformation into one in which ε^{ik} are replaced by arbitrary functions $\varepsilon^{ik}(u)$. Under this "generalized Lorentz transformation" we assume that ψ^A and $h^k{}_\mu$ transform as

$$\delta\psi^A = \tfrac{1}{2}\varepsilon^{k\ell}(u)\ T_{(k\ell)}{}^A{}_B\psi^B \quad ,$$
$$\delta h^k{}_\mu = \varepsilon^k{}_\ell(u)h^\ell{}_\mu \quad . \qquad (7.6)$$

Then in order that I remains invariant under (7.6), it is necessary to introduce a new field $A^{k\ell}{}_\mu(u) = -A^{\ell k}{}_\mu(u)$ with the following transformation law:

$$\delta A^{k\ell}{}_\mu = \varepsilon^k{}_m A^{m\ell}{}_\mu + \varepsilon^\ell{}_m A^{km}{}_\mu + \partial\varepsilon^{k\ell}/\ell u^\mu \quad . \tag{7.7}$$

The new Lagrangian density is then given by

$$\mathcal{L}(\psi^A, \nabla_\mu \psi^A, h^k{}_\mu) = hL(\psi^A, h_k{}^\mu \nabla_\mu \psi^A) \quad , \tag{7.8}$$

where

$$\nabla_\mu \psi^A = \partial\psi^A/\partial u^\mu - (1/2)A^{k\ell}{}_\mu T_{(k\ell)}{}^A{}_B \psi^B \quad . \tag{7.9}$$

We now take as our basic space-time, some Riemannian space whose metric is $g_{\mu\nu}(u) = h^k{}_\mu h_{k\nu}$ and whose affine connection is $\Gamma^\lambda{}_{\mu\nu} = (1/2)g^{\lambda\sigma}(\partial g_{\sigma\mu}/\partial u^\nu + \partial g_{\nu\sigma}/\partial u^\mu - \partial_{\mu\nu}/\partial u^\sigma)$. In order to obtain the relationship between $A^{k\ell}{}_\mu$ and $h^k{}_\mu$, let us take for ψ^A the local tensor $\psi^{k\ell}$. Then from Equation (7.9) we obtain

$$\nabla_\mu \psi^{k\ell} = \partial\psi^{k\ell}/\partial u^\mu - A^{km}{}_\mu \psi_m{}^\ell - A^{\ell m}{}_\mu \psi^k{}_m \quad .$$

Accordingly, replacing $\psi^{k\ell}$ by $\psi^{k\nu} = h_m{}^\nu \psi^{km}$, we obtain

$$\nabla_\mu \psi^{k\nu} = \partial\psi^{k\nu}/\partial u^\mu - A^{km}{}_\mu \psi_m{}^\nu + \Gamma'{}_{\rho\mu}{}^\nu \psi^{k\rho} \quad , \tag{7.10}$$

where

$$\Gamma'{}_{\nu\mu}{}^\rho = h_\ell{}^\rho(\partial h^\ell{}_\nu/\partial u^\mu) - h_k{}^\rho h_{\ell\nu} A^{k\ell}{}_\mu \quad . \tag{7.11}$$

Equation (7.10) shows that the covariant derivative obtained here is the usual one where for Greek indices the Γ' appear instead of the usual affinity Γ, and for Latin indices the $A^{k\ell}{}_\mu$ must be inserted instead of Γ. The relation (7.10) can be generalized (see Problem 20). One also finds, under the <u>ad hoc</u> assumption that Γ' is symmetric in its lower two indices (see Problem 21), that

$$\Gamma'{}_{\mu\nu}{}^\rho = \tfrac{1}{2}g^{\rho\sigma}\left(\frac{\partial g_{\sigma\mu}}{\partial u^\nu} + \frac{\partial g_{\nu\sigma}}{\partial u^\mu} - \frac{\partial g_{\mu\nu}}{\partial u^\sigma}\right) = \Gamma_{\mu\nu}{}^\rho \quad , \tag{7.12}$$

and

$$h_\ell{}^\rho \frac{\partial h^\ell{}_\nu}{\partial u^\mu} - A^\rho{}_{\nu\mu} \quad \Gamma_{\nu\mu}{}^\rho \quad , \tag{7.13}$$

where $A^\rho{}_{\nu\mu} = h_k{}^\rho h_{\ell\nu} A^{k\ell}{}_\mu$.

Accordingly, we have obtained a general expression for the covariant derivative. For example, if we take for the field ψ^A the spinor field ψ, we obtain

$$\nabla_\mu \psi = \partial\psi/\partial x^\mu - (i/4)A^{k\ell}{}_\mu [\gamma_k, \gamma_\ell]\psi \quad ,$$

where γ_k are the usual Dirac γ matrices.

7.4 Free field case

Let us now consider the Lagrangian density $\mathcal{L}_0$ for the free field, i.e. the case without the multiplet ψ^A. The Lagrangian density $\mathcal{L}_0$ is a function of the functions h and A, $\mathcal{L}_0(h^k{}_\mu,\ A^{k\ell}{}_\mu,\ \partial A^{k\ell}{}_\mu/\partial u^\nu)$. Since $\mathcal{L}_0$ should be invariant under the "generalized Lorents transformation", it follows that $\mathcal{L}_0$ must depend on the field A through the form $\mathcal{L}_0(h^k{}_\mu,\ F^{k\ell}{}_{\mu\nu})$, where F is defined by

$$F^{k\ell}{}_{\mu\nu} = \frac{\partial A^{k\ell}{}_\nu}{\partial u^\mu} - \frac{\partial A^{k\ell}{}_\mu}{\partial u^\nu} - \frac{1}{4} f_{ab}{}^{k\ell}{}_{mn}(A^{ab}{}_\mu A^{mn}{}_\nu - A^{ab}{}_\nu A^{mn}{}_\mu)$$

$$= \frac{\partial A^{k\ell}{}_\nu}{\partial u^\mu} - \frac{\partial A^{k\ell}{}_\mu}{\partial u^\nu} + A^{kb}{}_\mu A^\ell{}_{b\nu} - A^{kb}{}_\nu A^\ell{}_{b\mu} \quad . \tag{7.14}$$

One can then show that (Problem 22):

$$F^{k\ell}{}_{\mu\nu} = h^{\ell\lambda} h^k{}_\alpha R^\alpha{}_{\lambda\mu\nu} \quad , \tag{7.15}$$

where $R^\alpha{}_{\lambda\mu\nu}$ is the Riemann tensor.

The total Lagrangian density is given by $\mathcal{L}_t = \mathcal{L}(\psi^A,\ \nabla_\mu\psi^A,\ h^k{}_\mu) + \mathcal{L}_0(h^k{}_\mu,\ F^{k\ell}{}_{\mu\nu})$. The field equations for ψ and h are given by

$$\frac{\delta\mathcal{L}}{\delta\psi^A} = 0 \ , \quad \frac{\delta\mathcal{L}_t}{\delta h^i{}_\mu} = 0 \ .$$

The field equations of gravitation are usually obtained from a particular Lagrangian density, $\mathcal{L}_0 = hR$, where R is defined by $R = g^{\mu\nu}R_{\mu\nu} = h_\ell{}^\mu h_k{}^\nu F^{k\ell}{}_{\mu\nu}$, and $R_{\mu\nu} = R^\lambda{}_{\mu\lambda\nu}$. Taking the variation with respect to h gives

$$\frac{\delta\mathcal{L}_0}{\delta h^i{}_\mu} + \frac{\delta\mathcal{L}}{\delta h^i{}_\mu} = 0 \quad ,$$

one obtains

$$\frac{\delta\mathcal{L}_0}{\delta \quad \mu}\, \delta h^i{}_\mu = \frac{\delta\mathcal{L}_0}{\delta g_{\rho\mu}}\, h_{i\rho}\, \delta h^i{}_\mu + \frac{\partial}{\partial u^\mu}\left\{ \frac{\partial\mathcal{L}_0}{\partial g_{\rho\sigma,\mu}}\, h_{i\rho}\, \delta h^i{}_o \right\} ,$$

where

$$\frac{\delta\mathcal{L}_0}{\delta g_{\rho\sigma}} = -h(R^{\rho\sigma} - \tfrac{1}{2} g^{\rho\sigma} R) .$$

Accordingly, the action principal leads to the field equations

$$h(R^{\rho\sigma} - \tfrac{1}{2}g^{\rho\sigma}R) = \tau^{\rho\sigma} \quad , \tag{7.16}$$

where $\tau^{\rho\sigma} = \tau^{\rho}{}_{i}h^{i\sigma}$, and $\tau^{\rho}{}_{i}$ is given by

$$\tau^{\rho}{}_{i} = \delta\mathcal{L}/\delta h^{i}{}_{\rho} \quad .$$

Here $\tau^{\rho\sigma}$ is the symmetric energy-momentum tensor density of the original field ψ. The symmetry character of $\tau^{\rho\sigma}$ can be proved (see Problem 23).

7.5 Poincaré invariance and the gravitational field

Kibble [29] has pointed out that Utiyama's method, discussed above, is a rather unsatisfactory procedure since it is the purpose of the method to supply an argument for introducing the gravitational field variables, including the metric and the affine connections. To overcome these difficulties and make possible the introduction of the vierbein components $h^{i}{}_{\mu}$ as well as the local affine connections $A^{ij}{}_{\mu}$ as new field variables analogous to the electromagnetic potential A_{μ}, Kibble extended Utiyama's discussion and considered the 10-parameter inhomogeneous Lorentz group (Poincaré group) instead of the homogeneous 6-parameter Lorents group. He showed it is then unnecessary to introduce a priori curvilinear coordinates or a Riemannien metric, and that the new field variables introduced as a consequence of the argument include the vierbien components $h_{k}{}^{\mu}$ as well as the local affine connection $A^{ij}{}_{\mu}$. The extended transformations for which the 10 parameters become arbitrary functions of position may be interpreted as general coordinate transformations and rotations of the $h_{k}{}^{\mu}$ field. The Lagrangian density proposed, then yields the free space field equations $R_{\mu\nu} = 0$, but when matter is presented the resultant equations show that there is a difference from the theory of general relativity which arises from the fact that $A^{ij}{}_{\mu}$ appear in the matter field Lagrangian. As a consequence this means that, although the covariant derivative of the metric vanishes, the affine connections $\Gamma^{\lambda}_{\gamma\beta}$ is nonsymmetric.

We will not discuss in details this Poincaré invariant theory. Instead, we will return and formulate the gravitational field equations in a SL(2,C) invariant way, so as to exhibit the gauge aspects of the theory. This is done in the next section.

8. SL(2,C) INVARIANCE AND THE GRAVITATIONAL FIELD

In the last section we discussed the theories of Utiyama and Kibble of applying the Yang-Mills method in order to relate the gravitational field to a generalized gauge field associated with the

Lorentz group, where one starts with flat space and introduces at each point a curved space-time. On the other hand, we saw in Section 2 how spinors are introduced in a Riemannian space, at each space-time point in a tangent two-dimensional complex space. The two procedures are therefore the opposite of each other. It is thus, an attractive idea to relate the two approaches, the one that is based on Yang-Mills method and the other that is based on spinor formalism, to the gravitational field. In this section, it is shown how the theory of general relativity, given in the spinorial form, can be recast into a Yang-Mills-type theory by use of the group SL(2,C). To be more sure, we will not follow the prescription of Utiyama, thus not starting with a Dirac field and going into a curved space-time since, as has been pointed out by Weinberg [30], this is a somewhat arbitrary procedure. Instead, we will reverse Utiyama's procedure since we start with the curved space-time and subsequently introduce at each space-time point a tangent space in which a complex three-dimensional linear space is introduced. Another difference exists between the present theory and that of gauge fields since in the latter case it is the spin affinities that are considered as potentials, whereas the potential matrices here will be defined differently (see Equation (8.2) below). Obviously spin affinities are not space-time vectors whereas the potentials to be defined here are.

8.1 Spin frame gauge

The gravitational field dynamical variables of general relativity can be divided into the sets: (1) the Riemann tensor, decomposed into its irreducible components (the Weyl tensor, the trace-free parts of the Ricci tensor, and the Ricci scalar); (2) the spin coefficients; and (3) a tetrad system of vectors (from which one obtains the metric tensor). They are connected by three sets of first-order partial differential equations which describe the gravitational field.

We will represent the spin coefficients and the components of the Riemann tensor in the form of linear combinations of the infinitesimal generators of the group SL(2,C). This representation is very similar to the way Yang and Mills write their dynamical variables in terms of the Pauli spin matrices. The spin coefficients take the role of the Yang-Mills-like potentials, whereas the Riemann tensor components take the role of the fields.

There is an essential difference, however, between this representation and that of Yang and Mills. The group underlying the symmetry here is SL(2,C) whereas in the Yang-Mills case it is SU_2. The group SL(2,C) seems to fit in with general relativity in a remarkable and natural way, just as 2-component spinors do. This is not an unexpected result since spinors, as we have seen in Section 1, describe the finite-dimensional representation of the group SL(2,C).

We start by introducing at each point of space-time two 2-component spinors $\zeta^A{}_a$, where a = 0, 1, to define a spin frame. Each one of these two spinors might be considered as a complex wave function describing a spin ½ particle, but one assumes nothing as to whether they satisfy any dynamical wave equation. As was done in Section 4 the two spinors $\zeta_a{}^A$ are normalized such as $\zeta_a{}^B \varepsilon_{BA} \zeta_b{}^A = \zeta_{aA}\zeta_b{}^A = \varepsilon_{ab}$, where as usual ε's are the skew-symmetric Levi-Civita symbols defined by $\varepsilon_{01} = 1$. Such a frame has already been discussed in Section 4, where the two spinors were denoted by ℓ^A and n^A.

A spin frame gauge van be defined [31] as an arbitrary way of choosing the orientation of the spin frame axes at all space-time points, in analogy with the isotopic gauge which is an arbitrary way of choosing the orientation of the isotopic spin axes at all points. One then demands that all physical processes be invariant under the spin frame transformation

$$\zeta = S\zeta' \quad , \tag{8.1}$$

where ζ is a 2 x 2 complex matrix whose elements are $\zeta_a{}^A$, and S represents a spin frame rotation which is a 2 x 2 unimodular complex matrix whose elements $S_a{}^b$ are functions of space-time.

An arbitrary spinor $G^{AB'}$ can now be written in terms of the spin frame, $G^{AB'} = G^{ab'}\zeta_a{}^A\zeta_{b'}{}^{B'}$, where $G^{ab'}$ are the dyad components of the spinor $G^{AB'}$ and are given by $G^{ab'} = G_{AB'}\zeta^{aA}\zeta^{b'B'}$. As before, lower-case indices behave the same way algebraically as ordinary spinor indices except when covariant differentiation is applied in which case no term involving an affine connection appears for them. By the same token, the quantity $\nabla_\mu \xi^A$, obtained by taking the covariant derivative of a spinor ξ^A, can also be written in terms of the spin frame as $\nabla_\mu \xi^A = B^b{}_\mu \zeta_b{}^A$, where $B^b{}_\mu$, with b = 0, 1, are some space-time vectors. In particular the last formulae applies to the two spinors $\zeta_a{}^A$ defining the spin frame. This gives $\nabla_\mu \zeta_a{}^A = B_a{}^b{}_\mu \zeta_b$, where $B_a{}^b{}_\mu$, with a, b = 0, 1, are some vectors.

8.2 Potentials and fields

In the Yang-Mills theory, it is the spinor affine connection which are considered as potentials. However, these quantities are not space-time vectors, as is well known, in the Riemann space, and alternative quantities have to be found. Fortunately, such quantities are available. For instance, one can take the vectors $B_a{}^b{}_\mu$, obtained above from the covariant derivatives of $\zeta_a{}^A$, as the potentials. For convenience one rewrites the relation $\nabla_\mu \zeta_a{}^A = B_a{}^b{}_\mu \zeta_b{}^A$ as

$$\nabla_\mu \zeta = B_\mu \zeta \quad , \tag{8.2}$$

where B_μ and ζ are 2 x 2 complex matrices whose elements are $B_a{}^b{}_\mu$

and $\zeta_a{}^A$, respectively. The normalization condition that the two spinors $\zeta_a{}^A$ have to satisfy then implies that the matrix B_μ be traceless and the matrix ζ be unimodular.

The commutator of the covariant derivatives $(\nabla_\nu\nabla_\mu - \nabla_\mu\nabla_\nu)$, when applied on ζ gives $F_{\mu\nu}\zeta$, where

$$F_{\mu\nu} = \partial_\nu B_\mu - \partial_\mu B_\nu + [B_\mu, B_\nu] \tag{8.3}$$

is a 2 x 2 complex traceless matrix whose elements are skew-symmetric tensors. The commutator $[B_\mu, B_\nu] = B_\mu B_\nu - B_\nu B_\mu$. Hence the relation between the F- and B-matrices is identical to that of the Yang-Mills field, Equation (6.3), but with the exception that the potentials are now defined by Equation (8.2) rather than taken as the spinor affine connections as is done in that case. Furthermore, under a change of the spin frame (8.1) one easily finds that B_μ and $F_{\mu\nu}$ transform into

$$B'_\mu = S^{-1}B_\mu S - S^{-1}\partial_\mu S \quad , \tag{8.4}$$

$$F'_{\mu\nu} = S^{-1}F_{\mu\nu}S \quad , \tag{8.5}$$

identically to those of the Yang-Mills field, Equations (6.2) and (6.4), when subjected under an isotopic gauge transformation.

The matrix B_μ defines 12 complex functions, whereas the matrix $F_{\mu\nu}$ defines 18 complex functions. The latter is equivalent to the 20 real components of the Riemann tensor plus the 16 real components of the tetrad field σ^μ_{ab}, (see Equation (8.10) below).

8.3 Spin coefficients as potentials

Since the matrices B_μ and $F_{\mu\nu}$ are traceless, it follows that they both can be written as linear combinations of the infinitesimal generators of the group SL(2,C), similar to the way Yang and Mills write their dynamical variables in terms of the Pauli spin matrices. The infinitesimal generators of the group SL(2,C) are three traceless matrices that can be chosen as [32]:

$$g_1 = \begin{pmatrix} 0 & 0 \\ 1 & 0 \end{pmatrix}, \quad g_2 = \begin{pmatrix} 1 & 0 \\ 0 & -1 \end{pmatrix}, \quad g_3 = \begin{pmatrix} 0 & 1 \\ 0 & 0 \end{pmatrix} \tag{8.6}$$

The matrices g_1, g_2, and g_3 are tangent vectors to the one-parameter subgroups

$$g_1(z) = \begin{pmatrix} 1 & 0 \\ z & 1 \end{pmatrix}, \quad g_3(z) = \begin{pmatrix} e^z & 0 \\ 0 & e^{-z} \end{pmatrix}, \quad g_3(z) = \begin{pmatrix} 1 & z \\ 0 & 1 \end{pmatrix} \tag{8.7}$$

where z is a complex parameter, and satisfy

$$g_m(z_1 + z_2) = g_m(z_1)g_m(z_2); \quad \text{(no summation on m)} \quad ,$$

for m = 1, 2, 3. The matrices (8.7) are identical to those appeared in Equation (4.18) Section 4 (but with a slight change in the notation of $g_2(z)$), and every matrix of the group SL(2,C) can be presented uniquely as product of them. The infinitesimal generators g_1, g_2, and g_3 are obtained from $g_1(z)$, $g_2(z)$, and $g_3(z)$, as usual by

$$g_m = [dg_m(z)/dz]_{z=0} \quad ,$$

and conversely, the matrices $g_1(z)$, $g_2(z)$, and $g_3(z)$ can be expressed in terms of the infinitesimal generators g_1, g_2, and g_3 by

$$g_m(z) = \exp(zg_m) \quad .$$

Accordingly, one can write

$$B_\mu = \underline{b}_\mu \cdot \underline{g} \quad , \tag{8.8}$$

$$F_{\mu\nu} = \underline{f}_{\mu\nu} \cdot \underline{g} \quad , \tag{8.9}$$

where $\underline{g} = (g_1, g_2, g_3)$, and b_μ and $f_{\mu\nu}$ are vectors in the complex 3-dimensional space of SL(2,C).

We now define another set of Hermitian matrices related to the matrix $\tilde{\sigma}^\mu$ (see Section 2) by

$$\sigma^\mu = \zeta\tilde{\sigma}^\mu\zeta^\dagger \quad , \tag{8.10}$$

where $\zeta^\dagger$ is the Hermitian conjugate of ζ. Contrary to $\tilde{\sigma}^\mu$ whose covariant derivative vanishes by definition, the covariant derivative of the matrix σ^μ does not vanish and one has, using Equation (8.2),

$$\nabla_\alpha\sigma^\mu = B_\alpha\sigma^\mu + \sigma^\mu B_\alpha^\dagger \quad . \tag{8.11}$$

The geometrical metric can then be written as $g^{\mu\nu} = \tilde{\sigma}^\mu_{AB'}\tilde{\sigma}^{\nu AB'} = \sigma^\mu{}_{ab'}\sigma^{\nu ab'}$. The elements of the matrix σ^μ define a null tetrad of vectors where $\sigma^\mu_{00'}$ and $\sigma^\mu_{11'}$ are real, whereas $\sigma^\mu_{01'}$ and $\sigma^\mu_{10'}$ are complex, conjugate to each other. Moreover, they satisfy the orthogonality relation $\sigma^\mu{}_{ab'}\sigma_{\mu cd'} = \varepsilon_{ac}\varepsilon_{b'd'}$. These are the same null vectors ℓ^μ, m^μ, $\bar{m}^\mu$, and n^μ introduced in Section 4, where $\sigma^\mu_{00'} = \ell^\mu$, $\sigma^\mu_{01'} = m^\mu$, $\sigma^\mu_{10'} = \bar{m}^\mu$, and $\sigma^\mu_{11'} = n^\mu$.

The three sets of matrices B_μ, $F_{\mu\nu}$, and σ^α describe all of the dynamical variables of the gravitational field. From the B_μ and $F_{\mu\nu}$ one can obtain two new sets of matrices which are just new representations of the B and F matrices:

$$B_{ab'} = \sigma^\mu{}_{ab'}\, B_\mu \quad , \tag{8.12}$$

$$F_{ab'cd'} = \sigma^\mu{}_{ab'}\sigma^\nu{}_{cd'}F_{\mu\nu} \tag{8.13}$$

Again one may write the latter matrices as linear combinations of g:

$$B_{cd'} = \underline{b}_{cd'} \cdot \underline{g} \quad , \tag{8.14}$$

$$F_{ab'cd'} = \underline{f}_{ab'cd'} \cdot \underline{g} \quad , \tag{8.15}$$

where the new four 3-vectors $\underline{b}_{cd'}$ and the six 3-vectors $\underline{f}_{ab'cd'}$ are related to $\underline{b}_\mu$ and $\underline{f}_{\mu\nu}$ of Equations (8.8) and (8.9) by

$$\underline{b}_{cd'} = \sigma^\mu{}_{cd'}\underline{b}_\mu \quad , \quad \underline{f}_{ab'cd'} = \sigma^\mu{}_{ab}\,\sigma^\nu{}_{cd'}\underline{f}_{\mu\nu} \quad .$$

The four 3-vectors $b_{cd'}$ in the complex SL(2,C) space will be denoted by

$$\begin{aligned} \underline{b}_{00'} &= (-\kappa, \varepsilon, \mu) \quad , \quad \underline{b}_{01'} = (-\sigma, \beta, \mu) \quad , \\ \underline{b}_{10'} &= (-\rho, \alpha, \lambda) \quad , \quad \underline{b}_{11'} = (-\tau, \gamma, \nu) \quad . \end{aligned} \tag{8.16}$$

Using Equation (8.14), we see that the four matrices $B_{cd'}$ will then have the form

$$\begin{aligned} B_{00'} &= \begin{pmatrix} \varepsilon & -\kappa \\ \pi & -\varepsilon \end{pmatrix} , \quad B_{01'} = \begin{pmatrix} \beta & -\sigma \\ \mu & -\beta \end{pmatrix} , \\ B_{10'} &= \begin{pmatrix} \alpha & -\rho \\ \lambda & -\alpha \end{pmatrix} , \quad B_{11'} = \begin{pmatrix} \gamma & -\tau \\ \nu & -\gamma \end{pmatrix} \end{aligned} \tag{8.17}$$

when the representation (8.6) is used for g.

The twelve complex functions ε, κ, π, etc., were first introduced by Newman and Penrose [33] and are known in general relativity as spin coefficients. From the point of view of Yang-Mills field theory these same quantities are the potentials the field of which is given by the F-matrices according to Equation (8.3).

8.4 Symmetry of $F_{ab'cd'}$

It is convenient to introduce another matrix, $\tilde{B}_\mu$, connected to the matrix B_μ by a similarity transformation

$$\zeta\tilde{B}_\mu = B_\mu\zeta \quad . \tag{8.18}$$

The new matrix then satisfies

$$\nabla_\mu\zeta = \zeta\tilde{B}_\mu \quad . \tag{8.19}$$

The matrix elements of $\tilde{B}_\mu$ and B_μ are related as follows. If $B_a{}^b{}_\mu$

is the ab element of the matrix B_μ, then $B_A{}^B{}_\mu$ is the AB element of the matrix $\tilde{B}_\mu$. This fact can easily be seen by writing the matrix elements of both sides of Equation (8.18). The left-hand side gives

$$(\zeta\tilde{B}_\mu)_e{}^F = \zeta_e{}^D B_D{}^F{}_\mu \quad ,$$

whereas the right-hand side gives

$$(B_\mu\zeta)_e{}^F = B_e{}^d{}_\mu \zeta_d{}^F \quad .$$

As can be easily seen, both of these expressions are equal to $B_e{}^F{}_\mu$. Hence, while the matrix element indices of B_μ are spinorial.

In the same way we can define another matrix $\tilde{F}_{\mu\nu}$,

$$\zeta\tilde{F}_{\mu\nu} = F_{\mu\nu}\zeta \quad , \tag{8.20}$$

which satisfies

$$(\nabla_\nu\nabla_\mu - \nabla_\mu\nabla_\nu) = \zeta\tilde{F}_{\mu\nu} \tag{8.21}$$

and whose explicit expression is given by

$$\tilde{F}_{\mu\nu} = \nabla_\nu\tilde{B}_\mu - \nabla_\mu\tilde{B}_\nu - [\tilde{B}_\mu, \tilde{B}_\nu] \quad . \tag{8.22}$$

Similar to the potential matrix B_μ, the matrix elements of $F_{\mu\nu}$ and $\tilde{F}_{\mu\nu}$ will be $F_a{}^b{}_{\mu\nu}$ and $F_A{}^B{}_{\mu\nu}$, respectively.

To find the SL(2,C) structure of the matrices $F_{ab'cd'}$, we proceed as follows [34].

Let ξ^P be an arbitrary spinor. Then

$$(\nabla_\nu\nabla_\mu - \nabla_\mu\nabla_\nu)\xi^P = (\nabla_\nu\nabla_\mu - \nabla_\mu\nabla_\nu)\xi^g\zeta_g{}^P = \xi^g(\nabla_\nu\nabla_\mu - \nabla_\mu\nabla_\nu)\zeta_g{}^P \quad .$$

Now using Equation (8.21), we obtain

$$(\nabla_\nu\nabla_\nu - \nabla_\mu\nabla_\nu)\xi^P = \xi^g\zeta_g{}^H F_H{}^P{}_{\mu\nu} \quad .$$

Hence we have

$$(\nabla_\nu\nabla_\mu - \nabla_\mu\nabla_\nu)\xi_Q = F_{PQ\mu\nu}\xi^P \quad ,$$

or equivalently,

$$(\nabla_{AC'}\nabla_{BD'} - \nabla_{BD'}\nabla_{AC'})\xi_Q = F_{PQBD'AC'}\xi^P \quad . \tag{8.23}$$

By decomposing the commutator of differentation on the left-hand side of Equation (8.23), we obtain (see Problem 24):

$$\frac{1}{2}\varepsilon_{C'D'}(\nabla_{AF'}\nabla_B{}^{F'} + \nabla_{BF'}\nabla_A{}^{F'})\xi_Q +$$
$$+ \frac{1}{2}\varepsilon_{AB}(\nabla_{EC'}\nabla^E{}_{D'} + \nabla_{ED'}\nabla^E{}_{C'})\xi_Q = F_{PQBD'AC'}\xi^P \quad . \tag{8.24}$$

But the left-hand side of Equation (8.24) is equal to (see Problem 25):

$$\varepsilon_{C'D'}[\psi_{ABQP} - \Lambda(\varepsilon_{PA}\varepsilon_{BQ} + \varepsilon_{PB}\varepsilon_{AQ})]\xi^P + \varepsilon_{AB}\phi_{QPC'D'}\xi^P \qquad (8.25)$$

where ψ_{ABCD} is the totally symmetric spinor which represents the Weyl spinor, $\phi_{QPC'D'}$ represents the trace-free part of the Ricci spinor having the symmetry

$$\phi_{QPC'D'} = \phi_{PQC'D'} = \phi_{QPD'C} = \bar{\phi}_{C'D'QP} \quad ,$$

and $\Lambda = R/24$, where R is the Ricci scalar.

Accordingly, we obtain

$$F_{PQBD'AC'} = \varepsilon_{C'D'}[\psi_{ABQP} - \Lambda(\varepsilon_{PA}\varepsilon_{BQ} + \varepsilon_{PB}\varepsilon_{AQ})] + \varepsilon_{AB}\phi_{QPC'D'} \quad ,$$

and the same relation holds for lower-case indices:

$$F_p{}^q{}_{bd'ac'} = \varepsilon_{c'd'}[\psi_p{}^q{}_{ab} - \Lambda(\varepsilon_{pa}\delta_b{}^q + \varepsilon_{pb}\delta^q{}_a)] + \varepsilon_{ab}\phi_p{}^q{}_{c'd'} \quad .$$

Using the standard notation

$$\psi_{0000} = \psi_0 \quad , \quad \psi_{0001} = \psi_1 \quad , \quad \psi_{0011} = \psi_2 \quad ,$$

$$\psi_{0111} = \psi_3 \quad , \quad \psi_{1111} = \psi_4 \quad ,$$

and

$$\phi_{000'0'} = \phi_{00} \quad , \quad \phi_{010'1'} = \phi_{11} \quad , \quad \phi_{000'1'} = \phi_{01} \quad ,$$

$$\phi_{011'1'} = \phi_{12} \quad , \quad \phi_{010'0'} = \phi_{10} \quad , \quad \phi_{110'1'} = \phi_{21} \quad ,$$

$$\phi_{001'1'} = \phi_{02} \quad , \quad \phi_{111'1'} = \phi_{22} \quad , \quad \phi_{110'0'} = \phi_{20} \quad ,$$

we finally obtain for the 3-vector $f_{ab'cd'}$:

$$\begin{aligned}
f_{01'00'} &= (-\psi_0, \psi_1, \psi_2 + 2\Lambda) \quad , \\
f_{11'10'} &= (-\psi_2 - 2\Lambda, \psi_3, \psi_4) \quad , \\
f_{10'00'} &= (-\phi_{00}, \phi_{10}, \phi_{20}) \quad , \\
f_{11'01'} &= (-\phi_{02}, \phi_{12}, \phi_{22}) \quad , \\
f_{11'00'} &= (-\psi_1 - \phi_{01}, \psi_2 + \phi_{11} - \Lambda, \psi_3 + \phi_{21}) \quad , \\
f_{10'01'} &= (\psi_1 - \phi_{01}, -\psi_2 + \phi_{11} + \Lambda, -\psi_3 + \phi_{21}) \quad ,
\end{aligned} \qquad (8.26)$$

and for the six matrices $F_{ab'cd'}$:

$$F_{01'00'} = \begin{pmatrix} \psi_1 & -\psi_0 \\ \psi_2+2\Lambda & -\psi_1 \end{pmatrix},$$

$$F_{11'10'} = \begin{pmatrix} \psi_3 & -\psi_2-2\Lambda \\ \psi_4 & -\psi_3 \end{pmatrix},$$

$$F_{10'00'} = \begin{pmatrix} \phi_{10} & -\phi_{00} \\ \phi_{20} & -\phi_{10} \end{pmatrix},$$

$$F_{11'01'} = \begin{pmatrix} \phi_{12} & -\phi_{02} \\ \phi_{22} & -\phi_{12} \end{pmatrix}, \tag{8.27}$$

$$F_{11'00'} = \begin{pmatrix} \psi_2+\phi_{11}-\Lambda & -\psi_1-\phi_{01} \\ \psi_3+\phi_{21} & -\psi_2-\phi_{11}+\Lambda \end{pmatrix},$$

$$F_{10'01'} = \begin{pmatrix} -\psi_2+\phi_{11}+\Lambda & \psi_1-\phi_{01} \\ -\psi_3+\phi_{21} & \psi_2-\phi_{11}-\Lambda \end{pmatrix}.$$

9. GRAVITATIONAL FIELD EQUATIONS

Having defined the gravitational field dynamical variables, given by the components of the Riemann tensor (8.27), the spin coefficients (8.17), and the four null vectors $\sigma^{\mu}{}_{ab'}$, we are now in a position to find out the gravitational field equations that relate these quantities. We encounter a different situation from that of the Yang-Mills case. In the latter case, the field equations are obtained from a Lagrangian density that is postulated. In the present case, we have the Einstein equations which relate the Einstein tensor to matter. This latter condition, the Einstein condition, is imposed on the field equations already at our disposal, including the identities, which consequently ceased to be identities and became part of the field equations. The procedure of using the identities as part of the field equations is well known in general relativity [35]. Nevertheless, we will see that two out of three sets of these same equations can be derived from a Lagrangian density.

9.1 Identities

The matrices $F_{\mu\nu}$ and $\tilde{F}_{\mu\nu}$ defined in the last section satisfy some identities which can be found as follows:

We calculate the expression

$$\nabla_{\alpha}F_{\beta\gamma} + \nabla_{\beta}F_{\gamma\alpha} + \nabla_{\gamma}F_{\alpha\beta} \quad . \tag{9.1}$$

The curl part of F does not contribute to (9.1) as can easily be verified. The second part of F, the commutator $[B_\mu, B_\nu]$, contributes to (9.1) the expression

$$[(\nabla_{\gamma}B_{\alpha}-\nabla_{\alpha}B_{\gamma}), B_{\beta}] + [(\nabla_{\alpha}B_{\beta}-\nabla_{\beta}B_{\alpha}), B_{\gamma}] + [(\nabla_{\beta}B_{\gamma}-\nabla_{\gamma}B_{\beta}), B_{\alpha}] \quad .$$

Now add to this the expression

$$[[B_{\alpha}, B_{\gamma}], B_{\beta}] + [[B_{\beta}, B_{\alpha}], B_{\gamma}] + [[B_{\gamma}, B_{\beta}], B_{\alpha}] \quad ,$$

which is identically zero, we obtain

$$\nabla_{\alpha}F_{\beta\gamma}+\nabla_{\beta}F_{\gamma\alpha}+\nabla_{\gamma}F_{\alpha\beta} = [B_{\alpha}, F_{\beta\gamma}] + [B_{\beta}, F_{\gamma\alpha}] + [B_{\gamma}, F_{\alpha\beta}]. \tag{9.2}$$

Note that the covariant derivatives in Equation (9.2) can be replaced by partial derivatives without affecting that equation.

To find the identity the $\tilde{F}_{\mu\nu}$ satisfy, we express its covariant derivative in terms of those of $F_{\mu\nu}$. Since, by Equation (8.20), $\tilde{F}_{\beta\gamma} = \zeta^{-1}F_{\beta\gamma}\zeta$, one obtains

$$\nabla_{\alpha}\tilde{F}_{\beta\gamma} + (\nabla_{\alpha}\zeta^{-1})F_{\beta\gamma}\zeta + \zeta^{-1}(\nabla_{\zeta}F_{\beta\gamma})\zeta + \zeta^{-1}F_{\beta\gamma}\nabla_{\alpha}\zeta \quad .$$

Using Equation (8.2) and the fact that $\nabla_{\alpha}\zeta^{-1} = -\zeta^{-1}(\nabla_{\alpha}\zeta)\zeta^{-1}$, one obtains

$$\nabla_{\alpha}\tilde{F}_{\beta\gamma} = \zeta^{-1}(\nabla_{\alpha}F_{\beta\gamma} - [B_{\alpha}, F_{\beta\gamma}])\zeta \quad .$$

Using this equation in Equation (9.2) we obtain the identity that the matrix $\tilde{F}_{\alpha\beta}$ has to satisfy:

$$\nabla_{\alpha}\tilde{F}_{\beta\gamma} + \nabla_{\beta}\tilde{F}_{\gamma\alpha} + \nabla_{\gamma}\tilde{F}_{\alpha\beta} = 0 \quad . \tag{9.3}$$

9.2 Field equations

In Section 8, we have defined the matrix $F_{\mu\nu}$ in terms of the matrix B_μ by Equation (8.3), and in this section we showed that these matrices satisfy the identity (9.2).

By contracting Equations (8.3) and (9.2) with $\sigma^{\mu}{}_{ab'}\sigma^{\nu}{}_{cd'}$ and $\sigma^{\alpha}{}_{ab'}\sigma^{\beta}{}_{cd'}\sigma^{\gamma}{}_{ef'}$, respectively, and using Equation (8.11), one obtains two sets of first-order partial differentia equations that now connect the four matrices $B_{ab'}$ and the six matrices $F_{cd'ed'}$. A supplementary set of equations which connect the matrices σ^{μ} with the matrices $B_{ab'}$ and which is called the <u>metric equation</u> is, furthermore, obtained from Equation (8.11).

Multiplying Equation (8.3) by $\sigma^{\mu}{}_{ab'}\sigma^{\nu}{}_{cd'}$ and using Equation (8.12), we obtain

$$\partial_{cd'}B_{ab'}-\partial_{ab'}B_{cd'}-(\nabla_{cd'}\sigma^{\mu}{}_{ab'}-\nabla_{ab'}\sigma^{\mu}{}_{cd'})B_{\mu} + [B_{ab'}, B_{cd'}] = F_{ab'cd'} , \tag{9.4}$$

where the two differential operators ∂ and ∇ are defined by $\partial_{ab'} = \sigma^{\mu}{}_{ab'}\partial_{\mu}$ and $\nabla_{ab'} = \sigma^{\mu}{}_{ab'}\nabla_{\mu}$.

The third and fourth terms in Equation (9.4) can be found using Equation (8.11). Contracting the latter with $\sigma^{\alpha}{}_{cd'}$ we obtain

$$\nabla_{cd'}\sigma^{\mu} = B_{cd'}\sigma^{\mu} + \sigma^{\mu}B^{\dagger}{}_{\alpha}\bar{\sigma}^{\alpha}{}_{d'c} ,$$

where use has been made of the Hermiticity property of $\sigma^{\alpha}{}_{cd'}$. Hence we obtain

$$\nabla_{cd'}\sigma^{\mu} = B_{cd'}\sigma^{\mu} + \sigma^{\mu}B^{\dagger}{}_{d'c} . \tag{9.5}$$

Here the four matrices $B^{\dagger}{}_{d'c}$ are Hermitian conjugate to the matrices $B_{dc'}$ given in Equation (8.17). For completeness, and convenience of the reader, we list them below:

$$B^{\dagger}{}_{0'0} = \begin{pmatrix} \bar{\varepsilon} & \bar{\pi} \\ -\bar{\kappa} & -\bar{\varepsilon} \end{pmatrix}, \quad B^{\dagger}{}_{0'1} = \begin{pmatrix} \bar{\beta} & \bar{\mu} \\ -\bar{\sigma} & -\bar{\beta} \end{pmatrix},$$
$$B^{\dagger}{}_{1'0} = \begin{pmatrix} \bar{\alpha} & \bar{\lambda} \\ -\bar{\rho} & -\bar{\alpha} \end{pmatrix}, \quad B^{\dagger}{}_{1'1} = \begin{pmatrix} \bar{\gamma} & \bar{\nu} \\ -\bar{\tau} & -\bar{\gamma} \end{pmatrix} \tag{9.6}$$

Accordingly we obtain from Equation (9.5)

$$\nabla_{cd'}\sigma^{\mu}{}_{ab'} = (B_{cd'}\sigma^{\mu})_{ab'} + (\sigma^{\mu}B^{\dagger}{}_{d'c})_{ab'} ,$$

where $(\ )_{ab'}$ is the ab' element of the matrix ( ). Writing this last equation in terms of matrix elements we obtain

$$\nabla_{cd'}\sigma^{\mu}{}_{ab'} = (B_{cd'})_{a}{}^{f}\sigma^{\mu}{}_{fb'} + \sigma^{\mu}{}_{af'}(B^{\dagger}{}_{d'c})^{f'}{}_{b'} . \tag{9.7}$$

Here $(B_{cd'})_{a}{}^{f}$ and $(B^{\dagger}{}_{d'c})^{f'}{}_{b'}$ are the af and f'b' elements of the matrices $B_{cd'}$ and $B^{\dagger}{}_{d'c}$, respectively.

Using Equations (9.7) and (8.12) in Equation (9.4) one obtains

$$\begin{aligned}\partial_{cd'}B_{ab'} - \partial_{ab'}B_{cd'} - (B_{cd'})_{a}{}^{f}B_{fb'} - (B^{\dagger}{}_{d'c})^{f'}{}_{b'}B_{af'} + \\ + (B_{ab'})_{c}{}^{f}B_{fd'} + (B^{\dagger}{}_{b'a})^{f'}{}_{d'}B_{cf'} + [B_{ab'},B_{cd'}] = F_{ab'cd'}\end{aligned} \tag{9.8}$$

Similarly, one can reqrite Equation (9.2) in terms of the corresponding quantities with dyad indices to express relations between the matrices $B_{ab'}$ and $F_{ab'cd'}$ by using Equations (8.11) and (8.13). Multiplying Equation (9.2) by $\sigma^{\alpha}{}_{ab'}\sigma^{\beta}{}_{cd'}\sigma^{\gamma}{}_{ef'}$, and using Equation (9.7), one obtains

$$\partial_{ab'}F_{cd'ef'} + \partial_{cd'}F_{ef'ab'} + \partial_{ef'}F_{ab'cd'} - (B_{ab'})_c{}^g F_{gd'ef'} -$$
$$- (B^\dagger{}_{b'a})^{g'}{}_{d'}F_{cg'ef'} - (B_{ab'})_e{}^g F_{cd'gf'} - (B^\dagger{}_{b'a})^{g'}{}_{f'}F_{cd'eg'} -$$
$$- (B_{cd'})_e{}^g F_{gf'ab'} - (B^\dagger{}_{d'c})^{g'}{}_{f'}F_{eg'ab'} - (B_{cd'})_a{}^g F_{ef'gb'} -$$
$$- (B^\dagger{}_{d'c})^{g'}{}_{b'}F_{ef'ag'} - (B_{ef'})_a{}^g F_{gb'cd'} - (B^\dagger{}_{f'e})^{g'}{}_{b'}F_{ag'cd'} -$$
$$- (B_{ef'})_c{}^g F_{ab'gd'} - (B^\dagger{}_{f'e})^{g'}{}_{d'}F_{ab'cg'} = [B_{ab'}, F_{cd'ef'}] +$$
$$[B_{cd'}, F_{ef'ab'}] + [B_{ef'}, F_{ab'cd'}] \quad . \tag{9.9}$$

One can easily verify that Equations (9.8) and (9.9) are the usual field equations obtained using the formalism of Newman and Penrose [36].

We finally obtain the metric equation which connects $\sigma^\mu{}_{ab'}$ with $B_{ab'}$. This equation can easily be obtained from Equation (9.7) and is given by:

$$\partial_{ab'}\sigma^\mu{}_{cd'} - \partial_{cd'}\sigma^\mu{}_{ab'} = (B_{ab'}\sigma^\mu)_{cd'} + (\sigma^\mu B^\dagger{}_{b'a})_{cd'} -$$
$$-(B_{cd'}\sigma^\mu)_{ab'} - (\sigma^\mu B^\dagger{}_{d'c})_{ab'} \quad . \tag{9.10}$$

9.3 Gravitational Lagrangian

The gravitational field equations, connecting the dynamical variables of general relativity, were obtained from Equations (8.3), (8.11), and (9.2) by rewriting them in terms of the corresponding quantities with dyad indices, and substituting the desired expression in terms of the energy-momentum tensor for the Ricci tensor components in the curvature matrices (8.27). However, one can also obtain two sets of these field equations from an action principle that is based on the analogy of the present theory to that of the Yang-Mills theory.

The simplest Lagrangian density which is invariant under both general coordinate transformation and spin frame transformation was shown by Carmeli to be given by [37]:

$$- (1/4)(-g)^{\frac{1}{2}}\mathrm{Tr}(F_{\mu\nu}F^{\mu\nu}) \quad . \tag{9.11}$$

It is also a most natural generalization to the free-field Yang-Mills Lagrangian density (6.9). It follows, however, that the equation of motion obtained from such a Lagrangian density does not give the complete set of gravitational field equations but only the empty space ones.

Another Lagrangian density can be contructed, however, which gives the complete set of field Equations (9.8) and (9.9). This was shown by Carmeli and Fickler [38] to be given by $-(1/4)(-g)^{\frac{1}{2}}\mathrm{Tr}(H^{\mu\nu}F_{\mu\nu})$, where $H^{\mu\nu} = \tilde{\sigma}^{\mu CB'}\tilde{\sigma}^{\nu AD'}\tilde{\sigma}^\alpha{}_{AB'}\tilde{\sigma}^\beta{}_{CD'}F_{\alpha\beta}$. The first order form of this Lagrangian density is:

$$\mathcal{L} = -(-g)^{\frac{1}{2}}\mathrm{TR}\{H^{\mu\nu}(-\tfrac{1}{2}F_{\mu\nu} + \partial_\nu B_\mu - \partial_\mu B_\nu + [B_\mu, B_\nu])\} \quad . \tag{9.12}$$

A complex conjugate expression can be added to (9.12) so that the Lagrangian density becomes real. The matrix elements of B_μ and $F_{\mu\nu}$ are considered to be the independent field variables, and Equation (8.3) is assumed to be unknown. The matrices $\tilde{\sigma}^\mu$ are introduced in (9.12) as auxiliary quantities in order to accomplish invariance but they are not considered as part of the dynamical variables.

Application of the usual procedure of variational calculus then leads to the field Equation (8.3) and to the following equation of motion:

$$\partial_\nu((-g)^{\frac{1}{2}}H^{\mu\nu}) - [B_\nu, (-g)^{\frac{1}{2}}H^{\mu\nu}] = 0 \quad . \tag{9.13}$$

Equation (9.13) gives the dynamical equation of motion which the Riemann tensor has to satisfy, and accordingly we have a full description of the dynamical system. Equation (8.3) gives the Riemann tensor in terms of the spin coefficients, whereas Equation (8.11) gives the spin coefficients in terms of the tetrad of null vectors.

To recover the gravitational field Equations (9.8) - (9.10) one has merely to rewrite these equations in terms of the dynamical varables using Equations (8.12), (8.13), (8.17), and (8.27). One obtains the field Equations (9.8), (9.10), and the following:

$$\begin{aligned} &\partial^{cd'}F_{cb'ad'} - \{(B^{pd'})^{c}{}_{p} + (B^{\dagger q'c})_{q'}{}^{d'}\}F_{cb'ad'} - \\ &- \{\delta^{f'}{}_{b'}(B^{cd'})^{e}{}_{a} + \delta^{e}{}_{a}(B^{\dagger d'c})_{b'}{}^{f'}\}F_{cf'ed'} - \\ &- [B^{cd'}, F_{cb'ad'}] = 0 \quad . \end{aligned} \tag{9.14}$$

When written in details, Equations (9.9) and (9.14) follow to be identical.

The Lagrangian density (9.12) is a natural generalization of the free-field Lagrangian density (9.11), and reduces to the latter in the free field case. This can easily be seen since the expression in braces can be written as $H^{\mu\nu}F_{\mu\nu}$, and by Equation (8.13) this is equal to $F^{cb'ad'}F_{ab'cd'}$. In empty space (i.e., when all ϕ's and Λ are assumed to be zeros) this last expression can be seen, by Equation (8.27), to be equal to $F^{ab'cd'}F_{ab'cd'}$, or equal to $F^{\mu\nu}F_{\mu\nu}$, thus giving the expression

$$-\tfrac{1}{2}(-g)^{\frac{1}{2}}\,\mathrm{Tr}\{F^{\mu\nu}(-\tfrac{1}{2}F_{\mu\nu} + \partial_\nu B_\mu - \partial_\mu B_\nu + [B_\mu, B_\nu])\} \quad .$$

for the Lagrangian density (9.12) in free space, which is the Lagrangian density (9.11).

PROBLEMS

1. Use Equation (2.1) to show that the geometrical metric can be written as

$$g^{\mu\nu} = \tilde{\sigma}^{\mu}{}_{AB'}\tilde{\sigma}^{\nu AB'} .$$

Also show that

$$\tilde{\sigma}_{\mu}{}^{A}{}_{C'}\tilde{\sigma}_{\nu}{}^{BC'} + \tilde{\sigma}_{\nu}{}^{A}{}_{C'}\tilde{\sigma}_{\mu}{}^{BC'} = g_{\mu\nu}\varepsilon^{AB} .$$

2. Use the identity

$$\varepsilon_{AB}\varepsilon_{CD} + \varepsilon_{AC}\varepsilon_{DB} + \varepsilon_{AD}\varepsilon_{CD} = 0$$

to show that an arbitrary spinor with two indices, ξ^{AB}, will satisfy the equation

$$\xi_{AB} - \xi_{BA} = \varepsilon_{AB}\xi_{C}{}^{C} ,$$

where $\xi_{C}{}^{C} = \varepsilon^{CD}\xi_{CD}$.

3. Show that the spinor equivalent to the tensor

$$\varepsilon^{\alpha\beta}{}_{\mu\nu} = (-g)^{\frac{1}{2}}\varepsilon_{\rho\sigma\mu\nu}g^{\alpha\rho}g^{\beta\sigma}$$

is given by

$$\varepsilon^{AB'CD'}{}_{EF'GH'} = i(\delta^{A}_{E}\delta^{C}_{G}\delta^{B'}_{H'}\delta^{D'}_{F'} - \delta^{A}_{G}\delta^{C}_{E}\delta^{B'}_{F'}\delta^{D'}_{H'}) .$$

See R. Penrose, Ann. Phys. (N.Y.) **10**, 171 (1960).

4. The tensor $*F_{\mu\nu}$, defined by

$$*F_{\mu\nu} = \tfrac{1}{2}(-g)^{\frac{1}{2}} F^{\rho\sigma}\varepsilon_{\mu\nu\rho\sigma} ,$$

is called the dual to the tensor $F_{\mu\nu}$. Show that if the spinor equivalent of $F_{\mu\nu}$ is given by Equation (3.3), then its dual can be given as

$$*F_{AB'CD'} = i(\varepsilon_{AC}\bar{\phi}_{B'D'} - \phi_{AC}\varepsilon_{B'D'}) .$$

5. Prove Equations (3.5) - (3.7).

6. Show that the function $\lambda = \chi_{AB}{}^{AB}$ is real.

7. The right dual tensor of the Riemann tensor is defined by

$$S_{\mu\nu\rho\sigma} = \tfrac{1}{2}(-g)^{\frac{1}{2}} R_{\mu\nu}{}^{\alpha\beta}\varepsilon_{\alpha\beta\rho\sigma} .$$

Find its spinor equivalent in terms of the spinors χ_{ABCD} and $\phi_{ABC'D'}$.

8. Show that the Bianchi identity,

$$\nabla_\alpha R_{\mu\nu\beta\gamma} + \nabla_\beta R_{\mu\nu\gamma\alpha} + \nabla_\gamma R_{\mu\nu\alpha\beta} = 0 \quad ,$$

is equivalent to the equation

$$\nabla_{DE'}\chi_{ABC}{}^{D} = \nabla_{CF'}\phi_{ABE'}{}^{F'} \quad .$$

[R. Penrose, Ann. Phys. (N.Y.) 10 171 (1960).]

9. Show that the eigenvalue Equation (4.3) corresponds to the tensor equation

$$F^{+}{}_{\mu\nu}\alpha^{\nu} = -2\lambda\alpha_{\mu} \quad ,$$

or equivalently to

$$F_{\mu\nu}\alpha^{\nu} = -2(\mathrm{Re}\lambda)\alpha_{\mu} \quad ,$$

where the vector α_μ is null and given by $\alpha^\mu = \tilde{\sigma}^{\mu}{}_{AB'}\alpha^{A}\bar{\alpha}^{B'}$.

10. Show that the eigenvalue Equation (5.5) corresponds to the tensor equation

$$C^{+}{}_{\alpha\beta\gamma\delta}\, F^{+\gamma\delta} = 2\lambda F^{+}{}_{\alpha\beta}$$

or the equation

$$C_{\alpha\beta\gamma\delta}F^{+\gamma\delta} = \gamma F^{+}{}_{\alpha\beta} \quad ,$$

where $F^{+}{}_{\alpha\beta}$ and ϕ_{AG} are related by Equation (4.2).

11. Find the matrix R of Equation (5.17); show that it is orthogonal and its determinant equals +1.

12. Show that

$$\begin{aligned} \mathrm{Tr}\Psi^3 &= 3 \det \Psi \\ &= 3(\chi_0\chi_1^2/2 + \chi_0^2\chi_2/\sqrt{3} - \chi_0\chi_3^2/2 - \\ &\quad - \chi_1^2\chi_2/2\sqrt{3} - \chi_1\chi_3\chi_4 - \chi_2^3/3\sqrt{3} - \\ &\quad - \chi_2\chi_3^2/2\sqrt{3} + \chi_2\chi_4^2/\sqrt{3})/\sqrt{2} \quad . \end{aligned}$$

13. Show that the Hamiltonian derived from the Lagrangian density given in Equation (6.10) is positive definite in the ab-

sence of the field of isotopic spin.

14. To quantize the Yang-Mills field it is sometimes convenient to start with the Lagrangian density which is not obviously gauge invariant:

$$\mathcal{L} = - \frac{\partial \underline{b}_\mu}{\partial x^\nu} \frac{\partial \underline{b}_\mu}{\partial x^\nu} + 2(\underline{b}_\mu \times \underline{b}_\nu) \frac{\partial \underline{b}_\mu}{\partial \chi^\nu} - (\underline{b}_\mu \times \underline{b}_\nu) + $$
$$+ \underline{J}_\mu \cdot \underline{b}_\mu - \bar{\psi}(\gamma_\mu \partial_\mu + m)\psi \quad .$$

Show that the equations of motion obtained from this Lagrangian density are

$$\frac{\partial^2 \underline{a}}{\partial x^{\nu 2}} + 2\underline{b}_\nu \times \frac{\partial \underline{a}}{\partial x^\nu} = 0 \quad ,$$

where $\underline{a} = \partial \underline{b}_\mu / \partial x^\mu$.

15. Show that the Hamiltonian density derived from the Lagrangian density of Problem 14 is given by $H = H_o = H_{int}$, where

$$H_o = -\underline{\pi}_\mu \cdot \underline{\pi}_\mu + \frac{\partial \underline{b}_\mu}{\partial x^i} \cdot \frac{\partial \underline{b}_\mu}{\partial x^i} + \bar{\psi}(\gamma_j \partial_j + m)\psi \ ,$$
$$H_{int} = 2(\underline{b}_i \times \underline{b}_o) \cdot \underline{\pi}_i - 2(\underline{b}_\mu \times \underline{b}_j) \cdot (\partial \underline{b}_\mu / \partial x^j) +$$
$$+ (\underline{b}_i \times \underline{b}_j)^2 - \underline{J}_\mu \cdot \underline{b}_\mu \quad .$$

Here $\underline{\pi}_\mu$ is defined by

$$\underline{\pi}_\mu = - \partial \underline{b}_\mu / \partial x^0 + 2(\underline{b}_\mu \times \underline{b}_0) .$$

Show also that the equal-time commutation rule between b_μ and $\underline{\pi}_\mu$ is given by

$$[b_\mu{}^i(x), \pi_\nu{}^j(x')]_{t=t'} = -\delta_{ij}\delta_{\mu\nu}\delta^3(x-x') \quad .$$

16. Discuss the properties of the b quanta.

17. Show that if the field $F_{\mu\nu}$ defined by Equation (6.3) vanishes everywhere, then the potential B_μ can be transformed away by a gauge transformation.

18. Show that the internal holonomy groups at different points are isomorphic.

19. Prove that compatibility of the internal holonomy group $\mathcal{H}$ and the gauge group G requires that either (1) $\mathcal{H}$ be a subgroup of G, or (2) G be an invariant subgroup of $\mathcal{H}$.

20. Use Equation (7.9) to generalize the covariant derivative law

(7.10) for the mixed tensor $\psi^{k\nu}$ to a tensor $\psi^{k\ell\cdots\rho\sigma\cdots}$

21. Assuming $\Gamma'_{\mu\nu}{}^{\rho} = \Gamma'_{\nu\mu}{}^{\rho}$, prove Equations (7.12) and (7.13).

22. Prove Equation (7.15).

23. Show that $\tau^{\rho\sigma}$ of Equation (7.16) is symmetric.

24. Prove Equation (8.24).

25. Prove Equation (8.25).

26. Prove Equation (9.9).

REFERENCES AND FOOTNOTES

1. M.A. Naimark, Linear Representations of the Lorentz Group, Pergamon Press, Inc., New York, 1964.
2. I.M. Gelfand, R.A. Minlos and Z.Ya. Shapiro, Representations of the Rotation and Lorentz Groups and their Applications, Pergamon Press, Inc., New York, 1963.
3. I.M. Gelfand, M.I. Graev and N.Ya. Vilenkin, Generalized Functions, Vol. 5: Integral Geometry and Representation Theory, Academic Press, New York, 1966.
4. L. Infeld and B.L. van der Waerden, Sb. preuss.Akad. Wiss. Phys.-mat. Kl. **380** (1933).
5. R. Penrose, Ann. Phys.,(N.Y.) **10**, 171 (1960).
6. Note as a consequence of Equations (8.13) one has $R_{\mu\nu\rho\sigma}=R_{\rho\sigma\mu\nu}$.
7. L. Witten, Phys. Rev. **113**, 357 (1959).
8. From the Weyl spinor ψ_{ABCD} one can construct the tensor $T_{\alpha\beta\gamma\delta}$ equivalent is given by $T_{AB'CD'EF'GH'} = \psi_{ACEG}\psi_{B'D'F'H'}$. It is known as the Robinson-Bell tensor (sometimes, gravitational density or super energy). The tensor $T_{\alpha\beta\gamma\delta}$ is symmetric in all its indices, has vanishing traces, and vanishing covariant divergences when $\phi_{ABC'D} = 0$. See L. Bell, Compt. Rend. **247**, 1094 (1958) and **248**, 1297 (1959).
9. The following is essentially based on G. Ludwig, Amer. J. Phys. **37**, 1225 (1969).
10. A null bispinor could also be defined as one which is orthogonal to itself, or has a zero inner product in E_3, where inner products of two spinors ϕ_{AB} and ϕ_{AB} is defined by $\chi_m\chi'_m = \phi_{AB}\phi'^{AB}$, and ρ_m and ρ'_m are the dyad components in the orthogonal basis η_{mAB}.
11. For applications of the 3 x 3 complex orthogonal matrix representation of the homogeneous Lorentz group to other physical problems, see B. Kursunoglu, Modern Quantum Theory, W.H. Freeman, San Francisco, 1962.
12. The null tetrad ℓ^μ, m^μ, $\bar{m}^\mu$, and n^μ satisfies the normalization conditions:

$$\ell_\mu n^\mu = -m_\mu m^\mu = 1 \quad ,$$

$$\ell_\mu \ell^\mu = n_\mu n^\mu = m_\mu m^\mu = \bar{m}^\mu \bar{m}_\mu$$

$$= \ell_\mu m^\mu = \ell_\mu \bar{m}^\mu = n_\mu m^\mu = n_\mu \bar{m}^\mu = 0 \quad .$$

It will be noted that ℓ^μ and n^μ are real, whereas m^μ is complex.

13. The matrices P, Equation (4.16), corresponding to the three matrices $g_1(z)$, $g_2(z)$, and $g_3(z)$ can be obtained by putting the appropriate values in P. One can verify that the determinants of $P_1(z)$, $P_2(z)$, and $P_3(z)$ are all +1. Hence the determinant of P is also +1.
14. Just like in the 3-dimensional case for the Maxwell spinor, one can introduce an inner product in E_5 as follows. If two Weyl tensors $C_{\alpha\beta\gamma\delta}$ and $C'_{\alpha\beta\gamma\delta}$ have components χ_m and χ'_n, respectively, in the basis (5.10), then their inner product is defined by

$$\underline{\chi}\cdot\underline{\chi}' = \sum_{m=0}^{4} \chi_m \chi'_m = \psi_{ABCD}\psi'^{ABCD} = (1/16)C^{+}_{\alpha\beta\gamma\delta}C'^{+\alpha\beta\gamma\delta}$$

where ψ_{ABCD} and ψ'_{ABCD} are the spinor Weyl spinors associated with the Weyl tensors $C_{\alpha\beta\gamma\delta}$ and $C'_{\alpha\beta\gamma\delta}$ respectively. Two Weyl tensors or, equivalently, Weyl spinors are orthogonal if their inner product vanishes. A Weyl tensor which is orthogonal to itself is called null. A unit Weyl tensor is one for which the self-inner product is unity. As for bivectors, the existence of inner product allows the introduction of the notion of direction in E_5 in the usual manner.

15. A Weyl spinor of type N is often called null. This terminology is unfortunate, since Weyl spinors of types N, III, and some of type I, are null, as we shall see later on. In the following, the word "null" shall not be synonymous with a gravitations field of type N.

16. H. Weyl, The Theory of Groups and Quantum Mechanics, Dover, New York, 1931.
17. W. Heisenberg, Z. Physik **77** 1 (1932).
18. G. Breit, E. Condon and R. Present, Phys. Rev. **50** 825 (1936).
19. J. Schwinger [Phys. Rev. **78** 135 (1970)] pointed out that the small difference may be attributed to magnetic interactions.
20. The total isotopic spin was first introduced by E. Wigner, Phys. Rev. **51** 106 (1937); B. Cassen and E. Condon, Phys. Rev. **50** 846 (1936).
21. T. Lauritsen, Ann. Rev. Nucl. Sci. **1** 67 (1952); D.R. Inglis, Revs. Mod. Phys. **25** 390 (1953).
22. R.H. Hildebrand, Phys. Rev. **89** 1090 (1953).
23. C.N. Yang and R.L. Mills, Phys. Rev. **96**, 191 (1954).

24. The equations of motion (6.11) and (6.12) can be completed by the supplementary condition $\partial b_\mu/\partial x^\mu = 0$ which serves to eliminate the scalar part of the field in b_μ. This imposes a condi-

tion on the possible isotopic gauge transformations. The infinitesimal isotopic gauge transformation $S = 1 - i\sigma \cdot \delta\omega$ must satisfy the condition:

$$2\underline{b}_\mu \times \partial \underline{\delta\omega}/\partial x^\mu + \partial^2 \underline{\delta\omega}/\partial x^{\mu 2} = 0 \quad .$$

This equation is the analog of the equation $\partial^2\alpha/\partial\chi^{\mu 2} = 0$ that must be satisfied by the gauge transformation $A'_\mu = A_\mu + e^{-1}(\partial\alpha/\partial x^\mu)$ of the electromagnetic field.

25. S.L. Glashow and M. Gell-Mann, *Ann. Phys., (N.Y.)* **15** 437 (1961).
26. H.G. Loos, *J. Math. Phys.* **8** 2114 (1967).
27. J.F. Schell, *J. Math. Phys.* **2** 202 (1961); J.N. Goldberg and R.P. Kerr, *ibid*. 321 and 332 (1961).
28. R. Utiyama, *Phys. Rev.* **101** 1597 (1956).
29. T.W.B. Kibble, *J. Math. Phys.* **2** 212 (1961).
30. S. Weinberg, *Phys. Rev.* **138** B988 (1965).
31. M. Carmeli, *Ann. Phys., (N.Y.)* **71** 603 (1972).
32. Our matrices are related to those of Gelfand, Graev and Vilenkin by $g_1 = a_-$, $g_2 = 2a_o$, and $g_3 = a_+$. See I.M. Gelfand, M.I. Graev, and N.Ya. Vilenkin, *Integral Geometry and Representation Theory*, Academic, New York, 1966.
33. E.T. Newman and R. Penrose, *J. Math. Phys.* **3** 566 (1962).
34. M. Carmeli, *J. Math. Phys.* **11** 2728 (1970).
35. M. Carmeli, *Nuovo Cimento Lett.* **4** 40 (1970); *Nuovo Cimento* **7A** 9 (1972).
36. These equations are given in F.A.E. Pirani, *Lectures on General Relativity*, Prentice-Hall, Englewood Cliffs, New Jersey, 1964, p. 350.
37. M. Carmeli, *Nucl. Phys.,* **B38** 621 (1972).
38. M. Carmeli and S.I. Fickler, *Phys. Rev.* **D5** 290 (1972).

COORDINATE SYSTEMS IN RIEMANNIAN SPACE-TIME: CLASSIFICATIONS AND TRANSFORMATIONS; GENERALIZATION OF THE POINCARÉ GROUP*

M. Halpern and S. Malin

Department of Information Science, Temple University
Philadelphia, Pennsylvania

Department of Physics and Astronomy, Colgate University
Hamilton, New York

ABSTRACT. Coordinate conditions that can be specified prior to determination of the geometrical structure of space-time are investigated, and new methods of dividing all coordinate systems into ten parametric classes are introduced. They are based on finding the degree of freedom in specifying the metric tensor and its derivatives to arbitrary high order at one point. These methods are used to present an exact formulation of the following idea: the limitation of geometrical description to gravitation only induces a corresponding restriction on the degree of invariance of the laws of nature. Geodesic Fermi coordinates are studied in detail and the set of transformations between them is shown to form a "quasi-group" which is a natural generalization of the Poincaré group to Riemannian space-time.

1. INTRODUCTION

To what extent can coordinate conditions be specified without knowledge of the geometrical structure of space-time? In General Relativity the structure of space-time is not given a-priori, but is rather part of the solution to be determined. Therefore coordinate conditions that are not linked to specific geometrical structure are of special interest. We begin this series of lectures with an investigation of coordinate conditions of this kind in Riemannian space-times with Euclidean topology.

Mathematically, the simplest coordinates that can be specified without reference to the geometry of space-time are Riemannian,

---

* This work has been supported in part by the Sloan Foundation and the Colgate Research Council.

and, in particular, normal [1,2]. The definitions and some relevant properties of these coordinate systems are recalled in Section 2. Sections 2 and 3 include a study of the transformations between normal frames [3]: the explicit formula for the general infinitesimal transformation between normal frames is derived and it is shown that the set of all transformations between normal coordinate systems does not form a group. It forms a novel mathematical structure (a "quasi-group"), and contains the homogeneous Lorentz group as a subset.

These results form the mathematical basis for the derivations in Sections 4-7. In spite of their mathematical simplicity, the possibilities of direct usage of normal coordinates in physical problems are somewhat limited by the following feature: the coordinate systems which they induce on constant-time hypersurfaces $x^4 = c$ (c - any real constant) depend on the value of c and correspond, therefore, to an observer who changes his system of measurement continuously as time goes by.

Another kind of coordinate system, which does not share with normal frames the above-mentioned unphysical feature, is introduced and investigated in Sections 4 and 5. It is known as "geodesic Fermi coordinates" [4]. These coordinate systems are a natural generalization of the inertial frames of Special Relativity to curved space-time [5] in the sense of satisfying the following requirements:

(1) their spatial origins move on time-like geodesics.

(2) They are locally Cartesean around the spatial origin at all times.

(3) Together with a given system x^{μ} all systems $x^{\mu'}$, defined by $x'^i = x^i$, $x'^4 = x^4 + c$ (c - any real constant) are geodesic Fermi.

(4) All geodesics belonging to hyper-surfaces $x^4 = $ const. and passing through the time-axis are given by equations of the form $x^i = c^i s$, $x^4 = c$ (c^i, c - any real constants).

(5) In the limit of flat space-time they reduce to inertial frames in the sense of Special Relativity.

Like the set of inertial frames, the set of geodesic Fermi frames is tenparametric: a geodesic Fermi coordinate system is uniquely defined once an origin (4 parameters) and four mutually perpendicular directions of axes (6 parameters) are chosen.

Section 5 is devoted mainly to analytic characterization of these coordinate systems in terms of the metric tensor and its derivatives of various orders at the origin.

In Section 6 we proceed to study the general problem of classification of coordinate systems in curved space-time. Generalizing the results of Sections 2 and 5 it is shown that again, prior to determination of the geometrical structure, it is possible to divide all coordinate systems into classes according to the values of the metric tensor and certain combinations of its derivatives of an arbitrarily high order at a given point. The idea behind this division is the following: combinations of derivatives of the metric tensor can be broken up into two sets: (1) those that form components of tensor; (2) all the other independent combinations. Tensor components cannot, in general, be chosen arbitrarily: if all components vanish in one frame they vanish in all frames. The combinations belonging to group (2), however, can be specified at will and their values form a basis for divisions of all coordinate systems into classes. All such classes turn out to be ten-parametric - the set of geodesic Fermi frames is a particular example. It turns out that the separation of combinations of derivatives into sets (1) and (2) does not uniquely determine one division, but rather allows for a wide variety of divisions of all frames into ten-parametric classes.

The physical meaning of such classifications is explored in Section 7. They are shown to correspond to characterizations of classses of coordinate systems, in terms of pre-assigned results of sets of measurements. The word "pre-assigned" is used in the following sense: prior to determination of the geometrical structure of space-time each class can be characterized by the numerical results of a chosen set of measurements.

The significance of the results of Sections 4-7 in relation to the principle of covariance is studied in Sections 8 and 9. Let us briefly indicate the direction of the study: in flat space-time (Special Relativity) the laws of nature are assumed to be invariant only under transformations within certain ten-parametric classes of frames, e.g. Cartesean frames. Ordinarily, with the transition to Riemannian space-time, the principle of covariance, according to which the laws of nature take the same form in all frames, is postulated. The principle of covariance raises serious physical objections. It is shown in Section 8 that it can be replaced by a principle of invariance under transformation within ten-parametric classes of frames only; and that this can be done without an arbitrary preference of one coordinate system over another.

In Section 9 we proceed to discuss the transformations between geodesic Fermi coordinates in the context of this new invariance principle. These transformations are shown to form a "quasi-group", which can be considered as a generalization of the Poincaré group to Riemannian space-time. The search for such a generalization stems from the unique role of the Poincaré group in flat space-time: the representations of this group indicate a deep connection between the geometry of space-time and properties of particles, because the Poincaré group itself is the group of motion in flat space-time, and the eigenvalues of its Casimir operators can be interpreted in terms of masses and spins of particles.

All the theorems stated in this work will be subject to the same restrictions used by Veblen and Thomas in their classical paper [2,5]: only analytic transformations between coordinate systems will be considered and the metric tensor components are assumed to be analytic functions of the coordinates. These restrictions will not be re-stated in each theorem. For example, the expression "all coordinate systems", whenever used, refers only to coordinate systems satisfying these restrictions.

The Einstein summation convention is used. Repeated Greek indices imply summation over 1,2,3,4; repeated Latin indices imply summation over 1,2,3.

2. RIEMANNIAN AND NORMAL COORDINATES [1,2]

Within any system of coordinates x^μ geodesic lines can always be put into the form

$$\frac{d^2x^\mu}{ds^2} + \Gamma^\mu_{\alpha\beta} \frac{dx^\alpha}{ds} \frac{dx^\beta}{ds} = 0 \tag{2.1}$$

by an appropriate choice of the parameter s. For non-minimal geodesics s can be chosen as the invariant distance. $\Gamma^\mu_{\alpha\beta}$ are the Christoffel symbols of the second kind.

Differentiating Equation (2.1) successively with respect to s we get, for all

$$\frac{d^nx^\mu}{ds^n} + \Gamma^\mu_{\alpha_1 \ldots \alpha_n} \frac{dx^{\alpha_1}}{ds} \cdots \frac{dx^{\alpha_n}}{ds} = 0 \tag{2.2}$$

where the Γ-symbols are defined by

$$\Gamma^\alpha_{\beta\gamma\delta} = \frac{1}{3} P \left(\frac{\partial}{\partial x^\delta} \Gamma^\alpha_{\beta\gamma} - 2 \Gamma^\alpha_{\mu\beta} \Gamma^\mu_{\gamma\delta} \right) \tag{2.3}$$

and in general

$$\Gamma^\alpha_{\beta\gamma\delta\ldots\mu\nu} = \frac{1}{N} P \left[\frac{\partial}{\partial x^\nu} \Gamma^\alpha_{\beta\gamma\delta\ldots\mu} - (N-1) \Gamma^\alpha_{\xi\gamma\delta\ldots\mu} \Gamma^\xi_{\beta\nu} \right] \tag{2.4}$$

where P means that all terms obtained by cyclic permutations of the subscripts should be added together and N is the number of subscripts.

Consider now a geodesic through the origin of the coordinate system with direction given by

$$\xi^\mu = \left(\frac{dx^\mu}{ds}\right)_0 \tag{2.5}$$

(a quantity with subscript 0 denotes the value of the quantity at the origin). Expanding its equations in Taylor series, we get from Equations (2.1), (2.2)

$$\begin{aligned} x^\mu &= \xi^\mu s + \sum_{n=2}^{\infty} \frac{1}{n!} \left(\frac{d^nx^\mu}{ds^\mu} \right)_0 s^n = \\ &= \xi^\mu s - \sum_{n=2}^{\infty} \frac{1}{n!} (\Gamma^\mu_{\alpha_1\ldots\alpha_n})_0 \, \xi^{\alpha_1} \ldots \xi^{\alpha_n} s^n \end{aligned} \tag{2.6}$$

A system of coordinates y^μ is called Riemannian if Equation (2.1) for all the geodesics through the origin is of the form

$$y^\mu = \xi^\mu s \tag{2.7}$$

where ξ^μ are constants. The following theorem now follows from Equation (2.6).

Theorem 1. A coordinate system is Riemannian if and only if for all $n \geq 2$, $\mu, \alpha_1, \ldots, \alpha_n = 1, \ldots, 4$

$$(\Gamma^\mu_{\alpha_1 \ldots \alpha_n})_0 = 0 \tag{2.8}$$

Corollary. For any $n \geq 2$ the set of combinations of derivatives of the metric tensor

$$\{\Gamma^\mu_{\alpha_1 \ldots \alpha_n}\} \ \mu, \alpha_1, \ldots \alpha_n = 1, \ldots 4$$

does not form a tensor and does not contain any subset that forms a tensor.

Proof: A tensor that vanishes in one frame vanishes in all frames. None of the expressions $\Gamma^\mu_{\alpha_1 \ldots \alpha_n}$ is identically zero (i.e., vanishes irrespective of the geometrical structure and choice of coordinate system) and yet they all vanish in Riemannian frames (Q.E.D.).

Consider now the following transformation between the given coordinates x^μ and another set of coordinates y^μ:

$$x^\mu = y^\mu + \sum_{n=2}^{\infty} \frac{1}{n!} \left(\frac{\partial^n x^\mu}{\partial y^{\alpha_1} \ldots \partial y^{\alpha_n}}\right)_0 y^{\alpha_1} \ldots y^{\alpha_n}$$

$$= y^\mu - \sum_{n=2}^{\infty} \frac{1}{n!} (\Gamma^\mu_{\alpha_1 \ldots \alpha_n})_0 \, y^{\alpha_1} \ldots y^{\alpha_n} \tag{2.9}$$

The Jacobian of the transformation is non-vanishing at the origin, and, therefore, the transformation can be inverted. The inverse transformation is, in fact, given explicitely by [2]

$$y^\mu = x^\mu + \sum_{n=2}^{\infty} (\Lambda^\mu_{\alpha_1 \ldots \alpha_n})_0 \, x^{\alpha_1} \ldots x^{\alpha_n} \tag{2.10}$$

where the Λ-symbols are defined by

$$\begin{aligned} \Lambda^\alpha_{\beta\gamma} &= \Gamma^\alpha_{\beta\gamma} \\ \Lambda^\alpha_{\beta\gamma\delta} &= \Gamma^\alpha_{\beta\gamma\delta} + P\,(\Lambda^\alpha_{\mu\beta}\,\Gamma^\mu_{\gamma\delta}) \\ \Lambda^\alpha_{\beta\gamma\delta\varepsilon} &= \Gamma^\alpha_{\beta\gamma\delta\varepsilon} + P\,(\Lambda^\alpha_{\mu\beta}\,\Gamma^\mu_{\gamma\delta\varepsilon} + \Lambda^\alpha_{\mu\nu}\,\Gamma^\mu_{\beta\gamma}\,\Gamma^\nu_{\delta\varepsilon} + \Lambda^\alpha_{\mu\beta\gamma}\,\Gamma^\mu_{\delta\varepsilon}) \end{aligned} \tag{2.11}$$

etc. P means that all the terms obtained by cyclic permutation of the subscripts should be added together.

From Equations (2.6) and (2.9) it follows that in the coordinate system y^μ all geodesics through the origin have the form (2.7). The y^μ constitute, therefore, a system of Riemannian coordinates. Since the linear form of Equation (2.7) is conserved if and only if the coordinates undergo linear transformation, it follows that: (i) all Riemannian frames with common origin are connected with each other by linear transformations; (ii) given an arbitrary coordinate system x^μ there exists one and only one Riemannian frame y^μ such that the transformation between x^μ and y^μ reduce to the identity transformation in first order. We have, therefore,

Theorem 2. Corresponding to an arbitrary coordinate system x^μ there exists one and only one Riemannian frame y^μ having the same origin and directions of coordinate axes. The components of the metric tensor at the origin are the same in both frames.

Consider an arbitrary Riemannian frame and denote its metric tensor at the origin by $(g_{\mu\nu})_0$. Consider the quadratic form $(g_{\mu\nu})_0\, y^\mu y^\nu$. According to Sylvester's law of inertia there exists a linear transformation with real coefficients

$$y'^\mu = c^\mu{}_\alpha y^\alpha \qquad \det |c^\mu{}_\alpha| \neq 0 \tag{2.12}$$

such that in the primed system of coordinates the coefficients of the quadratic form are $\pm\,\delta_{\mu\nu}$; the difference between the number of + and - along the diagonal being equal to the signature, i.e., -2:

$$(g_{\mu\nu})_0\, y^\mu y^\nu = \eta_{\mu\nu}\, y'^\mu y'^\nu \tag{2.13}$$

where

$$\eta_{\mu\nu} = \begin{matrix} -1 & & & \\ & -1 & & \\ & & -1 & \\ & & & +1 \end{matrix} \tag{2.14}$$

The transformation (2.12) does not effect the form of Equation (2.7). The y'^μ coordinates are, therefore, Riemannian.

Riemannian coordinates that satisfy, in addition to Equation (2.8) also

$$(g_{\mu\nu})_0 = \eta_{\mu\nu} \tag{2.15}$$

are called <u>normal coordinates</u>.

Theorem 3. The set of all normal coordinate systems having the same origin is six parametric and the set of transformations between them is identical with the homogeneous Lorentz group.

Proof: Let y^μ be normal coordinates. The most general transformation y^μ to another system of Riemannian coordinates y'^μ is given by Equation (2.12). The coordinates y'^μ are normal if and only if

$$(g'_{\mu\nu})_0 = \eta_{\mu\nu} = (g_{\alpha\beta})_0 \frac{\partial y^\mu}{\partial y'^\alpha} \frac{\partial y^\nu}{\partial y'^\beta} = \eta_{\alpha\beta} \frac{\partial y^\mu}{\partial y'^\alpha} \frac{\partial y^\nu}{\partial y'^\beta} \tag{2.16}$$

These equations are equivalent to

$$(g_{\mu\nu})_0 = \eta_{\mu\nu} = (g'_{\alpha\beta})_0 \frac{\partial y'^\alpha}{\partial y^\mu} \frac{\partial y'^\beta}{\partial y^\nu} = \eta_{\alpha\beta}\, C^\alpha{}_\mu\, C^\beta{}_\nu \tag{2.17}$$

The set of 4 x 4 matrices C with elements $C^\alpha{}_\beta$ satisfying Equation (2.17) is precisely the homogeneous Lorentz group. This group is six-parametric and each matrix C corresponds, according to Equation (2.12) to one transformation between the given normal coordinates y^μ and another normal frame y'^μ having the same origin. (Q.E.D.)

Theorem 3 dealt with the set of normal coordinates having the same origin. Since any point in space can be chosen as origin, the choice of the four coordinates of the origin allows for four additional parameters. We thus have

Theorem 4. The set of all normal coordinates in space-time is ten-parametric.

Geometrically, after a choice of origin (4 parameters) has been made, the choice of the metric tensor at the origin specifies the relative directions of the four unit vectors in the directions of the coordinate axes. Equation (2.15), for example, means that this tetrad of unit vectors should be orthogonal (i.e., the directions of the coordinate axes in a normal coordinate systems are mutually perpendicular). The choice of $(g_{\mu\nu})_0$ does not specify, however, the absolute orientation of the tetrad in space: a rotation of the tetrad as a whole does not effect $(g_{\mu\nu})_0$. Since the group of rotations in fourdimensions is six parametric (considering Lorentz transformations as complex rotations), the total number of parameters is ten. Theorem 4 is, therefore, readily generalized as follows:

Theorem 5. Corresponding to any symmetric matrix $a_{\mu\nu}$ with signature -2 and non-vanishing determinant there exists a ten-parametric set of Riemannian coordinate systems such that

$$(g_{\mu\nu})_0 = a_{\mu\nu} \tag{2.18}$$

Proof: We have previously seen that corresponding to any set of coefficients of a quadratic form with signature -2 and non-vanishing determinant there exists a transformation (2.12) to a quadratic form with coefficients $\eta_{\mu\nu}$. Since the Jacobian of the transformation is non-vanishing the inverse transformation exists, and when applied to normal coordinates it transforms the metric tensor at the origin from $\eta_{\mu\nu}$ to $a_{\mu\nu}$.

In analogy with (2.17) the transformations (2.12) that conserve $(g_{\mu\nu})_0$ satisfy

$$a_{\mu\nu} = a_{\alpha\beta} C^{\alpha}{}_{\mu} C^{\beta}{}_{\nu} \tag{2.19}$$

This is a set of 10 equations for 16 unknowns. Therefore its real solutions are at most six parameters. Choose one particular transformation $\bar{C}^{\mu}{}_{\nu}$ between normal coordinates and Riemannian coordinates satisfying Equation (2.18). By successive application of an arbitrary homogeneous Lorentz transformation and the transformation $\bar{C}^{\mu}{}_{\nu}$ a correspondence between all normal frames and all Riemannian frames satisfying Equation (2.18) is established. Since the former is six parametric (Theorem 3) so is the latter. Allowance of 4 additional parameters for choice of origin completes the proof.

Theorem 5 defines a division of all Riemannian coordinate systems into ten-parametric classes according to the value of $(g_{\mu\nu})_0$. This is a special case of a general result that will be proved in Section 6.

In the following sections we shall make use of three-dimensional normal co-ordinates in a space-like hyper-surface. In particular we need the following:

Theorem 6. The set of all normal frames in a three-dimensional hypersurface is six-parametric. A normal frame is uniquely defined once an origin (3 parameters) and three mutually perpendicular directions of axes (3 parameters) have been chosen.

This is the analogue of Theorem 4 for three instead of four-dimensional space.

3. TRANSFORMATIONS BETWEEN NORMAL COORDINATE SYSTEMS

The general infinitesimal transformation between two normal coordinate systems will be derived in two stages:

(1) Let y^{μ}, y'^{μ} be two normal coordinate systems with origins at P, P', and such that

$$\left(\frac{\partial y'^{\mu}}{\partial y^{\alpha}}\right)_0 = \delta^{\mu}{}_{\alpha} \tag{3.1}$$

The transformation between them will be derived as follows: from the equations of transformation of Christoffel symbols

$$\Gamma'^{\mu}_{\sigma\rho} \frac{\partial y'^{\sigma}}{\partial y^{\alpha}} \frac{\partial y'^{\rho}}{\partial y^{\beta}} = \Gamma^{\sigma}_{\alpha\beta} \frac{\partial y'^{\mu}}{\partial y^{\sigma}} - \frac{\partial^2 y'^{\mu}}{\partial y^{\alpha} \partial y^{\beta}} \tag{3.2}$$

and from Equations (2.8), (3.1) it follows that

$$\left(\frac{\partial^2 y'^{\mu}}{\partial y^{\alpha} \partial y^{\beta}}\right)_{P} = -\left(\Gamma'^{\mu}_{\alpha\beta}\right)_{P} \tag{3.3}$$

By differentiation Equation (3.2) with respect to y^γ and summing over the cyclic permutations of α, β, γ it likewise follows from Equation (2.8) that

$$\left(\frac{\partial^3 y'^\mu}{\partial y^\alpha \partial y^\beta \partial y^\gamma}\right)_P = -(\Gamma'^\mu_{\alpha\beta\gamma})_P \tag{3.4}$$

(The Γ-symbols are defined by Equations (2.3), (2.4)). Proceeding by induction

$$\left(\frac{\partial^n y'^\mu}{\partial y^{\alpha_1} \ldots \partial y^{\alpha_n}}\right)_P = -(\Gamma'^\mu_{\alpha_1 \ldots \alpha_n})_P \tag{3.5}$$

Therefore the transformation $y'^\mu(y^\alpha)$ is given by

$$y'^\mu(y^\alpha) = y'^\mu(P) + y^\mu - \sum_{n=2}^{\infty} \frac{1}{n!} (\Gamma'^\mu_{\alpha_1 \ldots \alpha_n})_P \, y^{\alpha_1} \ldots y^{\alpha_n} \tag{3.6}$$

Let us now specialize to the case of infinitesimal transformations. Let

$$y'^\mu(P) = b^\mu \tag{3.7}$$

Then, to first order in b^μ, by Equation (2.8)

$$\begin{aligned}(\Gamma'^\mu_{\alpha_1 \ldots \alpha_n})_P &= (\Gamma'^\mu_{\alpha_1 \ldots \alpha_n})_{P'} + (\Gamma'^\mu_{\alpha_1 \ldots \alpha_n, \nu})_{P'} b^\nu \\ &= (\Gamma'^\mu_{\alpha_1 \ldots \alpha_n, \nu})_{P'} b^\nu \end{aligned} \tag{3.8}$$

where $,\nu$ denotes usual differentiation:

$$\Gamma'^\mu_{\alpha_1 \ldots \alpha_n, \nu} \equiv \frac{\partial \Gamma'^\mu_{\alpha_1 \ldots \alpha_n}}{\partial y'^\nu} \tag{3.9}$$

By continuity we have, again up to first order in b^ν

$$(\Gamma'^\mu_{\alpha_1 \ldots \alpha_n, \nu})_{P'} b^\nu = (\Gamma^\mu_{\alpha_1 \ldots \alpha_n, \nu})_P \, b^\nu \tag{3.10}$$

$$b^\mu = y'^\mu(P) = -y^\mu(P') \equiv -a^\mu \tag{3.11}$$

Equation (3.6) becomes, for infinitesimal transformations

$$y'^\mu(y^\alpha) = y^\mu - a^\mu + D^\mu{}_\nu(y^\alpha) a^\nu \tag{3.12}$$

where

$$D^\mu{}_\nu(y^\alpha) \equiv \sum_{n=2}^{\infty} \frac{1}{n!} (\Gamma^\mu_{\alpha_1 \ldots \alpha_n, \nu})_0 \, y^{\alpha_1} \ldots y^{\alpha_n} \tag{3.13}$$

(2) In Section 2 (Theorem 3) we have shown that the set of transformations between normal coordinates with a fixed origin is identical

with the homogeneous Lorentz group. An infinitesimal transformation between two normal frames y^μ, y'^μ having the same origin is, therefore, of the form

$$y'^\mu = y^\mu + \frac{1}{2} \omega^{\alpha\beta} (M\alpha\beta)^\mu{}_\rho\, y^\rho \tag{3.14}$$

where $M\alpha\beta$ are the 4 x 4 matrices which correspond to the infinitesimal transformations of the homogeneous Lorentz group and $\omega^{\alpha\beta}$ are the corresponding parameters [7].

A general infinitesimal transformation is now obtained by a direct combination of Equations (3.12), (3.14):

$$y'^\mu = y^\mu - a^\mu + \frac{1}{2} \omega^{\alpha\beta} (M\alpha\beta)^\mu{}_\rho\, y^\rho + D^\mu_\alpha (y^\alpha)\, a^\partial \tag{3.15}$$

We proceed now to discuss the set N of all transformations between normal frames. The set N contains the homogeneous Lorentz group as a subset (Theorem 2), but N itself is different from the inhomogeneous Lorentz group. In fact, N does not form a group at all because Equation (3.12) depends not only on the infinitesimal parameters but also on the geometrical structure (on the values of $(\Gamma^\mu_{\alpha_1 \ldots \alpha_n, \nu})_0$).

Let us denote by $n(P,\lambda)$ a normal coordinate system with origin at P and directions of axes denoted collectively by λ and by $n(P, \lambda \to P',\lambda') \in$'N the transformation from $n(P,\lambda)$ to $n(P',\lambda')$. Let us define multiplication of transformations in the usual way: if $n_1 \equiv n_1 (P_1, \lambda_1 \to P_1', \lambda_1')$ is the transformation $y^\mu = y^\mu (x^\alpha)$ and $n_2 \equiv n_2 (P_2, \lambda_2 \to P_2', \lambda_2')$ is the transformation $\zeta^\xi = \zeta^\xi(y^\mu)$, then

$$n_1 \cdot n_2 \equiv \zeta^\xi\, [y^\mu(x^\alpha)] \tag{3.16}$$

In contradistinction to the case of that space, it now follows from (3.12) that for curved space-time, if $n_1, n_2 \in N$, $n_1 \cdot n_2 \in N$ is not necessarily true. Because of Theorem 2 $n_1 \cdot n_2 \in N$ if $P_1' = P_2$. If, however, $P_1' \neq P_2$ then, in general, $n_1 \cdot n_2 \notin N$. It follows then that the set N is not a group. We call the mathematical structure exemplified by N a "quasi-group", which we define as follows [8]:

A set $A = \{a_\alpha\}$, where α stands for any number of discrete or continuous parameters, is a quasi-group if:

(1) Corresponding to every element $a_\alpha \in A$ $\exists B_\alpha \subseteq A$ and $B_\alpha' \subseteq A$ such that if $b \in B_\alpha$, the multiplication $a_\alpha \cdot b$ is defined and $a_\alpha \cdot b \in A$ and if $b \in B_\alpha'$ then $b \cdot a_\alpha$ is defined and $b \cdot a_\alpha \in A$.

(2) The associative law: for any a, b, c $\in$ A if $a \cdot b$ and $b \cdot c$ are defined then $(a \cdot b) \cdot c$ and $a \cdot (b \cdot c)$ are also defined and

$$(a \cdot b) \cdot c = a \cdot (b \cdot c)$$

(3) Existence of unit element: among the elements of A there is one and only one element e which has the property that $a \cdot e$ and

$e \cdot a$ are defined for all $a \in A$ and

$$a \cdot e = e \cdot a = a$$

(4) Existence of an inverse: corresponding to every element $a \in A$, $\exists a' \in B_\alpha, B'_\alpha$ such that

$$a \cdot a' = a' \cdot a = e$$

Thus a quasi-group is different from a group in that the product of two elements is not always defined. It resembles a group in the sense that the associative law is satisfied whenever the products are defined, and in having a unit element and an inverse to every element.

4. GENERALIZATION OF INERTIAL FRAMES TO CURVED SPACE-TIME

Mathematically, the simplest ten-parametric set of coordinate systems is the set of all four-dimensional normal frames of reference. Physically, however, four-dimensional normal coordinates are unsatisfactory as a generalization of the inertial frames of Special Relativity to curved space time: the time axis of a normal frame reference is a time-like geodesic which satisfies

$$(g_{\mu\nu})_0 = \eta_{\mu\nu} \tag{4.1}$$

$$[\alpha\beta,\mu]_0 = (\Gamma^\mu_{\alpha\beta})_0 = 0 \tag{4.2}$$

where $(g_{\mu\nu})_0$, $[\alpha\beta,\mu]_0$, $(\Gamma^\mu_{\alpha\beta})_0$ are the values of the metric tensor and the Christoffel symbols of the first and second kind at the origin (o,o,o,o). The corresponding equations are not satisfied, however, at the spatial origin at times other than t = o, i.e. in general for $t \neq o$

$$g_{\mu\nu}\ (o,o,o,t) \neq \eta_{\mu\nu} \tag{4.3}$$

$$\Gamma^\mu_{\alpha\beta}\ (o,o,o,t) \neq 0 \tag{4.4}$$

Thus, in general a normal coordinate system is not locally Cartesean at the spatial origin (except at time t = o); and, more important, its properties around the spatial origin change with time. Physically, it corresponds, therefore, to an observer whose system of measurement changes continuously as time goes by.

This difficulty comes about because time and space coordinates are treated in the same way in the definition of normal coordinates: Equations (2.8) and (2.15) are completely symmetrical in time and space variables. In reality, however, the nature of our measurements is such that the time axis is distinguished: the observer is contrained to move along it as he takes his measurements.

In this section we introduce a ten-parametric set of coordinate systems that takes this special role of the physical observer into consideration. Looking for a generalization of inertial frames to curved space-time we would like our frame x^μ to satisfy the following requirements:

(1) x^μ is locally Cartesean around the spatial origin at all times;

(2) its spatial origin moves on a time-like geodesic (i.e. the line $x^i = 0$ $(i = 1,2,3)$, $x^4 = s$ is a geodesic);

(3) together with the given system x^μ all systems x'^μ, defined by $x'^i = x^i$ $(i = 1,2,3)$, $x'^4 = x^4 + c$ (c - any real constant), belong to the set;

(4) in the limit of flat space-time x^μ reduces to an inertial frame in the sense of Special Relativity.

Requirement (2) is really a consequence of (1); by definition x^μ is locally Cartesean around the spatial origin at all times if and only if

$$g_{\mu\nu}(o,o,o,x^4) = \eta_{\mu\nu} \qquad -\infty < x^4 < +\infty \tag{4.5}$$

$$\Gamma^\mu_{\alpha\beta}(o,o,o,x^4) = 0 \qquad -\infty < x^4 < +\infty \tag{4.6}$$

Since the equations of a geodesic are (2.1) it follows from Equation (4.6) that the time axis

$$x^i = o \qquad x^4 = s \tag{4.7}$$

is a geodesic.

Definition: A geodesic Fermi frame [4] is a coordinate system (i) which is locally Cartesean along all points of its time axis and (ii) all its hypersurfaces $x^4 = c$ (for all real numbers c) are geodesic hypersurfaces [9] perpendicular to the time axis and the coordinates induced on them by setting $x^4 = c$ are three-dimensional normal coordinates.

As a corollary to Theorem 9 (Section 5) we will show that geodesic Fermi frames satisfy the above-stated requirements (1)-(4). In this section we prove two theorems concerning these frames.

Theorem 7. A geodesic Fermi frame is uniquely determined by choice of a point for its origin and of four mutually perpendicular directions at this point (three space-like and one time-like) for directions of its axes.

Proof: the following properties follow from the definition of geodesic Fermi frames:

(a) its time axis is a geodesic (this follows from the remark preceding the definition).

(b) the three-dimensional normal coordinates of any hypersurface $x^4 = c$ are such that their x^1, x^2, x^3 directions at (o,o,o,c) are parallel to the x^1, x^2, x^3 directions at (o,o,o,o) in Levi-Civita's sense of parallelism.

Indeed, by the definition of parallel transfer, the change in the components of any vector R^u when displaced parallel to itself along an elementary path dx^β is given by

$$dR^u = - \Gamma^u_{\alpha\beta} R^\alpha dx^\beta \tag{4.8}$$

It follows, therefore, from Equation (4.6) that the components of the unit vectors in the x^1, x^2 and x^3 directions do not change by a parallel displacement along the elementary path (o,o,o,dx^4).

Given an origin and four mutually perpendicular directions it follows from (a) that the time axis is uniquely determined as the time-like geodesic in the given time-like direction. By requirement (ii) of the definition all the space-like hypersurface $x^4 = c$ are uniquely determined. From Theorem 6 (Section 2) the three given space-like directions uniquely determine a normal frame of reference on the geodesic hypersurface $x^4 = o$ and by (b) and Theorem 6 once the normal coordinates on $x^4 = o$ are fixed the normal coordinates on all surfaces $x^4 = c$ are uniquely determined. (Q.E.D.)

Corollary. The set of geodesic Fermi frames in space-time is ten-parametric.

The following theorem amounts to an alternative definition of geodesic Fermi frames. It exhibits the similarities and differences between geodésic Fermi and normal coordinates.

Theorem 8. A coordinate system is geodesic Fermi if and only if it is locally Cartesean at all points of the time axis and for all real values of the numbers c^1, c^2, c^3, c the lines

$$x^1 = c^i s \quad (i = 1,2,3) \quad x^4 = c \tag{4.9}$$

where s is the invariant distance, are geodesics.

Proof: we have to show that requirement (ii) of the definition of geodesic Fermi frames is satisfied if and only if all lines of the form (4.9) are geodesics.

If (ii) is satisfied then the hypersurface $x^4 = c$ is geodesic and the three-dimensional coordinate systems x^1, x^2, x^3 induced on it by the geodesic Fermi system by setting $x^4 = c$ are normal. By definition of a three-dimensional normal frame all lines of the form

$$x^i = c^i s \quad (i = 1,2,3) \tag{4.10}$$

are geodesics in the hypersurface. The additional requirement $x^4 = c$ insures that the lines belong to the hypersurface and from the definition of geodesic hypersurface it follows that (4.9) are geodesics in the four-dimensional space-time.

Conversely, if in a given surface all lines satisfying Equation (4.10) are geodesics the coordinate system on that surface is normal. By Equation (4.9) all the lines generating any hypersurface $x^4 = c$ are geodesics and all such hypersurfaces are, therefore, geodesics. (Q.E.D.)

5. ANALYTIC CHARACTERIZATION OF GEODESIC FERMI FRAMES

The following theorem gives an analytic characterization of geodesic Fermi frames in terms of the metric tensor and certain combinations of its derivations at the origin. It amounts again to an alternative definition of these frames.

Theorem 9. A system of coordinates is geodesic Fermi if and only if the following conditions are met for all $n \geq 2$, $k \geq 1$, $\mu, \nu, \alpha = 1,\ldots,4$ and $i_1\ldots,i_n = 1,2,3$:

$$(g_{\mu\nu})_0 = \eta_{\mu\nu} \tag{5.1}$$

$$(\Gamma^{\mu}_{\alpha 4})_0 = 0 \tag{5.2}$$

$$(\Gamma^{\mu}_{i_1\ldots i_n})_0 = 0 \tag{5.3}$$

$$\left(\frac{\partial^k \Gamma^{\mu}_{\alpha 4}}{(\partial x^4)^k}\right)_0 = 0 \tag{5.4}$$

$$\left(\frac{\partial^k \Gamma^{\mu}_{i_1\ldots i_n}}{(\partial x^4)^k}\right)_0 = 0 \tag{5.5}$$

Proof: we divide the proof into 2 parts. In Part (1) we show that a system of coordinates is locally Cartesean at all points of the time axis if and only if

$$(g_{\mu\nu})_0 = \eta_{\mu\nu} \tag{5.6}$$

$$(\Gamma^{\mu}_{\alpha\beta})_0 = 0 \tag{5.7}$$

$$\left(\frac{\partial^k \Gamma^{\mu}_{\alpha\beta}}{(\partial x^4)^k}\right)_0 = 0 \tag{5.8}$$

for all $k \geq 1$, $\mu, \alpha, \beta = 1,\ldots,4$. In Part (2) we show that for all real values of c^1, c^2, c^3, c the lines (4.9) are geodesics if and only if Equations (5.3) and (5.5) are satisfied for all $k \geq 1$, $n \geq 2$, $\mu = 1,\ldots,4$; $i_1\ldots,i_n = 1,2,3$. Because of Theorem 8 this will complete the proof.

(1) By definition a system is locally Cartesean at all points of

the time axis if for all x^4

$$g_{\mu\nu}(o,o,o,x^4) = \eta_{\mu\nu} \tag{5.9}$$

and

$$\Gamma^{\mu}_{\alpha\beta}(o,o,o,x^4) = 0 \tag{5.10}$$

Let us show that Equations (5.9) and (5.10) are equivalent to Equations (5.6)-(5.8). First, expanding $\Gamma^{\mu}_{\alpha\beta}(o,o,o,x^4)$ around the origin,

$$\Gamma^{\mu}_{\alpha\beta}(o,o,o,x^4) = (\Gamma^{\mu}_{\alpha\beta})_0 + \sum_{k=1}^{\infty} \frac{1}{k!} \left(\frac{\partial^k \Gamma^{\mu}_{\alpha\beta}}{(\partial x^4)^k}\right)_0 (x^4)^k \tag{5.11}$$

we see that Equation (5.10) is equivalent to Equations (5.7) and (5.8). In the continuation we use the equivalence of Equations (5.7) and (5.8) to

$$[\alpha\beta,\mu]_0 = 0 \tag{5.12}$$

$$\left(\frac{\partial^k [\alpha\beta,\mu]}{(\partial x^4)^k}\right)_0 = 0 \tag{5.13}$$

This equivalence is proved as follows: at any point P all Christoffel symbols of the second kind vanish if and only if all Christoffel symbols of the first kind vanish (since any of these sets of symbols vanish if and only if all first order derivatives of the metric tensor vanish at the point). Therefore, Equation (5.10) hold if and only if

$$[\alpha\beta,\mu]\ (o,o,o,x^4) = [\alpha\beta,\mu]_0 + \sum_{n=1}^{\infty} \frac{1}{n!} \left(\frac{\partial^n [\alpha\beta,\mu]}{(\partial x^4)^n}\right)_0 (x^4)^n = 0 \tag{5.14}$$

and Equations (5.7) and (5.8) are equivalent to Equations (5.12) and (5.13) respectively.

We are ready now to show that Equations (5.9) and (5.10) follow from (5.6)-(5.8) and vice versa: expanding $g_{\mu\nu}(o,o,o,x^4)$ around the origin:

$$g_{\mu\nu}(o,o,o,x^4) = (g_{\mu\nu})_0 + \left(\frac{\partial g_{\mu\nu}}{\partial x^4}\right)_0 x^4 + \sum_{n=2}^{\infty} \frac{1}{n!} \left(\frac{\partial^n g_{\mu\nu}}{(\partial x^4)^n}\right)_0 (x^4)^n$$

$$= \eta_{\mu\nu} + ([\mu 4,\nu] + [\nu 4,\mu])_0\, x^4 + \tag{5.15}$$

$$+ \sum_{n=2}^{\infty} \frac{1}{n!} \left\{\frac{\partial^{n-1}}{(\partial x^4)^{n-1}} ([\mu 4,\nu] + [\nu 4,\mu])\right\}_0 (x^4)^n$$

If Equations (5.6)-(5.8) are satisfied it follows from Equations (5.11), (5.15) and the equivalence of (5.12), (5.13) to (5.7), (5.8) that Equations (5.9) and (5.10) are satisfied. Conversely, if Equations (5.9), (5.10) hold, it follows from Equation (5.11) that Equations (5.7) and (5.8) are true; Equation (5.6) too follows now because of the above-mentioned equivalence and Equation (5.15).

(2) Consider the equations of geodesics (2.1). From these equations it follows that all lines (4.9) are geodesics if and only if $\Gamma^{\mu}_{ij}c^i c^j$ vanish along them ($\mu = 1,\ldots,4$). Expanding $\Gamma^{\mu}_{ij}c^i c^j$ around the point (o,o,o,c) along a given line

$$\Gamma^{\mu}_{ij}(s)c^i c^j = \Gamma^{\mu}_{ij}(o,o,o,c)c^i c^j + \sum_{n=1}^{\infty} \frac{1}{n!} \frac{d^n \Gamma^{\mu}_{ij}}{ds^n}(o,o,o,c)c^i c^j s^n \tag{5.16}$$

Now,

$$\frac{d\Gamma^{\mu}_{ij}}{ds} = \Gamma^{\mu}_{ij,k}\, c^k \tag{5.17}$$

$$\frac{d\Gamma^{\mu}_{ij}}{ds}(o,o,o,c)c^i c^j = \Gamma^{\mu}_{ij,k}(o,o,o,c)c^i c^j c^k \tag{5.18}$$

i, j, k are dummy indices, so we can permute them and add up all the permutations to get [10]

$$\frac{d\Gamma^{\mu}_{ij}}{ds}(o,o,o,c)c^i c^j = \Gamma^{\mu}_{ijk}(o,o,o,c)c^i c^j c^k \tag{5.19}$$

Thus a necessary and sufficient condition for the first term in the sum to vanish is

$$\Gamma^{\mu}_{ijk}(o,o,o,c) = 0 \tag{5.20}$$

Similarly, for any other term

$$\frac{d^n \Gamma^{\mu}_{ij}}{ds^n} = \Gamma^{\mu}_{ij,k_1\ldots k_n}\, c^{k_1}\ldots c^{k_n} \tag{5.21}$$

$$\begin{aligned}\frac{d^n \Gamma^{\mu}_{ij}}{ds^n}(o,o,o,c)c^i c^j &= \Gamma^{\mu}_{ij,k_1\ldots k_n}(o,o,o,c)c^i c^j c^{k_1}\ldots c^{k_n} \\ &= \Gamma^{\mu}_{ijk_1\ldots k_n}(o,o,o,c)c^i c^j c^{k_1}\ldots c^{k_n}\end{aligned} \tag{5.22}$$

Thus all terms in the expression vanish if and only if, for all real values of c, $\mu = 1,\ldots,4$, $n \geq 2$, $i_1,\ldots,i_n = 1,2,3$

$$\Gamma^{\mu}_{i_1\ldots i_n}(o,o,o,c) = 0 \tag{5.23}$$

Expanding $\Gamma^{\mu}_{i_1\ldots i_n}(o,o,o,c)$ around the origin we have

$$\Gamma^{\mu}_{i_1\ldots i_n}(o,o,o,c) = (\Gamma^{\mu}_{i_1\ldots i_n})_0 + \sum_{k=1}^{\infty} \frac{1}{k!} \left(\frac{\partial^k \Gamma^{\mu}_{i_1\ldots i_n}}{(\partial x^4)^k}\right)_0 c^n \tag{5.24}$$

Thus Equation (5.23) is equivalent to Equations (5.3) and (5.5). (Q.E.D.)

Corollaries. (1) for any $n \geq 2$ set of combinations of derivations of the metric tensor

$$\{ \Gamma^{\mu}_{i_1 \dots i_n}, \frac{\partial^k \Gamma^{\mu}_{i_1 \dots i_{n-k}}}{(\partial x^4)^k} \} \quad \mu = 1, \dots, 4;\ i_1, \dots, i_n = 1,2,3; \quad k = 1, \dots, n-2$$

does not form a tensor and does not contain any subset that forms a tensor.

This assertion is proved in complete analogy to the corollary to Theorem 1 (Section 2).

(2) Geodesic Fermi frames satisfy requirements (1)-(4) of Section 4.

Proof: requirement (1) is included in the definition and requirement (2) follows from it (see Section 4). In the course of proving Theorem 9, Equations (5.1)-(5.5) were shown to be equivalent to Equations (5.9), (5.10) and (5.23). If these equations are true for given coordinates x , they are also true for the coordinates x'^i defined by

$$x'^i = x^i \ (i = 1,2,3) \qquad x'^4 = x^4 + c \tag{5.25}$$

Therefore requirement (3) of Section 4 is satisfied too. Finally, concerning requirement (4), since inertial frames in flat space-time are defined by

$$g_{\mu\nu} = \eta_{\mu\nu} \tag{5.26}$$

at all points, Equations (5.1)-(5.5) are satisfied.

An alternative analytic characterization of geodesic Fermi frames is given in the following theorem.

Theorem 10. Equations (5.3) and (5.5) are satisfied if and only if at all points

$$\Gamma^{\mu}_{ij}(x^1, x^2, x^3, x^4)\, x^i x^j = 0 \qquad \mu = 1, \dots, 4 \tag{5.27}$$

Proof: Expand (5.26) around (o,o,o,x^4):

$$\Gamma^{\mu}_{ij}(x^1,x^2,x^3,x^4)x^i x^j = \sum_{k=0}^{\infty} \frac{1}{k!} \Gamma^{\mu}_{ij,i_1 \dots i_k}(o,o,o,x^4)x^i x^j x^{i_1} \dots$$

$$\dots x^{i_k} = \sum_{n=2}^{\infty} \frac{1}{(n-2)!} \Gamma^{\mu}_{i_1 \dots i_n}(o,o,o,x^4)x^{i_1} \dots x^{i_n} \tag{5.28}$$

The last step follows from the definition of the Γ-symbols (Equation (2.4)), in analogy with the derivation of Equation (5.21).

Thus, Equation (5.27) is satisfied if and only if for all $n \geq 2$, and all values of x^4

$$\Gamma^{\mu}_{i_1 \ldots i_n}(o,o,o,x^4) = 0 \quad \mu = 1,\ldots,4;\ i_1,\ldots,i_n = 1,2,3 \quad (5.29)$$

Expanding now $\Gamma^{\mu}_{i_1 \ldots i_n}(o,o,o,x^4)$ in Taylor's series around the origin Equation (4.29) is equivalent to

$$(\Gamma^{\mu}_{i_1 \ldots i_n})_0 = 0 \quad (5.30)$$

$$\left(\frac{\partial^k \Gamma^{\mu}_{i_1 \ldots i_n}}{(\partial x^4)^k}\right)_0 = 0 \quad (5.31)$$

for all $k \geq 1$, $\mu = 1,\ldots,4$, $i_1 \ldots, i_n = 1,2,3$ (Q.E.D.).

We conclude this section with a discussion of the natural one-to-one correspondence between normal and geodesic Fermi frames.

Both types of coordinate systems are uniquely determined by a choice of origin P and of four mutually perpendicular directions (three space-like and one time-like) denoted collectively by λ (see the discussion following Theorem 4 in Section 2 and Theorem 7 in Section 4). In analogy with the notation $n(P,\lambda)$ introduced in Section 3, let us use the notation $l(P,\lambda)$ to specify a geodesic Fermi frame with origin at P and directions of axes λ.

We define the natural one-to-one correspondence between the elements of the ten-parametric set of normal frames and the elements of the ten-parametric set of geodesic Fermi frames as

$$l(P,\lambda) \longleftrightarrow n(P,\lambda) \quad (5.32)$$

where P and λ are the same for l and n; i.e. the normal and geodesic Fermi frame that correspond to each other have the same origin and direction of axes.

The transformation between the corresponding frames are obtained as a special case of the transformation between any frame x^{μ} and the Riemannian frame y^{μ} with the same origin and direction of axes. The relevant formulas are given in Section 2: the transformations between $l(P,\lambda)$ and $n(P,\lambda)$ are given by (2.9) and (2.10) if x^{μ} are the geodesic Fermi coordinates and y^{μ} are the corresponding normal coordinates.

6. CLASSIFICATIONS OF COORDINATE SYSTEMS

Consider the following remarkable aspect of the results of the previous sections: without any knowledge of the geometrical structure of space-time it is always possible to set the metric tensor and certain combinations of its derivatives of an arbitrarily high order at a given point to zero by choice of coordinate system; and if the number of these combinations is big enough, their vanishing fixes the coordinate system up to a ten-parametric set.

In the present section we show this result to be a special case of a general theorem: again without knowledge of the geometrical structure of space-time, it is possible to divide all coordinate systems into classes according to the value of the metric tensor and some combinations of its derivatives of an arbitrarily high order at a given point; if the number of these combinations is big enough, particular numerical values fix the coordinate system up to a ten-parametric set.

The choice of combinations of derivatives is by no means unique. Different choices lead to different divisions. In the present section we present two possibilities: the "normal division" which contains as a class the ten-parametric set of normal coordinates and the "geodesic Fermi division" which contains the ten-parametric class of geodesic Fermi frames. The physical implications of the theorems presented here will be discussed in Section 7.

Theorem 11. All coordinate systems can be divided into ten-parametric classes as follows: each class is characterized by a set of numbers $a_{\nu\mu}$ $(\mu,\nu = 1,\ldots,4)$ and $b^{\mu}_{\alpha_1\ldots\alpha_n}$ $(\mu,\alpha_1,\ldots,\alpha_n = 1,\ldots,4;$ $n \geq 2)$ such that for coordinate systems belonging to the class

$$(g_{\mu\nu})_0 = a_{\mu\nu} \tag{6.1}$$

$$(\Gamma^{\mu}_{\alpha_1\ldots\alpha_n})_0 = b^{\mu}_{\alpha_1\ldots\alpha_n} \tag{6.2}$$

Remark: from Equations (6.1) and (6.2) it follows that the numbers are completely symmetric in the subscripts and that

$$\det a_{\mu\nu} \neq 0 \tag{6.3}$$

$$\mathrm{syg}\ a_{\mu\nu} = -2 \tag{6.4}$$

Proof: According to Theorem 2 (Section 2) any frame x^{μ} has one and only one corresponding Riemannian frame y^{μ} having the same origin and directions of axes. The transformation between x^{μ} and y^{μ} is given by

$$x^{\mu} = y^{\mu} + \sum_{n=2}^{\infty} \frac{1}{n!} \left(\frac{\partial^n x^{\mu}}{\partial y^{\alpha_1}\ldots\partial y^{\alpha_n}}\right)_0 y^{\alpha_1}\ldots y^{\alpha_n} \tag{6.5}$$

where, because of Equation (2.9)

$$\left(\frac{\partial^n x^{\mu}}{\partial y^{\alpha_1}\ldots\partial y^{\alpha_n}}\right)_0 = -\left(\Gamma^{\mu}_{\alpha_1\ldots\alpha_n}\right)_0 \tag{6.6}$$

According to Theorem 2 x^{μ} and y^{μ} have the same metric tensor at their common origin. From Equations (6.5), (6.6) it follows that x^{μ} satisfies Equations (6.1) and (6.2) if and only if it is connected with its corresponding Riemannian frame by

$$x^\mu = y^\mu - \sum_{n=2}^{\infty} \frac{1}{n!} b^\mu_{\alpha_1 \ldots \alpha_n} y^{\alpha_1} \ldots y^{\alpha_n} \tag{6.7}$$

The y^μ can be any Riemannian system. Therefore, if numerical values are assigned to all the b's Equation (6.7) establishes a one-to-one correspondence between all Riemannian frames y^μ and all coordinate systems x^μ that satisfy Equation (6.2). By Theorem 5 (Section 2) the set of Riemannian frames satisfying Equation (6.1) for any given numerical values of the $a_{\mu\nu}$ which satisfy Equations (6.3) and (6.4) is ten-parametric. Since the correspondence defined by Equation (6.7) is one-to-one, the set of frames characterized by Equations (6.1) and (6.2) with given numerical values of the a's and the b's is also ten-parametric (Q.E.D.).

We have thus defined a division of all frames into ten-parametric classes according to the values of $(g_{\mu\nu})_0$, $(\Gamma^\mu_{\alpha_1 \ldots \alpha_n})_0$. This particular division will be called the "normal division", because normal coordinates are obtained from Equations (6.1) and (6.2) by setting $a_{\mu\nu} = \eta_{\mu\nu}$ and $b^\mu_{\alpha_1 \ldots \alpha_n} = 0$; the set of all normal frames is, therefore, one of the classes defined by this division.

The normal division is by no means the only one. It depends on particular choice of combinations of derivations of $g_{\mu\nu}$ to an arbitrarily high order, namely, on the expressions $\Gamma^\mu_{\alpha_1 \ldots \alpha_n}$. A different division, called the "geodesic Fermi" division, which depends on the set of expression on the left hand side of Equations (5.2) - (5.5) is introduced in the following theorem.

Theorem 12. All coordinate systems can be divided into ten-parametric classes as follows: each class is characterized by a set of numbers $a_{\mu\nu}$, $b^{\mu(k)}_{\alpha\beta}$ and $b^{\mu(k)}_{i_1,\ldots i_n}$ ($k \geq 0$; $n \geq 3$; $\mu,\alpha,\beta = 1,\ldots,4$; $i_1,\ldots,i_n = 1,2,3$) such that for coordinate systems belonging to the class

$$(g_{\mu\nu})_0 = a_{\mu\nu} \tag{6.8}$$

$$(\Gamma^\mu_{\alpha 4})_0 = b^{\mu(0)}_{\alpha 4} \tag{6.9}$$

$$(\Gamma^\mu_{i_1 \ldots i_n})_0 = b^{\mu(0)}_{i_1 \ldots i_n} \tag{6.10}$$

$$\left(\frac{\partial^k \Gamma^\mu_{\alpha 4}}{(\partial x^4)^k}\right)_0 = b^{\mu(k)}_{\alpha 4} \tag{6.11}$$

$$\left(\frac{\partial^k \Gamma^\mu_{i_1 \ldots i_n}}{(\partial x^4)^k}\right)_0 = b^{\mu(k)}_{i_1 \ldots i_n} \tag{6.12}$$

Remarks (1): From Equations (6+8)-(6.12) it follows that all the a's and b's are completely symmetric in the subscripts and the a's satisfy Equations (6.3) and (6.4).

(2) The order of a b-symbol is defined as the number of

its subscripts plus k. For example, $b^{\mu(0)}_{\alpha\beta}$ are second order b-symbols, $b^{\mu(0)}_{i_1 i_2 i_3}$ and $b^{\mu(1)}_{\alpha\beta}$ are third order b-symbols, etc.

Proof: In analogy with the proof of Theorem 12, the present theorem will be proven by establishing a one-to-one correspondence between all coordinate systems that satisfy Equations (6.8)-(6.12) with particular values of the a's and b's and Riemannian frames satisfying Equations (6.8) with the same values of $a_{\mu\nu}$.

According to Theorem 2 (Section 2) any frame x^μ has one and only one corresponding Riemannian frame y^μ having the same origin and direction of axes. The tranformation between them is the identity transformation up to terms of the first order:

$$x^\mu = y^\mu + \sum_{n=2}^{\infty} \frac{1}{n!} \left(\frac{\partial^n x^\mu}{\partial y^{\alpha_1} \ldots \partial y^{\alpha_n}}\right)_0 y^{\alpha_1} \ldots y^{\alpha_n} \tag{6.13}$$

Because of (6.6), Equations (6.9) and (6.10) for n = 2 will be satisfied if and only if the second order b-symbols are

$$b^{\mu(0)}_{\alpha\beta} = -\left(\frac{\partial^2 x^\mu}{\partial y^\alpha \partial y^\beta}\right)_0 \tag{6.14}$$

The third order b-symbols will now be determined by the coefficients of the n = 3 term of Equation (6.3), namely

$$\left(\frac{\partial^3 x^\mu}{\partial y^{\alpha_1} \partial y^{\alpha_2} \partial y^{\alpha_3}}\right)_0 :$$

the $b^{\mu(0)}_{i_1 i_2 i_3}$ are obtained simply from Equation (6.6): Equation (6.10) for n = 3 is satisfied if and only if

$$b^{\mu(0)}_{i_1 i_2 i_3} = -\left(\frac{\partial^3 x^\mu}{\partial y^{i_1} \partial y^{i_2} \partial y^{i_3}}\right)_0 \tag{6.15}$$

To obtain the $b^{\mu(1)}_{\alpha\beta}$ symbols consider the transformation of the Christoffel symbols of the second kind:

$$\Gamma^\mu_{\sigma\rho} \frac{\partial x^\sigma}{\partial y^\alpha} \frac{\partial x^\rho}{\partial y^\beta} = \Gamma'^\sigma_{\alpha\beta} \frac{\partial x^\mu}{\partial y^\sigma} - \frac{\partial^2 x^\mu}{\partial y^\alpha \partial y^\beta} \tag{6.16}$$

(primed quantities referred to the y^μ coordinates; unprimed - to the x^μ).

Differentiating Equation (6.16) with respect to y^4 we obtain

$$\Gamma^\mu_{\sigma\rho,\nu} \frac{\partial x^\sigma}{\partial y^\alpha} \frac{\partial x^\rho}{\partial y^\beta} \frac{\partial x^\nu}{\partial y^4} + 2\,\Gamma^\mu_{\sigma\rho} \frac{\partial^2 x^\sigma}{\partial y^\alpha \partial y^4} \frac{\partial x^\rho}{\partial y^\beta} =$$
$$= \Gamma'^\sigma_{\alpha\beta,4} \frac{\partial x^\mu}{\partial y^\sigma} + \Gamma'^\sigma_{\alpha\beta} \frac{\partial^2 x^\mu}{\partial y^\sigma \partial y^4} - \frac{\partial^3 x^\mu}{\partial y^\alpha \partial y^\beta \partial y^4} \tag{6.17}$$

at the origin, since

$$\left(\frac{\partial x^\mu}{\partial y^\alpha}\right)_0 = \delta^\mu{}_\alpha \tag{6.18}$$

and using Equations (2.8) and (2.9)

$$(\Gamma^\mu_{\alpha\beta,4})_0 - 2\left(\frac{\partial^2 x^\mu}{\partial y^\sigma \partial y^\beta}\right)_0 \left(\frac{\partial^2 x^\sigma}{\partial y^\alpha \partial y^4}\right)_0 = (\Gamma'^\mu_{\alpha\beta,4})_0 - \left(\frac{\partial^3 x^\mu}{\partial y^\alpha \partial y^\beta \partial y^4}\right)_0 \tag{6.19}$$

Thus Equations (6.11) and (6.12) will be satisfied for k = 1 if and only if

$$b^{\mu(1)}_{\alpha\beta} = -\left(\frac{\partial^3 x^\mu}{\partial y^\alpha \partial y^\beta \partial y^4}\right)_0 + (\Gamma'^\mu_{\alpha\beta,4})_0 + 2\left(\frac{\partial^2 x^\mu}{\partial y^\sigma \partial y^\beta}\right)_0 \left(\frac{\partial^2 x^\sigma}{\partial y^\alpha \partial y^4}\right)_0 \tag{6.19}$$

Proceeding by induction one can express in this way all b-symbols of the nth order in terms of the

$$\left(\frac{\partial^k x^\mu}{\partial y^{\alpha_1} \ldots \partial y^{\alpha_k}}\right)_0$$

for $k \leq n$. For the $b^{\mu(0)}_{i_1 \ldots i_n}$ the result is simply (6.6):

$$b^{\mu(0)}_{i_1 \ldots i_n} = -\left(\frac{\partial^n x^\mu}{\partial y^{i_1} \ldots \partial y^{i_n}}\right)_0 \tag{6.21}$$

For the $b^{\mu(k)}_{i_1 \ldots i_n}$ with $k \geq 1$ the expression is more complicated. It is obtained by differentiating Equation (6.16) n-2 times and using the result at the origin. In analogy with (6.20) we obtain unique expressions. The one-to-one correspondence between the frames x^μ satisfying (6.8)-(6.12) and the Riemannian coordinates satisfying (6.8) with the same numbers $a_{\mu\nu}$ is thus established (Q.E.D.).

One difference between normal and geodesic Fermi divisions is apparent by comparing Equations (6.6) and (6.20): for the normal division the relation between the b-symbols and the coefficients of the transformation

$$\left(\frac{\partial^n x^\mu}{\partial y^{\alpha_1} \ldots \partial y^{\alpha_n}}\right)_0$$

is independent of the geometrical structure; for the geodesic Fermi division, however, the relation involves terms like $(\Gamma'^\mu_{\alpha\beta\,4})_0$ which will depend on the geometrical structure. This difference is of no real significance. It follows from the method of proof of Theorem 11, 12: in both cases a one-to-one correspondence with Riemannian frames was established. The above-mentioned difference stems from the fact that all Riemannian frames with a given value of $(g_{\mu\nu})_0$ belong to the same class according to the normal division but to different classes according to the geodesic Fermi division.

The unique determination of the b-symbols by the coefficients

$$\left(\frac{\partial^n x^\mu}{\partial y^{\alpha_1}\ldots\partial y^{\alpha_n}}\right)_0$$

both for the normal and geodesic Fermi divisions is not accidental. A general analysis of this relation is outside the scope of the present work. We confine ourselves here to the following remarks: define as nth order quantity an expression which involves derivatives of $g_{\mu\nu}$ up to nth order, and is linear in the nth order derivatives. A set of nth order quantities is defined as independent if no lower order quantity can be formed from them by a linear combination. Any set of independent quantities can be broken up into two sets: (1) those that differ from components of tensors only by lower order quantities; (2) all the other independent combinations. Tensor components cannot, in general, be chosen arbitrarily: if all the components vanish in one frame they vanish in all frames. The combinations belonging to set (2) however, can be specified at will and their values form a basis for a division of all coordinate systems into classes. The b-symbols are such quantities. (It was pointed out in corollary to Theorem 1 (Section 2) and corollary 1 to Theorem 9 (Section 4) that the b-symbols of the normal and geodesic Fermi division do not contain subsets that form tensors.

The crucial point is this: the number of independent nth order quantities of set (2) is equal to the number of n+2 order derivatives

$$\frac{\partial^{n+2}x^\mu}{\partial y^{\alpha_1}\ldots y^{\alpha_{n+2}}} \quad \text{which is} \quad \frac{4}{6}\frac{(n+4)!}{(n+1)!}.$$

Now, the total number of nth order derivatives of $g_{\mu\nu}$ is

$10\,\frac{(n+3)!}{6\cdot n!}$; and indeed, the difference,

$$\frac{10}{6}\,\frac{(n+3)!}{n!} - \frac{4}{6}\,\frac{(n+4)!}{(n+1)!} = (n-1)(n+2)(n+3) \tag{6.22}$$

is the number of independent quantities in set (1), i.e. the number of components in nth order tensors. Examples:
(i) $n = 1: (n-1)n+2)n+3) = 0$: no tensors can be built from $g_{\mu\nu}$ and its first order derivatives; (ii) $n = 2$: $(n-1)(n+2)(n+3) = 20$: the number of independent components of the Riemann tensor; (iii) $n= 3: (n-1)n+2)n+3) = 60$; the number of independent covariant derivatives of the Riemann tensor components (the Bianci identities reduce this number from 80 to 60!) etc.

Let us finally note the following feature of the divisions: in general, if the same transformations formula is applied to all elements of a given class, they will be mapped thereby into elements of several distinct classes. This is a consequence of the fact that the set of all transformations between elements of any class is a quasi-group rather than a group (see Section 3). Indeed, in non-homogeneous spaces application of the same transformation formulas to two coordinate systems with, say different origins, have in general different geometrical significance.

7. THE PRINCIPLE OF PRE-ASSIGNED MEASUREMENTS

A coordinate system is a correspondence between events in space-time and sets of four numbers (x^1,x^2,x^3,x^4). Once such a correspondence is set up, the metric tensor at all points of space-time is, in principle, measurable by a system of measurements that utilizes clocks and light signals. Therefore, the metric tensor and all its derivatives to an arbitrarily high order at one point can be determined by an appropriate set of measurements. Particular choices of division, as discussed in the previous section, correspond, therefore, to particular choices of sets of measurements. The classification of coordinate systems according to the values of a chosen set of expressions (the metric tensor and certain combinations of the derivatives to arbitrarily high order) corresponds to a classification of coordinate systems according to the results of the chosen set of measurements.

Within a given space-time structure any method of choosing coordinate conditions implies, of course, a corresponding set of measurements. In General Relativity space-time structure itself is to be determined by the equations. Therefore, a choice of coordinates that depends on the geometrical structure means that the set of measurements corresponding to it is decided upon only after the problem has been solved. However, the coordinate conditions defined as a choice of class according to either division introduced in Section 5 do not depend on the geometrical structure and can be made, therefore, while the problem is set.

A particular class of frames of reference is thus defined according to pre-assigned values of the chosen set of expressions. For example, the class of geodesic Fermi frames is defined by pre-assignment of the metric tensor at a particular point as $\eta_{\mu\nu}$ and all expressions

$$\Gamma^{\mu}_{i_1\ldots i_n} \quad \text{and} \quad \frac{\partial^k \Gamma^{\mu}_{i_1\ldots i_n}}{(\partial x^4)^k}$$

at this point as zero (Theorem 9). We use the term "pre-assigned" because this assignment is made prior to determination of the geometrical structure.

The physical significance of the results of Section 5 is now formulated as the principle of pre-determined measurements.

Without reference to the geometrical structure of space-time it is possible to choose sets of measurements which define divisions of all coordinate systems into ten-parametric classes. Having chosen a particular set, each class is defined by the pre-determined results of the measurements (i.e., coordinate systems belonging to a given class are such that if the chosen measurements are carried out, their results will be equal to the pre-assigned values).

In the flat space-time of Special Relativity it is possible to divide coordinate systems into ten-parametric classes by the direct metric significance of the coordinates. This is a consequence of the homogeneity of flat space-time; in an inhomogeneous curved space-time coordinates do not possess simple metric meaning. It is, however, possible to define the physical meaning to systems of coordinates as a whole, rather than the numberical values of the coordinates, in terms of the above-mentioned sets of measurements. In the limit of flat space-time the two ways of characterizing frames of reference are equivalent (e.g. second corollary to Theorem 9). Since, however, the second way makes no reference to the geometry of space it is directly generalizable to curved space-time.

The results of Section 5 show that if the set of measurements is big enough the corresponding set of frames are ten parametric. In flat space time the occurrence of ten-parametric sets of frames is linked with the degree of symmetry of space. Here we realize that such ten-parametric sets arise naturally when the maximum freedom in pre-assigning results of measurements is used.

In the derivation of the principle no mention was made of the equations which describe gravitation. The principle is, therefore, independent of any particular set of equations and can serve as a basis for a general treatment of the question of degree of invariance of the laws of nature.

8. THE DEGREE OF INVARIANCE OF THE LAWS OF NATURE

Laws of nature are relations between quantities which describe physical properties of systems, and means of predicting future change on the basis of initial conditions. If we consider Maxwell's equations as an example, two of them are relations between quantities referring to the same time and the other two describe the change in time of the electromagnetic field. In this section we shall use the phrase "The laws of nature are the same for two observers" in the following sense:

Definition. The laws of nature are the same in two frames of reference if the procedure which uniquely specifies all physical quantities at initial and later times in terms of given initial conditions is the same in both.

If one considers, for example, the gravitational field, then this procedure involves not only the equations for the metric tensor but also the expressions for gravitational energy, momentum and angular momentum. All such expressions must be the same for two observers if the laws of nature are the same in their respective systems.

Our definition is incompatible with the principle of covariance. Although the results of the preceding sections are obviously independent of whether or not one accepts this principle, our motivation for undertaking the present work lay in the following reservations concerning the principle of covariance:

(1) The vision behind the creation of the General Theory of Relativity included not only a formulation of the laws of nature that is the same for all observers, but also reduction of dynamics to geometry by showing that forces are nothing but geometrical manifestations of space-time. When the theory was proposed in 1915 the only known forces were gravitational and electromagnetic. Of these two only gravitation is described geometrically, but Einstein considered his theory as a first step towards a unified field theory that will describe both kinds of forces geometrically [12]. Since then strong and weak forces were discovered and the attainment of this aim seems hardly within grasp. Without taking a stand on whether or not this aim will ever be achieved, we believe that the limitation of geometrical description to gravitation only imposes a natural corresponding limitation on the class of observers for which the laws of nature are the same. Indeed, if all forces except for gravitation were switched off, and matter were treated as singularities in space-time, then all matter would have moved on time-like geodesics. All measurements would have been taken, therefore, by observers moving on time-like geodesics, and invariance with respect to such observers would have been the only physically meaningful question. Thus, as long as gravitation only is treated geometrically, the class of observers moving along geodesics is of special significance. Systems of coordinates in four-dimensional space-time, the spatial origin of which moves along a geodesic are singled out of the set of all systems of coordinates.

(2) The principle of covariance attempts to put on equal footing observers whose coordinates have, physically, hardly anything in common, e.g. whose coordinates mean distances and observers whose coordinates mean angles. What was the motivation for such a far-reaching generalization? In Einstein's own account it was the realization that in curved space-time (in contradiction to the flat space of Special Relativity) it is impossible to ascribe simple metric meaning to the coordinates themselves [13].

The emphasis here is on the search for metric meaning of the numerical values of the coordinates. If one looks instead for a physical meaning of the coordinate systems as a whole, new possibilities arise. Once it is recognized that frames can be naturally divided into ten-parametric sets, one can postulate that the laws of nature are the same for all members of the same set only and can take different forms when expressed in different ten-parametric sets. The choice of division as well as the choice of a ten-parametric set within the division becomes then a matter of convenience, not of principle [14].

Let us compare this approach with the situation of Special Relativity. The Principle of Relativity states that the laws of nature are the same for two observers in relative uniform motion using similarly constructed yardsticks and clocks. Stated mathematically, it prescribes a way of dividing all frames in flat space-time into ten-parametric classes and states that the laws of nature are the same for all members of the same class. It turns out that the laws (e.g. Maxwell's Equations) take the simplest form in the ten-parametric class of Cartesean frames; but this is a point of convenience, not of principle. In principle one is allowed to formulate everything in e.g. cylindrical frames whose origin are accelerated uniformly with a given magnitude and direction. The set of all these frames is ten-parametric and the laws of nature are the same in all of them.

Einstein pointed out the major difference between coordinate systems in flat and curved space-time. In the former the metric tensor can have a simple metric meaning while in the latter it does not [13].

In terms of classifications this difference is reflected as follows: in flat space-time there exists a unique division of frames to ten-parametric classes such that the laws of nature are invariant under transformations within each class. In curved space-time there are numerous divisions, depending on the choice of the linear combinations of the high derivatives of the metric tensor (see Section 6). In comparison with Special Relativity one has now a double freedom (a) choosing a classification; b) choosing a ten-parametric set within the classification. Let us emphasize that this situation comes about not only because space-time is curved, but also because its geometrical structure is no longer given a-priori.

The following, therefore, is our suggested formulation of the principle of invariance in the general case:

> Corresponding to any division of all frames into ten-parametric classes according to the values of the matric tensor and linear combinations of its derivatives of arbitrarily high order at one point, and corresponding to any choice of a ten-parametric class within such a division [14], it is always possible to formulate the laws of nature in a form which is invariant under transformations between members of the same class.

Remarks:

(i) This principle allows for the possibility that the laws of nature take a different form according to the choices made in the aforementioned double freedom, i.e. according to (a) which classification is chosen and (b) which class is chosen within a classification.

(ii) The principle avoids the difficulty of assuming the laws to take the same form for observers whose coordinates have entirely different physical meaning, e.g. lengths versus angles; or for observers who are freely falling (moving on geodesics) versus observers moving under the influence of electromagnetic forces.

(iii) Unlike Special relativity, since the metric tensor is now an unknown, not an a-priori given quantity (a "relative" rather than "absolute" element of the theory in Anderson's terminology [11]), the choice of class is not linked to a particular geometrical structure; it allows an infinite variety of structures according to the initial conditions. Two classes belonging to different divisions may be identical in a particular structure (e.g. normal and flat space-time) but will be different in different structures that will arise from different initial conditions.

9. GENERALIZATION OF THE POINCARE GROUP TO CURVED SPACE-TIME; AND CONCLUDING REMARKS

We pointed out in Section 4 that geodesic Fermi coordinates are the natural generalization of inertial frames to Riemannian space-time. Let us choose, therefore, the ten-parametric set of geodesic Fermi coordinate systems for a specific formulation of the principle of invariance. It follows from the general formulation of Section 8 that this set is one of the many ten-parametric sets which can be chosen for this purpose.

Let us consider therefore, the following principle: the laws of nature are the same in all geodesic Fermi frames.

We begin by pointing out that this principle, as well as the general one (Section 8), has nothing to do with particular choices of boundary values for the metric tensor at infinity. Thus, if one considers the problem of an isolated sun-earth system, observers at the center of the sun and at the center of the earth both move on geodesics, and are, therefore, geodesic Fermi if they use the appropriate system of measurement. Nevertheless, their metric tensors do not approach the same limit at large distances from the system.

Let us also emphasize that this principle is independent of a particular theory of gravitation, such as Einstein's equations with or without the cosmological constant, or the scalar-tensor theory. It is more in the nature of a "super-law" in Wigner's sense of the term: a limitation on acceptable theories.

From the formulation of the principle in terms of geodesic Fermi frames it follows that the laws of nature are unchanged under transformations between such frames. The set L of all transformations between such coordinate systems can be regarded as a generalization of the Poincaré group to curved space-time.

Explicit formulas for the elements of L can be obtained by a combination of the results of Section 3 and the natural correspondance between geodesic Fermi and normal coordinates defined at the end of Section 5. Given any two geodesic Fermi frames $l(P,\lambda)$ and $l(P',\lambda')$, with origins at P, P' and directions of axes denoted collectively by λ and λ', we carry out the transformations from $l(P,\lambda)$ to $l(P',\lambda')$ in three stages:

$$l(P,\lambda) \to n(P,\lambda) \to n(P',\lambda') \to l(P',\lambda') \tag{9.1}$$

The first and last stages are given by Equations (2.9) and (2.10). The second stage is the subject matter of Section 3. In analogy with the set N of all transformations between normal coordinates, the set L is a quasi-group, not a group.

A few remarks concerning transformation properties of physical entities and a preliminary consideration of the problem of energy-momentum complex within the framework of our principle may help clarify its meaning.

A quantity which describes an intrinsic property of a physical system is independent of the coordinate system in which it is expressed; its definition is incomplete unless, together with its expression is one frame, the transformation law to all frames is given. This understanding is maintained, of course, in the present approach. We do not assume, however, that the mathematical relations between the expressions for physical entities in all coordinate systems are always the same. A consideration of a non-tensorial physical entity, the gravitational energy-momentum complex will illustrate the difference.

The problem of definition of conserved gravitational energy and momentum complex within the framework of Einstein's equations has been a subject of research since 1915. According to the principle of covariance, the goal has always been to reach an expression for an arbitrary frame of reference. Recently Møller [15] has shown that no expression containing the metric tensor only can serve as a satisfactory definition. He introduced accordingly an expression in terms of certain unobservable quantities. In the present approach, however, Einstein's equations and the coordinate conditions which determine a particular class, say, the class of geodesic Fermi frames, are taken as a unified set of equations. Energy-momentum complexes could be constructed by application of Neother's theorem to a Lagrangian that generates the whole set. Different Lagrangians will be associated with different classes. Energy-momentum expressions will differ accordingly; none will involve unobservable quantities. Since within any particular class the complex is uniquely defined by the metric tensor, and since the division into classes involves all coordinate systems, the transformation of the energy-momentum complex between arbitrary frames will be uniquely defined.

Our principle implies a similar change of approach concerning the problem of quantization of the gravitational field.

Difficulties connected with this problem are well-known [16]. A major difficulty is the non-uniqueness of the Hamiltonian which reflects the fact that Einstein's equations (or any other set of tensorial equations) do not specify a unique solution for the metric tensor with given initial conditions on a space-like surface. This non-uniqueness is eliminated if quantization within a geodesic Fermi frame of reference, instead of an arbitrary frame, is aimed at. This is similar to the situation in the familiar quantum electrodynamics: the usual quantization of, say, the electromagnetic field is carried out in a Lorentz frame, not in an arbitrary system of coordinates.

ACKNOWLEDGEMENTS

We are grateful to Professor Asim O. Barut and Professor Robert H. Richtmayer for very stimulating discussions.

REFERENCES AND NOTES

1. L.P. Eisenhart, Riemannian Geometry, Princeton University Press, Princeton, 53 (1949).
2. Veblen and Thomas, Trans. Am. Math. Soc. 25, 551 (1923).
3. The terms "coordinate system", "frame of reference", "frame" will be used interchangeably.
4. These coordinate systems were introduced e.g. by J.L. Synge, Relativity, the General Theory, North-Holland, Amsterdam, 83 (1966). Synge calls these frames simply "Fermi coordinates". This name usually refers, however, to a much bigger class of coordinate systems, namely those that satisfy

$$g_{\mu\nu} = \eta_{\mu\nu} \text{ and } \Gamma^{\alpha}_{\beta\gamma} = 0$$

along a given curve (E. Fermi, Atti Accad. Naz. Lincei Rend. Classe Sci. Fis. Mat. Nat. 31, 21, 51 (1922); T. Levi-Civita, Math. Ann. 92, 291 (1926); L. O'Raifeartaigh, Proc. Roy. Irish Acad. A59, 15 (1958); L. O'Raifeartaigh and J.L. Synge, Proc. Roy. Soc. (London) A246, 299 (1958)). We prefer, therefore, to use the term "geodesic Fermi coordinates" suggested by A. Schield (Relativity Theory and Astrophysics, I. Relativity and Cosmology, Lectures of Applied Mathematics, American Mathematical Society, Providence, R.I., 1967, Vol. 8). Geodesic Fermi coordinates were recently used by J.L. Anderson, "Maximal Covariance Conditions and Kretschmann's Relativity Group", in Prospectives in Geometry and Relativity, B. Hoffman (ed.), Indian U.P., Bloomington, 1966.
5. The problem of generalizing inertial frames to curved space-time was discussed in a different context by N. Rosen, Proceedings of the Israel Academy of Sciences and Humanities, Sec. of Sciences, 12, 1 (1968). Professor Rosen investigates

coordinate systems within given models of the universe.

6. Problems which involve gravitational radiation are not included, therefore, in the present treatment.
7. S.S. Schwever, An Introduction to Relativistic Quantum Mechanics, Harper and Row Pub., New York, 40 (1962).
8. To the best of our knowledge such a structure has never been defined before.
9. For definition and discussion of geodesic manifold, see T. Levi-Civita, The Absolute Differential Calculus, Blackie & Son Ltd., London and Glasgow, 162 (1954).
10. Having proved part (1) we can assume Equation (5.11) too holds. Equations (2.3), (2.4) reduce then to the form

$$\Gamma^{\alpha}_{\beta\gamma\delta} = \frac{1}{3} P \left(\frac{\partial}{\partial x^{\delta}} \Gamma^{\alpha}_{\beta\gamma}\right)$$

$$\Gamma^{\alpha}_{\beta\gamma\delta\ldots\mu\nu} = \frac{1}{N} P \left(\frac{\partial}{\partial x^{\nu}} \Gamma^{\alpha}_{\beta\gamma\delta\ldots\mu}\right)$$

This form is used in deriving Equations (5.19) and (5.23).

11. The principle of covariance has been a subject of discussion ever since Krechmann's objection (Am. Physik. 53, 575 (1917)) to Einstein's original formulation. More recently, J.L. Anderson (Relativity Principles and the Role of Coordinates in Physics, in Gravitation and Relativity, Hong-Yee Chin and W. Hoffman (eds.), W.A. Benjamin, 175 (1964)) presented a new analysis of the problem and a new definition according to which the principle of general covariance holds. His definition is different from ours.
12. A. Einstein, The Meaning of Relativity, 3rd Ed., Princeton University Press, 133 (1960).
13. A. Einstein, Autobiographical Notes in Albert Einstein, Philosopher-Scientist, P.A. Schlipp (ed.), Vol. 1, Harper and Brothers, New York, 69 (1959).
14. Let us recall (Section 6) that choice of division physically means choice of a set of measurements, and choice of class within a division physically means choice of numerical values for the results of these measurements.
15. C. Møller, Mat. Fys. Medd. Dan. Vid. Selsk. 35, No. 3 (1966).
16. J.L. Anderson, Quantization of General Relativity in Gravitation and Relativity, Hong-Yee Chiu and W.H. Hoffman (eds.), W.A. Benjamin, Inc., 1964.

SYMMETRIC SPACES IN RELATIVITY AND QUANTUM THEORIES*

Hans Tilgner
Institut für theoretische Physik der TU
3392 Clausthal-Zellerfeld

1. INTRODUCTION

The following is an attempt to fill in the gap between some modern mathematical concepts and physics. The mathematics basically used is the differential geometry of symmetric spaces, which was formulated in a new way by O. Loos in his two books. He has shown that the old definition of a symmetric space coincides with his axioms (S1) to (S4) for a symmetric space as a manifold with multiplication. This definition makes the analogy with Lie groups (more general with arbitrary groups) obvious, where only the multiplication is changed to a (Lie) group multiplication. One can ask nearly all questions, which are solved for Lie groups, in the same way for symmetric spaces, and one can answer most of them! The most important concepts, related to a Lie group is its Lie algebra. In the same sense, there is a linear structure on the tangent space of a symmetric space, the "Lie triple system". A third example of this kind is discussed, where the manifold is a "domain of positivity", its tangent space carrying a "formal real Jordan algebra" structure. In the diagram

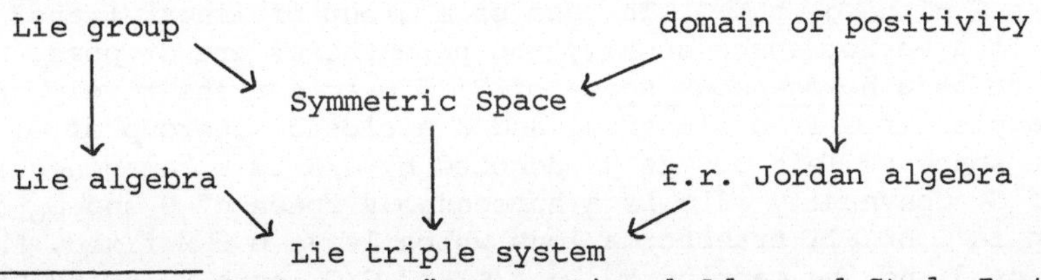

* Lecture given during the "International Advanced Study Institute on Mathematical Physics, Group Theory in Nonlinear Problems" in Istanbul, Turkey, 7-18 August 1972.

the vertical arrows are the tangent functors, i.e. mappings from the manifold onto its tangent space, carrying over the nonlinear structure on the manifold onto a multilinear structure in the tangent space. The other arrows mean that the spaces carry in a natural way the structures of the spaces onto which they point. However, there are anomalies in the diagram. A domain of positivity carries no (known) multiplication, different from the symmetric one, like groups do. A Lie triple system has a trilinear composition, whereas Lie and Jordan algebras have bilinear ones. Domains of positivity are defined as open subspaces of vector spaces, whereas the other two kinds of manifolds in general cannot be embedded into their tangent structures. However, there are cases, where such an embedding exists as well; for instance the set of invertible matrices is embedded into its tangent space, the Lie algebra $gl(n, \mathbb{R})$ of all matrices.

In Section 4 symmetric spaces are traced back to Lie groups via "homogeneous" spaces. Hence for every symmetric space in physics, there is a group theory. The converse is not true because condition (4.2) is not fulfilled in all possible cases. However, there seem to exist only few homogeneous spaces in physics which are not symmetric. Keeping those exceptions in mind, one can state, that the theory of symmetric spaces may be helpful for a more intuitive description of physical problems, which are described so far in group theoretical terms only.

2. LIE TRANSFORMATION GROUPS, LIE ALGEBRAS, COVERING AND PSEUDO-ORTHOGONAL GROUPS

2.1 Lie groups

A Lie group is a group G which is also an analytical manifold such that the mapping $(g,f) \to gf^{-1}$ of $G \times G$ into G is analytic.

Remark: It suffices to demand G to be a topological manifold only, since then there is unique differentiable and even analytical structure in which the above mapping is differentiable and even analytic.

A Lie group G with identity element e is said to act on a manifold M as a Lie transformation group if there is a mapping $\mu : G \times M \to M$, $\mu(g,p) = g(p)$ such that $e(p) = p$ and $g(h(p)) = (gh)(p)$ for all $p \in M$ and $g, h \in G$. In case of a group of linear transformations in a vector space usually the parentheses are dropped. M is said to be a homogeneous space of G if G acts transitively on M.

Example: If G is a Lie group and K a closed subgroup of G, then the space of left cosets gK denoted by G/K is a homogeneous space of G. Conversely if M is a homogeneous space of G and G_p the subgroup in G of all transformations which leave $p \in M$ fixed, then G_p is closed and the mapping $gG_p \to g(p)$ of G/G_p onto M is analytic.

2.2 Lie algebras

Every one-parameter subgroup θ of G, i.e. every Lie group homomorphism θ of the additive abelian group $\mathbb{R}$ into G, $\theta : \tau \to \theta_\tau$, defines a vector field X^θ on M, i.e. a derivation X^θ of the (commutative, associative) algebra F(M) of real valued, continuous functions on M (pointwise multiplication), by

$$(X^\theta \phi)(p) = \lim_{\tau \to 0} \frac{d}{d\tau} \phi(\theta_\tau(p)) \qquad p \in M, \ \phi \in F(M) \ , \tag{2.1}$$

The vector space of all vector fields is a Lie algebra, and the vector space of all vector fields (2.1) is a subalgebra with the dimension of G.

In mathematics the Lie algebra of a Lie group G is given as a special case of this definition: take M = G; then G operates on G as Lie transformation group by left multiplications l(G), l(g)h = gh and the set of all left invariant vector fields X, i.e.

$$(X\phi) o l(g) = X(\phi o l(g)) \qquad g \in G, \ \phi \in F(M) \quad , \tag{2.2}$$

is the Lie algebra of G. Because of specialisation of M to G is of no importance in mathematical physics, this concept is never used. In physics for instance M is the configuation space of a dynamical system or the space-time manifold (see, however, the work of Oszváth and Schücking, where the space-time manifold is a Lie group). Vector fields on M, defined by (2.1), with respect to a group G in physics are called generators of G. If M is a vector space there is another concept of Lie algebra, given by the "infinitesimal" transformations of G in M [Ti 71a]. Here the elements of the Lie algebra are rational transformations (not necessarily linear) of M rather than (linear) transformations of F(M). Example (linear): Lie group Gl(M,$\mathbb{R}$), Lie algebra of infinitesimal transformations gl(M,$\mathbb{R}$). Due to the definition by one-parameter subgroups, all these concepts define isomorphic Lie algebras if they exist simultaneously.

Tangent vectors X_p^θ of M in the point p are defined by

$$X_p^\theta \phi = (X^\theta \phi)(p) \qquad \phi \in F(M), \ p \in M \tag{2.3}$$

They are linear transformations $X_p : F(M) \to \mathbb{R}$ subject to

$$X_p(\phi\psi) = \phi(p)X_p(\psi) + \psi(p)X_p(\phi) \qquad p \in M \quad \phi,\psi \in F(M) \tag{2.4}$$

This relation can be used to transport the Lie algebra structure from the vector fields to the tangent vectors.

2.3 Coverings

A covering $(\tilde{M},\pi)$ of a manifold M is a pair of a manifold $\tilde{M}$ and an onto mapping $\pi : \tilde{M} \to M$ such that the fibre $\pi^{-1}(p) = \{\tilde{p} \in \tilde{M} / \pi(\tilde{p}) = p,\ p \in M\}$ over every point p is a discrete space. The study of coverings is particularly simple if the manifolds carry a multiplication, since the fibres of the coverings then become kernels of the covering homomorphisms π and therefore can be determined algebraically. If the left multiplication acts transitively, the number of points in the fibres over all points is the same, since the fibres of different points are transformed into each other by suitable left multiplications. The best known example in physics is the onto Lie group homomorphism $Sl(2,C) \to SO(1,3;\mathbb{R})$ where the kernel is $\pm$ the identity transformation in $Sl(2,C)$.

Covering homomorphisms induce isomorphisms of the local algebraic structures in the tangent spaces if these structures are related to the algebraic composition on the manifolds in a functorial way.

Lie groups are related to matrix Lie groups exactly by covering. This follows from the fact that every Lie group has a Lie algebra and that every finite-dimensional Lie algebra has a faithful finite-dimensional representation (theorem of Ado). However, there are Lie groups which have no faithful finite-dimensional representation.

2.4 Pseudo-orthogonal vector spaces

A pseudo-orthogonal vector space $(V,\langle,\rangle)$ is a pair of a vector space V of finite dimension n and a non-degenerate symmetric bilinear form $\langle,\rangle$. The invertible and symmetric matrix of $\langle,\rangle$ in some given basis will be denoted by I. There is a unique basis in V such that $I = \mathrm{diag}(id_{n_1}, -id_{n_2})$, $n_1+n_2 = n$, where (n_1,n_2) is the signature of $\langle,\rangle$.

The pseudo-orthogonal group of $(V,\langle,\rangle)$ is defined by

$$\mathrm{Aut}(V,\langle,\rangle) = \{A \in Gl(V,\mathbb{R}) / \langle Ax,Ay\rangle = \langle x,y\rangle,\ \forall x,y \in V\} \tag{2.5}$$

which is in matrix form in the basis defined by I

$$\mathrm{Aut}(V,\langle,\rangle) = \{A \in gl(V,\mathbb{R}) / A^t I A = I\} \tag{2.6}$$

If I is diagonal we write $\mathrm{Aut}(V,\langle,\rangle) = O(n_1,n_2)$ (since the ground field will be only $\mathbb{R}$ in the following this letter is dropped in $O(\ )$, $SO(\ )$ etc.).

The pseudo-orthogonal Lie algebra of $(V,\langle,\rangle)$ is the Lie algebra of infinitesimal transformations of $\mathrm{Aut}(V,\langle,\rangle)$, which can be calculated by inserting $A = \exp\tau B$ into (2.5) resp. (2.6) and taking $\tau \to 0$

$$\begin{aligned} \mathrm{der}(V,\langle,\rangle) &= \{B \in gl(V,\mathbb{R}) / \langle Bx,y\rangle + \langle x,By\rangle = 0,\ \forall x,y \in V\} \\ &= \{B \in gl(V,\mathbb{R}) / B^t I + IB = 0\}\ . \end{aligned} \tag{2.7}$$

If I is diagonal we write $der(V,<,>) = so(n_1,n_2)$. Every pseudo-orthogonal group is conjugate (but not equal since the underlying subsets of $Gl(V,\mathbb{R})$ are different) to a group $O(n_1,n_2)$ and hence globally isomorphic. The number of connectivity components and the connectivity of every component therefore are the same. As a consequence the Lie algebras are conjugate as well.

The connectivity component of the identity of a group G will be denoted by G_0 in the following.

PART I / SYMMETRIC SPACES AND LIE TRIPLE SYSTEMS

The idea of symmetric multiplication is due to O. Loos, although symmetric spaces are known much longer. The material of Section 3 is taken from his books. Example (c) with multiplication (3.3) has not been treated before. In Section 4, which is entirely taken from Loos' books, it is shown that symmetric spaces are exactly a certain class of factor spaces of (Lie) groups. Section 5 gives the tangent functor onto Lie triple systems. Again it is taken from Loos' books although this relation was known before [He p.189]. In Section 6 a class of symmetric spaces of the pseudo-orthogonal groups is classified such that those of the Lorentz group are included. Among them there are spaces which cannot be realized as orbits in the selfrepresentation space (= Minkowski space). Therefore the higher dimensional space of the adjoint representation, which is the Lie algebra, probably is more appropriate. Physically the program of this section for SU(3) in the adjoint representation is more interesting, since it is related to hadrons and strong interaction. Section 7 is devoted to the conformal group in order to give in Section 8 an example of a homogeneous space (of the conformal group) which is not symmetric. Section 9 gives a formulation of the problem of symmetric spaces in general relativity and cosmology.

3. SYMMETRIC SPACES

3.1 Definition

A symmetric space is a manifold M with a differentiable multiplication $\mu : M \times M \to M$, $\mu(p,q) = p\cdot q$ subject to

(S1) $p\cdot p = p$
(S2) $p\cdot(p\cdot q) = q$
(S3) $p\cdot(q\cdot r) = (p\cdot q)\cdot(p\cdot r)$
(S4) every p has a neighborhood U such that $p\cdot q = q$ implies $p = q$ for all $q \in U$.

A connected symmetric space has a natural real analytic structure such that the multiplication is real analytic. The left multiplication with p in M is denoted by S(p), i.e. $S(p)q = p\cdot q$. It is called the symmetry around p. The following statements are immediate:

(S5) p is an isolated fixpoint of S(p)
(S6) S(p) is an involutive automorphism of M.

A pointed symmetric space is a pair (M,o) of a symmetric space M and a point $o \in M$, called the base point. A homomorphism of pointed symmetric spaces is a homomorphism of symmetric spaces which maps base point onto base point. Isomorphisms and automorphisms are defined as usual. Every point of a symmetric space is a symmetric subspace.

3.2 Examples

Standard examples of symmetric spaces are:

(a) Lie groups with the symmetric multiplication

$$p\cdot q = pq^{-1}p \qquad p,q \in G. \tag{3.1}$$

these symmetric spaces are denoted by G^s;

(b) vector spaces with the symmetric multiplication

$$x\cdot y = 2x - y \tag{3.2}$$

are a particular case of (a) if one specializes the group G to the additive abelian group of the vector space;

(c) the set of elements outside the null cone $\{y \in V / \langle y,y\rangle = 0\}$ in a pseudo-orthogonal vector space $(V,\langle,\rangle)$ with the symmetric multiplication

$$x\cdot y = 2\,\frac{\langle x,y\rangle}{\langle y,y\rangle}\,x - \frac{\langle x,x\rangle}{\langle y,y\rangle}\,y \tag{3.3}$$

because of the identity (read z = y)

$$\langle x\cdot y, x\cdot z\rangle = \frac{\langle x,x\rangle^2\langle y,z\rangle}{\langle y,y\rangle\langle z,z\rangle} \quad ; \tag{3.4}$$

if we introduce the reflections S_y and inversions R by

$$S_y z = z - 2\,\frac{\langle y,z\rangle}{\langle y,y\rangle}\,y \ , \quad R(z) = \frac{z}{\langle z,z\rangle} \tag{3.5}$$

with $S_y^{\,2} = id_V$, $S_y \in Aut(V,\langle,\rangle)$ and $R^2 = id_V$, then

$$S(y) = -\langle y,y\rangle S_y \circ R = -\langle y,y\rangle R \circ S_y \ ; \tag{3.6}$$

hence it is homogeneous of degree -1;

this example has two important kinds of symmetric subspaces:

(d) every connectivity component of (c) is a symmetric subspace; for instance the interior of the null cone $\{y \in V / \langle y,y\rangle > 0\}$ and the exterior $\{y \in V / \langle y,y\rangle < 0\}$; we have the following four cases for the number of connectivity components of (c):
one if sign $\langle,\rangle = (0, n_2 \geqslant 1)$ or $= (n_1 \geqslant 1, 0)$; the interior or exterior of the null cone $\{0\}$ shrinks to the point $\{0\}$; $V \setminus \{0\}$ obviously is a cone (see Section 10 below);
two if sign $\langle,\rangle = (n_1 \geqslant 2, n_2 \geqslant 2)$ where interior and exterior of the null cone are cones, but not convex;
three components if sign $\langle,\rangle = (1, n_2 \geqslant 2)$, which is called the Lorentz signature, or sign $\langle,\rangle = (n_1 \geqslant 2, 1)$; in the Lorentz case the interior of the null cone (= light cone) decomposes into two convex cones, the interior of the forward light cone $\{y \in V / \langle t,y\rangle > 0,\ \langle y,y\rangle > 0\} = Y$ for a $t \in V$ with $\langle t,t\rangle = 1$ (the definition does not depend on the choice of such a t) and the interior of the backward light cone $\{y \in V / \langle t,y\rangle < 0,\ \langle y,y\rangle > 0\} = -Y$.
four components if sign $\langle,\rangle = (1,1)$ all of which are convex cones;

(e) the hyperboloids or mass shells of radius $\sqrt{\kappa}$ in $(V, \langle,\rangle)$

$$M^{\kappa} = \{y \in V / 0 \neq \langle y,y\rangle = \kappa \in \mathbb{C}\} \tag{3.7}$$

since (3.4) reduces to $\langle x \cdot y, x \cdot y\rangle = \kappa$ they become symmetric subspaces with

$$x \cdot y = \frac{2}{\kappa} \langle x,y\rangle\, x - y \ . \tag{3.8}$$

The null cone itself is excluded from the examples (e). It is an interesting question whether it can be made a symmetric space (see Section 8 below), and whether the union of the interior (or exterior) of the null cone together with the latter (its boundary) can be given a symmetric multiplication, the restriction of which to the interior is (3.3).

3.3 Special concepts

By a one-parameter symmetric subspace of a pointed symmetric space (M,o) we mean a homomorphism $\theta : (\mathbb{R},0) \to (M,o)$ of pointed symmetric spaces. Let us give some examples: (a) One-parameter subgroups of G induce one-parameter symmetric subspaces of G^s. (b) The semidirect product of $Gl(V,\mathbb{R})$ with the (normal subgroup of) translations is a subgroup of the automorphism group of the symmetric space $(V,\cdot)$. The one-parameter subgroups of the translations induce one-parameter symmetric subspaces $\mu \to o + \mu a$ for any base point $o \in V$. In $Gl(V,\mathbb{R})$ there are two classes of one-parameter subgroups [Ti 71a]: The degenerate one induces one-parameter subspaces by application to any base point, whereas the non-degenerate class does not for any base point $0 \neq o \in V$. This shows that not every one-parameter subgroup of the automorphism group of a symmetric space gives a one-parameter symmetric subspace when acting on a base point. For (c) and (d) we get one-parameter symmetric subspaces by applying one-parameter subgroups of the dilatations (which are in the automorphism group) to any base point. Clearly they are lines through the base point approaching zero. A further class of one-parameter symmetric subspaces of (c), (d) and (e) were described in [Ti 72].

A homomorphism of pointed symmetric spaces $\phi : (M,o) \to (C \setminus \{0\},1)$, where the multiplication of the latter is given by (3.1), is called a symmetric function on M. Examples: Characters of a Lie group G are symmetric functions of G^S. Symmetric functions for (b) are given by the functions f_a with $f_a(x) = e^{<a,x>}$ for any bilinear form <,> on V. They seem to be naturally related (but not identical) to the spherical functions of Lie groups, whence to the Gelfand-Neumark-Segal construction of unitary group representations, see [He Ch. X Section 4].

Spherical functions on symmetric spaces seem to be the easiest approach to the "special functions" of mathematical physics.

3.4 The square realization of symmetric spaces

The group generated by all $S(p)S(q)$ with $p,q \in M$ is called the group of displacements and is always denoted by Dis(M). From $S(\phi(p)) = \phi S(p)\phi^{-1}$ for $p \in M$ and $\phi \in \mathrm{Aut}(M)$, we see that Dis(M) is a normal subgroup of Aut(M). The square realization of (M,o) is the map $Q : M \to \mathrm{Dis}(M)$ defined by $Q(p) = S(p)S(o)$. From $Q(p)Q(q)^{-1} = S(p)S(q)$ we see that Q(M) generates Dis(M).

Theorem. $Q : (M,o) \to (\mathrm{Dis}(M)^S, \mathrm{id}_M)$ is a homomorphism of pointed symmetric spaces, i.e.

$$Q(p\cdot q) = Q(p)Q(q)^{-1}Q(p) \quad . \tag{3.9}$$

The proof is straightforward.
For example (b) Q is an isomorphism onto the translations of V. For example (c), where M is the outside of the null cone, we get for $n \geqslant 3$ and SO(4) excluded.

Theorem.

$$\mathrm{Dil}(V) \otimes \mathrm{Aut}_0(V,<,>) \subset \mathrm{Dis}(M) \subset \mathrm{Dil}(V) \otimes \mathrm{Aut}(V,<,>) \quad . \tag{3.10}$$

Remark: This implies (a) in the compact case $\mathrm{Dis}(M) = \mathrm{Dil}(V) \otimes \mathrm{Aut}(V,<,>)$ and (b) that Dis(M) contains at least two components in the non-compact case. In Minkowski space the second is the PT-component.

Proof: (3.6) gives $Q(x) = <x,x>S_x S_t$; hence Dis(M) is the restriction to M of a linear transformation group of V. Moreover $\langle Q(x)y,Q(x)z\rangle = \langle y,z\rangle$. It is easy to verify that $Q(\langle x,x\rangle^{-1/2}x) = S_xS_t$ for $\langle x,x\rangle > 0$ and for $\langle x,x\rangle < 0$ $Q((-\langle x,x\rangle)^{-1/2}x) = S_xS_t$. From this $Q(x)Q(\langle x,x\rangle^{-1/2}x)^{-1} = D_{\langle x,x\rangle}$ in the first case and $Q(x)Q((-\langle x,x\rangle)^{-1/2}x)^{-1} = D_{\langle x,x\rangle}$ in the second case. Hence $\mathrm{Dil}(V) \subset \mathrm{Dis}(M)$ and $\mathrm{Dis}(M) \subset \mathrm{Dil}(V) \otimes \mathrm{Aut}(V,\langle,\rangle)$. Every $A \in \mathrm{Aut}(V,\langle,\rangle)$ with $A \neq \mathrm{id}_V$, can be written as a product of at most n S_x, hence the group which is generated by the S_xS_y, $\mathrm{Dis}(M)\cap\mathrm{Aut}(V,\langle,\rangle)$ is a normal subgroup in $\mathrm{Aut}(V,\langle,\rangle)$. The theorem follows from the simplicity of $\mathrm{Aut}_0(V,\langle,\rangle)$.

Since $Q(x) = Q(y)$ implies $x = \pm y$ the square realization is a twofold covering homomorphism onto $Q(M)$.

For example (d) with the Lorentz signature, $y \in Y$ implies $\det Q(y) > 0$ since $\det S(x) = -1$ for all $x \in V$ [Ti 71a]. Hence the upper boundary in (3.10) shrinks to the two components with positive determinant. $y \in Y$ implies $S_t y \in -Y$ hence $S_t \notin \mathrm{Dis}(M)$. In Minkowski space this implies $\mathrm{Dis}(Y) = \mathrm{Dil}_0(V) \otimes \mathrm{Aut}_0(V,\langle,\rangle)$. Clearly there is a basis in V in which S_t is time reflection and $-S_t$ space reflection.

Q is an isomorphism of symmetric spaces from Y onto $Q(Y) = \langle Y,Y\rangle S_YS_t$. From this, using (3.4) one proves

$$S_{x\cdot y} = S_xS_yS_x \qquad x,y \in M \tag{3.11}$$

i.e. the map $M \to \mathrm{Aut}(V,\langle,\rangle)^s$ defined by $x \to S_x$ is a homomorphism of pointed symmetric spaces as well. $S_x = S_y$ iff $x = \lambda y$ with non-vanishing λ shows that S is an onto homomorphism of Y (which can be seen as a fibre bundle with base space M^κ, the fibres being the rays in Y from the origin) onto $S(M^\kappa) \approx M^\kappa$.

4. SYMMETRIC SPACES AS HOMOGENEOUS SPACES OF GROUPS

Let G be a connected Lie group with an involutive automorphism σ and

$$G^\sigma = \{g \in G \,/\, \sigma(g) = g\}$$

$$G_\sigma = \{g\sigma(g)^{-1} \,/\, g \in G\}$$

G^σ is the set of <u>fixed points</u> of σ. G_σ is called the <u>space of symmetric elements of G</u>. $g \in G\sigma$ implies $\sigma(g) = g^{-1}$.

$$G_\sigma \text{ is a symmetric sub-space of } G^s. \tag{4.1}$$

Proof: $g\sigma(g)^{-1}[h\sigma(h)^{-1}]^{-1} = g\sigma(g)^{-1}\sigma(h)\sigma(g\sigma(g)^{-1}\sigma(h))^{-1}$ since $\sigma(g)^{-1} = \sigma(g^{-1})$.

Let K be a sub-group of the Lie group G such that

$$(G^\sigma)_0 \subset K \subset G^\sigma \tag{4.2}$$

Then K is closed and $M = G/K$ is an (analytic) manifold.

M is a pointed symmetric space with the multiplication

$$gK \cdot hK = g\sigma(g)^{-1}\sigma(h)K \tag{4.3}$$

and base point eK.

A verification gives (S1), (S2) and (S3). For the proof of (S4) one needs the exponential mapping of manifolds.

G acts transitively on M by left multiplications $l(G)$, $l(g)hK = ghK$ and $l(h)$ is an automorphism of M for all $h \in G$. (4.4)

The proof is straightforward.

Let $q : M \to G^s$ be defined by $q(gK) = g\sigma(g)^{-1}$.

q is a homomorphism of symmetric spaces of M onto G_σ. (4.5)

Proof: $q(gK \cdot hK) = g\sigma(g)^{-1}\sigma(h)\sigma(g\sigma(g)^{-1}\sigma(h))^{-1} = g\sigma(g)^{-1}[h\sigma(h)^{-1}]^{-1}g\sigma(g)^{-1} = q(gK)q(hK)^{-1}q(gK) = q(gK)\cdot q(hK)$.

The set $\{g \in G \,/\, \sigma(g) = g^{-1}\}$ is a symmetric space: given g and h in this set $\sigma(g\cdot h) = \sigma(g)\sigma(h^{-1})\sigma(g) = (gh^{-1}g)^{-1} = (g\cdot h)^{-1}$.

Lemma. G_σ is the connectivity component of this set which contains e. (4.6)

The proof implies the exponential mapping again.

Lemma. The map $q : M = G/K \to G_\sigma$ is a covering with fibre G/K. (4.7)

Proof: $q(eK) = e$ iff $q(gK) = g\sigma(g)^{-1} = e$ iff $g \in G^\sigma$ iff $gK \in G^\sigma/K$ and this is discrete from (4.2).

Lemma. q induces an isomorphism $G/G^\sigma \to G_\sigma$ of symmetric spaces. (4.8)

Proof: The kernel of q is the set of all elements in the same fibre: $q(gK) = q(hK)$ iff $g\sigma(g)^{-1} = h\sigma(h)^{-1}$ iff $h^{-1}g = \sigma(h)^{-1}\sigma(g) = \sigma(h^{-1}g)$ iff $h^{-1}g \in G^\sigma$ iff $gG^\sigma = hG^\sigma$.

Thus G/K with (4.2) is a symmetric space. Conversely, every symmetric space has such a realization: let M be a connected symmetric space with base point o. Let Iso(o) be the isotropy group of o in Dis(M), i.e. $\mathrm{Iso}(o) = \{g \in \mathrm{Dis}(M)/g(o) = o\}$. It can be shown that Dis(M) is a connected Lie transformation group of M which acts transitively.

Theorem. $g \to \sigma(g) = S(o)gS(o)$ is an involutive automorphism of Dis(M) such that $(\mathrm{Dis}(M)^\sigma)_0 \subset \mathrm{Iso}(o) \subset \mathrm{Dis}(M)^\sigma$ and M is isomorphic to Dis(M)/Iso(o). Dis(M) is the smallest subgroup of Aut(M) which acts transitively on M. (4.9)

This theorem has a natural generalization to non-connected symmetric spaces. It shows that every symmetric space is a homogeneous space of its group of displacements.

5. LIE TRIPLE SYSTEMS AS THE LOCAL ALGEBRAIC STRUCTURES OF SYMMETRIC SPACES

A vector space V with a trilinear composition [ ] is called a Lie triple system if the following identities are fulfilled

(LT1) $[xxy] = 0$
(LT2) $[xyz] + [zxy] + [yzx] = 0$ (Jacobi-identity)
(LT3) $[xy[uvw]] = [[xyu]vw] + [u[xyv]w] + [uv[xyw]]$.

A sub-space V_0 of V is called a sub-system of V (resp. ideal) if $[V_0V_0V_0] \subset V_0$ (resp. $[V_0VV] \subset V_0$).

Standard examples of Lie triple systems are:

(a) Lie algebras with the triple composition

$$[xyz] = [[x,y], z] \tag{5.1}$$

where [,] is the Lie bracket;

(b) vector spaces with the zero composition, called the trivial Lie triple system;

(c) pseudo-orthogonal vector spaces with the composition

$$[xyz] = \pm \langle x,z\rangle y \mp \langle y,z\rangle x \quad . \tag{5.2}$$

These examples are related to the examples of Section 3 (where the Lie triple system (c) corresponds to the symmetric spaces (c), (e) (d) with a by one smaller dimension of the latter) by the following construction:

Given M = G/K as in Section 4. An involutive automorphism σ of G induces an involutive automorphism $d\sigma_e$ of the Lie algebra of G by the "functorial" relation

$$\sigma(\exp x) = \exp d\sigma_e(x) \quad x \in \mathfrak{g} \ , \tag{5.3}$$

where exp is the exponential mapping of $\mathfrak{g}$ into G [He p. 100]. In the special cases of matrix groups and the corresponding Lie algebras of infinitesimal (linear) transformations the involutive automorphisms σ and $d\sigma_e$ coincide since the exponential mapping becomes the exponential series and σ usually is of the form $\sigma(A) = BAB^{-1}$ with some invertible matrix B. Writing A = exp C we get

$$\sigma(A) = B(\exp C)B^{-1} = \exp BCB^{-1} = \exp d\sigma_e(C) \quad . \tag{5.4}$$

Let $\mathfrak{g}_\pm$ be the eigenspaces of eigenvalue ± 1 of $d\sigma_e$, i.e. $\mathfrak{g}_\pm = \{x \in \mathfrak{g}/d\sigma_e(x) = \pm x\}$. From $x = \frac{1}{2}(x+d\sigma_e(x)) \oplus \frac{1}{2}(x-d\sigma_e(x))$ for all $x \in \mathfrak{g}$ we have the direct sum decomposition $\mathfrak{g} = \mathfrak{g}_+ \oplus \mathfrak{g}_-$. A verification gives the Lie bracket relations of

$$[\mathfrak{g}_+, \mathfrak{g}_+] \subset \mathfrak{g}_+ \ , \quad [\mathfrak{g}_-, \mathfrak{g}_-] \subset \mathfrak{g}_+ \ , \quad [\mathfrak{g}_+, \mathfrak{g}_-] \subset \mathfrak{g}_- \ . \tag{5.5}$$

(Conversely, given a decomposition $\mathfrak{G} = \mathfrak{G}_+ \oplus \mathfrak{G}_-$ with (5.5) of a Lie algebra $\mathfrak{G}$, the mapping $x \to x$ for $x \in \mathfrak{G}_+$ and $x \to -x$ for $x \in \mathfrak{G}_-$ is an involutive automorphism of $\mathfrak{G}$ such that $\mathfrak{G}_\pm$ are its eigenspaces of eigenvalue ±1.) Hence the vector space $\mathfrak{G}_-$ becomes a Lie triple system with the composition (5.1). Since it can be identified with the tangential space of M in the point eK [He p. 113] it is proved that every symmetric space carries a Lie triple system in the tangential space of the base point. That (5.5) yields already all Lie triple systems is seem from proposition 2.3 [Lo I p. 78] which is the corresponding result to (4.9).

6. ON SYMMETRIC SPACES OF PSEUDO-ORTHOGONAL GROUPS

In the following $(V,<,>)$ is a pseudo-orthogonal vector space of dimension n and signature (n_1,n_2), I is the matrix of $<,>$ in some given basis and $\mathbb{R} \oplus V = \underset{\sim}{V}$, $\mathbb{R} \oplus V \oplus \mathbb{R} = \tilde{V}$. We restrict ourselves to the connectivity components $\mathrm{Aut}_0(\ ,\ )$. Every diagonal matrix I_{pq} with p-times 1 and q-times -1, p + q = n, induces an involutive automorphism

$$\sigma(A) = \mathrm{Ad}(I_{pq})A = I_{pq}\, A\, I_{pq}^{-1} \qquad (6.1)$$

of any pseudo-orthogonal group in p + q = n dimensions. In the following we discuss three types of I_{pq}, namely

$$I_1 = \begin{pmatrix} 1 & 0 \\ 0 & -id_n \end{pmatrix}, \quad I_2 = \begin{pmatrix} -1 & 0 & 0 \\ 0 & id_n & 0 \\ 0 & 0 & -1 \end{pmatrix}, \quad I_3 = \begin{pmatrix} 1 & 0 & 0 \\ 0 & id_n & 0 \\ 0 & 0 & -1 \end{pmatrix}$$

where the pseudo-orthogonal bilinear form in the first case is given by $\underset{\sim}{I} = \mathrm{diag}(1,I)$ and in the two other cases by $\langle\!\langle,\rangle\!\rangle$ with matrix $\tilde{I} = \mathrm{diag}(1,I,-1)$. Hence $\langle\!\langle,\rangle\!\rangle$ necessarily is indefinite whereas in the first case the positive-definite cases are included.

Type I_1 : For $\underset{\sim}{A} = \begin{pmatrix} \alpha & b^tI \\ a & A \end{pmatrix} \in \mathrm{Aut}_0(\underset{\sim}{V},\underset{\sim}{I})$, $\alpha \in \mathbb{R}$, $a,b \in V$, $A \in gl(V,\mathbb{R})$ we get

$$\mathrm{Aut}_0(V,I)^\sigma = \{\begin{pmatrix} \alpha & 0 \\ 0 & A \end{pmatrix} \in \mathrm{Aut}_0(\underset{\sim}{V},\underset{\sim}{I}) / A^tI\,A = I,\ \alpha^2 = 1\} \qquad (6.2)$$

The space of symmetric elements becomes

$$\mathrm{Aut}_0(V,I)_\sigma = \{\begin{pmatrix} 1-2<a,a> & -2\alpha a^tI \\ 2\alpha a & id_n - 2a \otimes a^tI \end{pmatrix} / \alpha^2 + <a,a> = 1\} \qquad (6.3)$$

Proof: From $id_{n+1} = \underset{\sim}{A}\underset{\sim}{A}^{-1} = \underset{\sim}{A}\underset{\sim}{I}^{-1}\underset{\sim}{A}^t\underset{\sim}{I}$ respectively $id_{n+1} = \underset{\sim}{A}^{-1}\underset{\sim}{A} = \underset{\sim}{I}^{-1}\underset{\sim}{A}^t\underset{\sim}{I}\,\underset{\sim}{A}$ we get the identities

$$\alpha^2 = 1 - <b,b>, \quad Ab = -\alpha a, \quad a \otimes a^tI + AI^{-1}A^tI = id_n$$
$$\text{resp. } \alpha^2 = 1 - <a,a> \qquad (6.4)$$

Hence $\underset{\sim}{A}\sigma(\underset{\sim}{A})^{-1} = \underset{\sim}{A}I_1\underset{\sim}{I}^{-1}\underset{\sim}{A}^t\underset{\sim}{I}I_1$ has the form (6.3).

The Lie algebra of (6.2) becomes

$$\left\{\begin{pmatrix} 0 & 0 \\ 0 & B \end{pmatrix} / B^t I + IB = 0 \right\} . \tag{6.5}$$

It is given by the set of n+1 square matrices of the form $\frac{1}{2}(\underset{\sim}{B} + I_1\underset{\sim}{B}\, I_1)$ with $\underset{\sim}{B} \in \mathrm{der}(\underset{\sim}{V},\underset{\sim}{I})$. The set of matrices $\frac{1}{2}(\underset{\sim}{B} - I_1\underset{\sim}{B}\, I_1)$ gives the Lie triple system

$$\left\{\begin{pmatrix} 0 & a^t I \\ -a & 0 \end{pmatrix} / a \in V \right\} \tag{6.6}$$

with the double commutation relations (5.2) (upper signs) and the bilinear form

$$-\langle a,b\rangle = \frac{1}{2}\,\mathrm{trace}\begin{pmatrix} 0 & a^t I \\ -a & 0 \end{pmatrix}\begin{pmatrix} 0 & b^t I \\ -b & 0 \end{pmatrix} .$$

The hyperboloid M_0^κ in $\underset{\sim}{V}$ in the interior of the "forward" light cone is given by (3.5) for $\tilde{0} \neq \kappa \in \mathbb{R}$. For base point take $k = \kappa \oplus 0$. Then $M_0^\kappa = \mathrm{Aut}_0(\underset{\sim}{V},\underset{\sim}{I})k$ since $\mathrm{Aut}_0(\underset{\sim}{V},\underset{\sim}{I})$ acts transitively on M_0^κ. Obviously $S(k) = \mathrm{diag}(1,-\mathrm{id}_n)$ and

$$\mathrm{Iso}(k) = \left\{\begin{pmatrix} 1 & 0 \\ 0 & A \end{pmatrix} \in \mathrm{Aut}_0(\underset{\sim}{V},\underset{\sim}{I}) \;/\; A^t I\, A = I\right\} \tag{6.7}$$

which has two connectivity components if I is indefinite and one if I is positive definite. The automorphism group of the symmetric space $(M_0^\kappa,.)$ is easily seen to be $\mathrm{Aut}_0(\underset{\sim}{V},\underset{\sim}{I})$. If we restrict $\underset{\sim}{I}$ to be indefinite this group is simple and therefore coincides with its non-trivial normal sub-group $\mathrm{Dis}(M_0^\kappa)$. Applying (4.9) we get the isomorphism $\phi : \mathrm{Aut}_0(\underset{\sim}{V},\underset{\sim}{I})/\mathrm{Iso}(k) \to M_0^\kappa$ defined by

$$\phi : \begin{pmatrix} \alpha & b^t I \\ a & A \end{pmatrix}\mathrm{Iso}(k) \to \begin{pmatrix} \alpha & b^t I \\ a & A \end{pmatrix}\begin{pmatrix} \kappa \\ 0 \end{pmatrix} = \kappa\begin{pmatrix} \alpha \\ a \end{pmatrix} \tag{6.8}$$

where the symmetrix multiplications are given by (4.3) and (3.6).

Proof: From the last equation in (6.4) $\kappa\alpha$ $\kappa a \in M_0^\kappa$. The monomorphism is easy to check. The surjectivity follows from

$$\begin{pmatrix} \beta \\ b \end{pmatrix} = \begin{pmatrix} \kappa^{-1}\beta & a^t I \\ \kappa^{-} b & A \end{pmatrix}\begin{pmatrix} \kappa \\ 0 \end{pmatrix}$$

where the n+1 square matrix is in $Aut_0(\underset{\sim}{V},\underset{\sim}{I})$.

Since $Iso(k) \subset Aut_0(\underset{\sim}{V},\underset{\sim}{I})$ from (6.8) follows that M_0^{κ} is among the possible n-dimensional symmetric spaces $Aut_0(\underset{\sim}{V},\underset{\sim}{I})/K$ with (4.2).

<u>The connected Lorentz group $SO_+^{\uparrow}(1,3)$</u>: $\underset{\sim}{I} = diag(1,-id_3)$ and $A^tA = id_3$, $\alpha = 1$ in (6.2). Hence $\det(\underset{\sim}{A}) = 1$ and $SO_+^{\uparrow}(1,3)^{\sigma} \cong SO(3)$. From (4.2) there is only one three-dimensional symmetric space $SO_+^{\uparrow}(1,3)/SO(3)$, which from (4.8) is isomorphic to the space of symmetric elements

$$SO_+^{\uparrow}(1,3)_{\sigma} = \{ \begin{pmatrix} 1+2\alpha^{\mu}\alpha^{\mu} & 2\alpha\vec{\alpha}^{\mu t} \\ 2\alpha\alpha^{\mu} & id_3+2\vec{\alpha}^{\mu}\otimes\vec{\alpha}^{\mu t} \end{pmatrix} / \alpha^2 - \alpha^{\mu}\alpha^{\mu} = 1 \} \qquad (6.9)$$

and from (6.8) to the mass shell in the forward light cone.

<u>The connected de-Sitter group $SO_+^{\uparrow}(2,3)$</u>: Here $\underset{\sim}{I} = diag(id_2,-id_3)$. In the case of positive α in (6.2) the upper left element of A must be positive, for negative α it must be negative. Since the whole matrix must have positive determinant we get the two components

$$\begin{aligned} SO_+^{\uparrow}(2,3) &= \{diag(1,A)/A \in SO_+^{\uparrow}(1,3)\} \cup \\ &\cup \{diag(-1,A)/A \in PTSO_+^{\uparrow}(1,3)\} \end{aligned} \qquad (6.10)$$

(P space reflection T time reflection). Hence there are

$$(SO_+^{\uparrow}(2,3)/SO_+^{\uparrow}(2,3)^{\sigma} \quad \text{and} \quad SO_+^{\uparrow}(2,3)/SO_+^{\uparrow}(1,3) \qquad (6.11)$$

as symmetric spaces. From (4.8) the first one is isomorphic to the space of symmetric elements (6.3) with the Lorentz metric I. The second one is the de-Sitter space M_0^{κ}, since $Iso(k) = SO_+^{\uparrow}(1,3)$. It covers the space of symmetric elements twice. There is a second pair of spaces given by the metric $\underset{\sim}{I} = diag(-1,I)$ which is discussed in [Ti 72].

Type I_2. In the following we write

$$\tilde{A} = \begin{pmatrix} \alpha & a^tI & \beta \\ b & A & c \\ \gamma & d^tI & \delta \end{pmatrix} \qquad (6.12)$$

Then the group of fixed points of σ is

$$\begin{aligned} Aut_0(\tilde{V},\langle,\rangle)^{\sigma} = \{\tilde{A} \in Aut_0(\tilde{V},\langle,\rangle)/a = b = c = d = 0 \\ \text{and } A^tI\,A = I, \begin{pmatrix} \alpha & \beta \\ \gamma & \delta \end{pmatrix} \in O(1,1)\} \end{aligned} \qquad (6.13)$$

has $\frac{1}{2}n(n-1)+1$ dimensions since $O(1,1)$ is one-dimensional. The expression for $Aut_0(\tilde{V},\langle\!\langle,\rangle\!\rangle)_\sigma$ is rather complicated. The Lie algebra of $Aut(\tilde{V},\langle\!\langle,\rangle\!\rangle)$ is

$$der(\tilde{V},\langle\!\langle,\rangle\!\rangle) = \{\begin{pmatrix} 0 & b^tI & \beta \\ -b & B & b' \\ \beta & b'^tI & 0 \end{pmatrix} / \; B^tI + IB = 0,\ b,\ b' \in V\} \quad (6.14)$$

From (5.4) the Lie algebra of (6.13) is given by the set of matrices (6.14) with $b = b' = 0$ and the Lie triple system is given by the set of these matrices with $B = 0$ and $\beta = 0$. Writing

$$o(a \otimes b)x = \langle a,c\rangle b - \langle b,c\rangle a \ , \quad o(a \otimes b) \in der(V,\langle,\rangle) \quad , \quad (6.15)$$

the double commutation relations of this Lie triple system become

$$[\tilde{a}\ \tilde{b}\ \tilde{c}] = (\langle a,b'\rangle - \langle a',b\rangle)\tilde{c} + [(o(a\otimes b) - o(a'\otimes b'))c]^{\sim}. \quad (6.16)$$

Since no element of $\tilde{V}$ is left invariant by (6.13) no symmetric space of this type can be identified to some orbit in $\tilde{V}$.

<u>The connected Lorentz group</u>: $I = -id_2$, i.e.

$$SO_+^\uparrow(1,3)^\sigma = \{\begin{pmatrix} \alpha & 0 & \beta \\ 0 & A & 0 \\ \gamma & 0 & \delta \end{pmatrix} \in SO_+^\uparrow(1,3) / \det A > 0,\ \det\begin{pmatrix} \alpha & \beta \\ \gamma & \delta \end{pmatrix} > 0\}$$

$$\cup \{\begin{pmatrix} \alpha & 0 & \beta \\ 0 & A & 0 \\ \gamma & 0 & \delta \end{pmatrix} \in SO_+^\uparrow(1,3) / \det A < 0,\ \det\begin{pmatrix} \alpha & \beta \\ \gamma & \delta \end{pmatrix} < 0\} \quad (6.17)$$

Proof: $\det\begin{pmatrix} \alpha & 0 & \beta \\ 0 & A & 0 \\ \gamma & 0 & \delta \end{pmatrix} = \det A \det\begin{pmatrix} \alpha & \beta \\ \gamma & \delta \end{pmatrix}$ with $A \in O(2)$ and $\begin{pmatrix} \alpha & \beta \\ \gamma & \delta \end{pmatrix} \in O(1,1)$. Hence $\alpha > 0$ gives $\det\begin{pmatrix} \alpha & \beta \\ \gamma & \delta \end{pmatrix} \gtrless 0$, from which $\det A \gtrless 0$.

The corresponding 4-dimensional symmetric spaces are

$$SO_+^\uparrow(1,3)/SO_+^\uparrow(1,1) \otimes SO(2) \quad \text{and} \quad SO_+^\uparrow(1,3)/SO_+^\uparrow(1,3)^\sigma \quad (6.18)$$

the second being isomorphic to the space of symmetric elements and covered twice by the first one. Both cannot be identified to some orbit in Minkowski space. From (6.16) we see that in the tangent space $V \otimes V$ there is only a natural positive definite metric so that there seems to be no application in general relativity.

Type I_3. The group of fixed points of σ is given by

$$\mathrm{Aut}_0(\tilde{V},\sphericalangle,\rangle)^\sigma = \{\mathrm{diag}(A,1)/\underset{\sim}{A} \in \mathrm{Aut}(\underset{\sim}{V},\underset{\sim}{I}),\ \det \underset{\sim}{A} > 0\} \cup \cup \{\mathrm{diag}(A,-1)/\underset{\sim}{A} \in \mathrm{Aut}(\underset{\sim}{V},\underset{\sim}{I}),\ \det \underset{\sim}{A} > 0\}\ . \tag{6.19}$$

The connected Lorentz group: $\alpha > 0$ imples $\underset{\sim}{A} \in SO_+^\uparrow(1,2)$ or $\underset{\sim}{A} \in R_3SO_+^\uparrow(1,2)$, hence

$$SO_+^\uparrow(1,3)^\sigma = \mathrm{diag}(SO_+^\uparrow(1,2),1) \cup \mathrm{diag}(R_3SO_+^\uparrow(1,2),-1)\ . \tag{6.20}$$

The space $SO_+^\uparrow(1,3)/SO_+^\uparrow(1,3)^\sigma$ is isomorphic to the space of symmetric elements and covered twice by $SO_+^\updownarrow(1,3)/SO_+^\updownarrow(1,2)$, which is isomorphic to the (space-like) mass shell of non-vanishing pure imaginary mass κ.

The above three types were discussed since they already give all symmetric spaces of the Lorentz group, induced by diagonal I_{pq} via (6.1). This is true up to a trivial change of sign of I_{pq} and trivial permutations of the three space dimensions. For certain pseudo-orthogonal groups there are other involutive automorphisms induced by non-diagonal matrices [Lo II p. 104]. Hence it remains to show that the above five symmetric spaces are all symmetric spaces of the Lorentz group.

7. CONFORMAL GROUPS OF PSEUDO-ORTHOGONAL VECTOR SPACES

The translation group of V is given by the set of non-linear transformations T_a with $T_a(x) = x + a$. The transformation

$$K_a(x) = \frac{x + \langle x,x\rangle b}{1+2\langle x,a\rangle + \langle x,x\rangle\langle a,a\rangle}\ , \tag{7.1}$$

defined on the open sub-set $\mathrm{Dom}\, K_a = \{x \in V/1+2\langle x,a\rangle+\langle x,x\rangle\langle a,a\rangle \neq 0\}$ of V, is called special conformal transformation. Together with $\mathrm{Aut}_0(V,\langle,\rangle)$ and the dilatations D_λ with $D_\lambda x = \lambda x$ for $0 < \lambda \in \mathbb{R}$ these transformations generate a sub-group of dimension $\frac{1}{2}(n+2)(n+1)$ in the group of birational transformations of V [Ti 71a], [Koe 69], called the full connected conformal group of $(V,\langle,\rangle)$ and written $\mathrm{Kon}_0(V,\langle,\rangle)$.

Given $\tilde{x} = \xi_o \oplus x \oplus \xi_{n+1} \in \mathbb{R} \oplus V \oplus \mathbb{R} = \tilde{V}$, we consider the subspace $D = \{\tilde{x} \in \tilde{V}^o/ \sphericalangle \tilde{x},\tilde{x} \rangle = 0,\ \xi_{n+1} \neq \xi_o\}$ of the light cone in $(\tilde{V},\sphericalangle,\rangle)$. The mapping

$$\Gamma : D \to V \qquad \Gamma : \tilde{x} \to \frac{x}{\xi_{n+1}-\xi_o} \tag{7.2}$$

is surjective. We define a mapping from $\mathrm{Aut}_0(\tilde{V},\sphericalangle,\rangle)$ into the group of birational transformations of V by

$$\Gamma : \tilde{A} \to \Gamma(\tilde{A})\ , \qquad \Gamma(\tilde{A})\Gamma(\tilde{x}) := \Gamma(\tilde{A}\tilde{x}) \tag{7.3}$$

It was proven by [Cl] that (7.3) is well defined and that its ker-

nel consists of the multiples of id_{n+2} only. To identify the image $\Gamma(\mathrm{Aut}_0(\tilde{V},\langle\!\langle,\rangle\!\rangle))$ consider the Lie algebra $\mathrm{der}(\tilde{V},\langle\!\langle,\rangle\!\rangle)$, (6.14). Every element of it can be written uniquely in the form
$\delta\tilde{K}_a \oplus \delta\tilde{T}_b \oplus \mathrm{diag}(0,B,0) \oplus \delta\tilde{D}_\beta =$

$$\begin{pmatrix} 0 & -a^tI & 0 \\ a & 0 & a \\ 0 & a^tI & 0 \end{pmatrix} \oplus \begin{pmatrix} 0 & b^tI & 0 \\ -b & 0 & b \\ 0 & b^tI & 0 \end{pmatrix} \oplus \begin{pmatrix} 0 & 0 & 0 \\ 0 & B & 0 \\ 0 & 0 & 0 \end{pmatrix} \oplus \begin{pmatrix} 0 & 0 & \beta \\ 0 & 0 & 0 \\ \beta & 0 & 0 \end{pmatrix} \tag{7.4}$$

Calculating the exponentials of these matrices we get

$$\tilde{K}_{\mu a} = \begin{pmatrix} 1-\frac{\mu^2}{2}\langle a,a\rangle & -\mu a^tI & -\frac{\mu^2}{2}\langle a,a\rangle \\ \mu a & id_n & \mu a \\ \frac{\mu^2}{2}\langle a,a\rangle & \mu a^tI & 1+\frac{\mu^2}{2}\langle a,a\rangle \end{pmatrix} \tag{7.5}$$

$$\tilde{T}_{\mu b} = \begin{pmatrix} 1-\frac{\mu^2}{2}\langle b,b\rangle & \mu b^tI & \frac{\mu^2}{2}\langle b,b\rangle \\ -\mu b & id_n & \mu b \\ -\frac{\mu^2}{2}\langle b,b\rangle & \mu b^tI & 1+\frac{\mu^2}{2}\langle b,b\rangle \end{pmatrix} \tag{7.6}$$

[Pi], [Hi p. 415]

$$\mathrm{diag}(1,\exp\mu B,1) \tag{7.7}$$

$$D_{\exp\beta\mu} = \begin{pmatrix} \cosh\beta\mu & 0 & \sinh\beta\mu \\ 0 & id_n & 0 \\ \sinh\beta\mu & 0 & \cosh\beta\mu \end{pmatrix} \tag{7.8}$$

respectively. Clearly these curves of linear transformations are one-parameter sub-groups of $\mathrm{Aut}_0(\tilde{V},\langle\!\langle,\rangle\!\rangle)$. Using

$$\xi_{n+1} + \xi_0 = (\xi_{n+1} - \xi_0)\langle x,x\rangle \quad \text{for } x \in D , \tag{7.9}$$

a straightforward calculation gives

$$\Gamma(\tilde{K}_a)(x) = K_a(x) \qquad \Gamma(\tilde{T}_b)(x) = T_b(x)$$
$$\Gamma(\mathrm{diag}(1,\exp B,1)) = \exp Bx \qquad \Gamma(\tilde{D}_\lambda)(x) = D_\lambda x \tag{7.10}$$

Since every element of $\mathrm{Aut}_0(\tilde{V},\langle\!\langle,\rangle\!\rangle)$ is a product of elements of the form (7.5)-(7.8), Γ maps $\mathrm{Aut}_0(\tilde{V},\langle\!\langle,\rangle\!\rangle)$ onto $\mathrm{Kon}_0(V,\langle,\rangle)$. The only multiples of id_{n+2} in $\mathrm{Aut}_0(\tilde{V},\langle\!\langle,\rangle\!\rangle)$ (the kernel of Γ) are id_{n+2} and in some cases (depending on dimension and signature of $(\tilde{V},\langle\!\langle,\rangle\!\rangle)$) $-id_{n+2}$. Let us summarize the relations between the various covering groups of the Lie algebra $der(\tilde{V},\langle\!\langle,\rangle\!\rangle)$ in the commutative diagram of short exact sequences (where we drop the trivial parts of the sequences)

$$\begin{array}{ccccccc}
 & & Fund(Aut_0(\tilde{V},\langle\!\langle,\rangle\!\rangle)) & & & & \pm\, id_C \\
 & & \searrow & & & \swarrow & \\
Ker(\Delta) & \longrightarrow & \widetilde{Aut_0}(\tilde{V},\langle\!\langle,\rangle\!\rangle) & \xrightarrow{\ \Delta\ } & Spin(\tilde{V},\langle\!\langle,\rangle\!\rangle) & & \\
 & & \searrow & & \swarrow & & \\
\pm\, id_{n+2} & \longrightarrow & Aut_0(\tilde{V},\langle\!\langle,\rangle\!\rangle) & \xrightarrow{\ \Gamma\ } & Kon_0(V,<,>) & &
\end{array}$$

Here $Spin(\tilde{V},\langle\!\langle,\rangle\!\rangle)$ is the group belonging to the skew elements of second degree in the Clifford algebra [Ch p. 66], id_C the identity in the Clifford algebra, Fund(G) the <u>funamental group</u> of a group G and $\tilde{G}$ the universal covering group of G.

In the special case of the conformal group of Minkowski space the diagram becomes

$$\begin{array}{ccccccc}
 & & Z & & & & \pm\, id_C \\
 & & \searrow & & & \swarrow & \\
Z & \longrightarrow & \widetilde{SO_+^{\uparrow}}(2,4) & \longrightarrow & Spin(2,4) \cong SU(2,2) & & \\
 & & \searrow & & \swarrow & & \\
\pm\, id_6 & \longrightarrow & SO_+^{\uparrow}(2,4) & \longrightarrow & Kon_0(1,3) & &
\end{array}$$

Here we used that $SO_+^{\uparrow}(2,4)$ is infinitely connected.

8. LIGHT CONES AS HOMOGENEOUS BUT NOT SYMMETRIC SPACES OF THE PSEUDO-ORTHOGONAL GROUPS

The matrix $\tilde{A} \in Aut_0(\tilde{V},\langle\!\langle,\rangle\!\rangle)$, see (6.12), is in the isotropy group of the point $\tilde{z} = 1 \oplus 0 \oplus 1$ of the light cone in $(\tilde{V},\langle\!\langle,\rangle\!\rangle)$ iff $\beta = 1-\alpha$, $c = -b$, $\delta = 1-\gamma$. From $id_{n+2} = \tilde{A}\tilde{A}^{-1} = \tilde{a}\tilde{I}^{-1}\tilde{A}^t\tilde{I}$ follows

$$Iso(\tilde{z}) = \{\tilde{T}_{-b} diag(1,A,1)/b \in V,\ A \in Aut_0(V,<,>)\} \qquad (8.1)$$

Here we used det(exp C) = exp(trace C) for $\delta\tilde{T}_{-b}$ and its restriction to $1+n_1$ dimensions, whence $\det\tilde{T}_{-b} = 1$ and $\tilde{T}_{-b} \in Aut_0(\tilde{V},\langle\!\langle,\rangle\!\rangle)$ (8.1) is the Pinski representation of the inhomogeneous connected pseudo-orthogonal group on $(V,<,>)$.

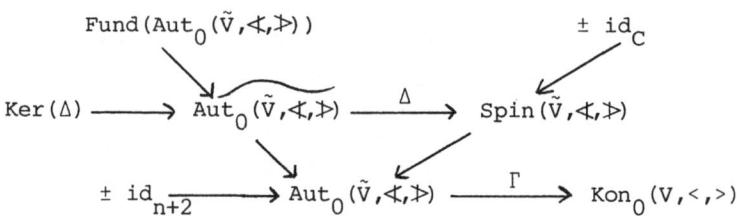

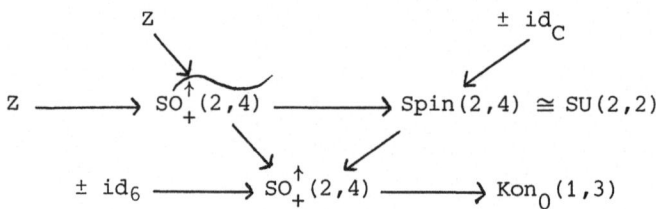

The space $Aut_0(\tilde{V},\langle\!\langle,\rangle\!\rangle)/Iso(\tilde{z})$ is a homogeneous space of $Aut_0(\tilde{V},\langle\!\langle,\rangle\!\rangle)$ which is diffeomorphic to the light cone in $(\tilde{V},\langle\!\langle,\rangle\!\rangle)$. However, it is not a symmetric space of $Aut_0(\tilde{V},\langle\!\langle,\rangle\!\rangle)$: the Lie algebra of (8.1) is spanned by the matrices $\delta\tilde{T}_b$ and diag(0,B,0) in (7.4). It is well known that the complementary sub-space in der$(\tilde{V}, \sphericalangle, \rangle)$, spanned by the matrices $\delta\tilde{K}_a$ and $\delta\tilde{D}_\beta$ in (7.4), is a sub-algebra [Ti 71b]. Hence the commutation relations (5.5) are not satisfied and the decomposition does not originate in a symmetric one.

The Pinski representation of the Poincaré group may be useful for the construction of its symmetric spaces.

9. APPLICATIONS IN GENERAL RELATIVITY

9.1 Pseudo-Riemannian globally symmetric spaces

Given the tangent bundle T(M) of the manifold M, M is called pseudo-riemannian if there is a real valued function on T(M) x T(M) such that its restriction to every tangent space is a symmetric non-denegerate bilinear form. This bilinear form in the tangent space $T_p(M)$ in the point p of M will be denoted by $<,>_p$. A symmetric space is called pseudo-riemannian if every symmetry S(p) around p induces a pseudo-orthogonal transformation of the tangent spaces; this means that for the induced linear transformation $dS(p) : T_q(M) \to T_{p\cdot q}(M)$, called the differential of S(p) [He p.22] and defined by

$$dS(p) : X_q \to X_{p\cdot q} \quad , \quad X_{p\cdot q}\phi = X_q(\phi o S(p)) \tag{9.1}$$

for the tangent vector X_q, $q \in M$ and the real valued function ϕ on M, we have

$$<dS(p)X_q, dS(p)Y_q>_{p\cdot q} = <X_q , Y_q>_q \tag{9.2}$$

9.2 Curvature and gravitational equations

The above pseudo-orthogonal structure can be transported from the tangent spaces to the vector spaces of vector fields on M by the definition

$$<X, Y>(p) = <X_p , Y_p>_p \tag{9.3}$$

Here $p \to <X,Y>(p)$ is a real valued function on M, which will be denoted by <X,Y> and which induces a tensor field g of type (0,2); i.e. $g : D^1(M) \times D^1(M) \to F(M)$, $g : (X,Y) \to <X,Y>$ is bilinear with respect to coefficients in F(M).

Following Kulkarni, a curvature tensor field C is a bilinear (with respect to F(M)) mapping from $D^1(M) \times D^1(M)$ into the modul of linear (with respect to F(M)) transformations of $D^1(M)$, such that for all vector fields $X,Y,Z,W \in D^1(M)$

(C1) $C(X,Y) = -C(Y,X)$

(C2) $\langle C(X,Y)Z,W\rangle = \langle C(Z,W)X,Y\rangle$

(C3) $C(X,Y)Z + C(Z,X)Y + C(Y,Z)X = 0$

Let $D_1(M)$ be the dual space of 1-forms on $D^1(M)$. For $\omega \in D_1(M)$ the mapping $(\omega,X,Y,Z) \to \omega(C(X,Y)Z)$, $D_1(M) \times D^1(M) \times D^1(M) \times D^1(M) \to F(M)$ is multilinear (with respect to $F(M)$), hence a tensor field of type (1,3).

Examples: (a) The trivial curvature structure is given by

$$\langle X,Z\rangle Y - \langle Y,Z\rangle X \tag{9.4}$$

(b) The canonical curvature structure is the structure defined with respect to an affine connection [GKM] ∇ by

$$R(X,Y)Z = \nabla_X\nabla_Y Z - \nabla_Y\nabla_X Z - \nabla_{XY-YX}Z \tag{9.5}$$

If ∇ is the unique Riemannian connection of $(M,\langle,\rangle)$ [GKM p. 78], then R is called the riemannian curvature structure. A canonical curvature R defines the Ricci tensor field by

$$\mathrm{Ric}(X,Y) = \mathrm{trace} : Z \to R(X,Z)Y \quad . \tag{9.6}$$

Obviously Ric is a tensor field of type (0,2). Writing $\langle \mathrm{Ric}.X,Y\rangle = \mathrm{Ric}(X,Y)$, Ric. is a linear (with respect to $F(M)$) transformation of $D^1(M)$. The curvature scalar Sc is defined by

$$\mathrm{Sc} = \mathrm{trace\ Ric}. \tag{9.6}$$

(c) For the definition of the Ricci and the conformal (or Weyl) curvature structures, see [Ku].

Given a pseudo-riemannian manifold $(M,\langle,\rangle)$ with $\dim M = n$ and equipped with its riemannian curvature, the equations

$$\mathrm{Ric}(X,Y) - \frac{1}{n}\,\mathrm{Sc}\,\langle X,Y\rangle = \varkappa E(X,Y) \tag{9.8}$$

are called the einsteinian gravitational equations. Here $E(X,Y)$ is the (0,2) energy-momentum tensor field and $\varkappa$ is the gravitational constant (we omit the possibility of a cosmological constant). In physics, usually E is given and the problem is to determine the topological structure of M and $\langle,\rangle$. If E vanishes, $(M,\langle,\rangle)$ is called einsteinian. Using trace: $Z \to \langle Y,Z\rangle X = \langle X,Y\rangle$, the trivial curvature has $\mathrm{Ric}(X,Y) = (n-1)\langle X,Y\rangle$ and $\mathrm{Sc} = n(n-1)$. Hence every $(M,\langle,\rangle)$ is einsteinian in its trivial curvature structure.

9.3 Remarks on classification

Since the tangent space of the base point of a pseudo-riemannian symmetric space of signature (1,3) physically is Minkowski space the classification of all such four-dimensional spaces reduces to two steps: (a) find all Lie triple systems in Minkowski space (this is a Lie algebraical problem from (5.5)), and (b) find all covering spaces for a given Lie triple system.

The canonical curvature on a symmetric space M is related to the Lie triple structure by

$$R(X,Y)Z = -[X\ Y\ Z] \tag{9.9}$$

[Lo I p. 84]. If one introduces a basis $X_1,\dots,X_n$ in $D^1(M)$,

$$R(X_i,X_k)X_l = \sum_{m=1}^{n} R^m{}_{ikl} X_m \tag{9.10}$$

gives the relation of R to the functions $R^m{}_{ikl}$. Comparison with (9.9) shows that the corresponding components of the tensor field R on the tangent space in the base point are the structure constants of the Lie triple system. Thus the classification of Lie triple systems is nothing but the classification of curvature tensors for symmetric spaces.

From (4.9), (5.5) and the theorem of Ado for Lie algebras (which states that every finite-dimensional Lie algebra has a faithful finite-dimensional representation) follows that every Lie triple system has such a representation as well. This can be used as a starting point for the solution of the second step: the exponential series of matrices leads to a symmetric space which is realized in the form G/K with matrix groups. The space of symmetric elements then gives a second symmetric space which in general does not coincide with the first one.

Note that from the above and the gravitational equations follows that symmetric spaces which are related to each other by covering have the same energy-momentum tensors.

9.4 Examples

A class of possible symmetric spaces in general relativity can be described directly: topological they are direct product of vector spaces and hyperboloids $\mathbb{R}^k \times M_o^\kappa$. By

$$(x,q)\cdot(y,b) = \left(2x-y,\ 2\,\frac{\langle a,b\rangle}{\kappa}\,a-b\right) \tag{9.11}$$

for $x,y \in \mathbb{R}^k$, $a,b \in M_o^\kappa$, they become symmetric spaces. When equipped with a physically acceptable pseudo-riemannian structure $\langle,\rangle$ of signature (1,3) and for $k+\dim M_o^\kappa = 4$, this class contains actual space-times: besides the Minkowski space for k = 4 and the two de-Sitter spaces for $\dim M_o^\kappa = 4$, there is the Oszváth and Schücking space-time, whose underlying manifold is the Lie group $\mathbb{R} \times S^3$, the three-dimensional sphere S^3 carrying its SU(2) multiplication.

It is shown in [Ti 72] that for the de-Sitter spaces, the riemannian and the trivial curvature structures coincide, hence they are einsteinian, reflecting the well known result that they have vanishing energy-momentum tensors. The same happens to the interior of the forward light cone Y in example (d) of Section 3, which is a pseudo-riemannian symmetric space as well.

PART II / DOMAINS OF POSITIVITY AND FORMAL REAL JORDAN ALGEBRAS

A special class of symmetric spaces is given by the domains of positivity which are a certain type of convex cones in pseudo-orthogonal vector spaces. They are described in Section 10, the material of which is taken from [Br], [Koe 68] and mainly [Koe 62]. Section 11 gives a sketchy description of those properties of Jordan algebras which are needed in the sequel. A general reference for Jordan algebras is [BK] and again [Koe 62]. In Section 12 the various concepts are related to each other. The tangent functor from domains of positivity onto Jordan algebras together with its inverse is due to M. Koecher [Koe 62], [Koe 68]. The proof that every domain of positivity is a symmetric space is due to O. Loos [Lo I]. In Section 13 the results of the preceeding sections are applied to a Jordan algebra which is defined on every pseudo-orthogonal vector space. A more detailed treatment of this Jordan algebra can be found in [BK]. The domain of positivity of the Jordan algebra of Minkowski space is the interior of the forward light cone. Koecher's result on the automorphism group of this domain of positivity coincides with the Zeeman theorem on the causal automorphisms of Minkowski space.

A second possible application of Jordan algebras and domains of positivity is in axiomatic quantum mechanics and quantum field theory. Actually the application in axiomatic quantum mechanics has led to the discovery of Jordan algebras [Jo]. The main idea is the same as in C*-algebra theory. The observables of a dynamical system are given by a Jordan algebra instead of the symmetric elements of a C*-algebra. The advantage is that there are no non-physical elements like the non-symmetric elements in a C*-algebra. Another advantage is the functorial relationship to domains of positivity, which leads to the statistical operators, representing the "states" of the dynamical system in question. More exactly, the "pure" states are given by the (primitive) idempotents [Hu] which lie on the boundary of the domain of positivity. The mixed states then are given by linear combinations of the pure states with coefficients in (0,1); graphically they are a hypersphere inside the domain of positivity with boundary on the convex cone.

According to this every dynamical system should have a description by a (not necessarily formal real) Jordan algebra which is to be represented as a Jordan algebra of self-adjoint operators in a Hilbert space. If the Jordan algebra is formal real ("compact"), the Hilbert space is finite-dimensional, otherwise necessarily infinite-dimensional.

For this approach to dynamics one needs a description of the Jordan algebra of self-adjoint operators in infinite-dimensional Hilbert spaces, and especially of the corresponding domain of positivity. A first step to a generalization to the infinite-dimensional case of Koechers results has been made in [Ja].

In Section 14 a Jordan algebraic description of non-relativistic spin is given following [Ka].

10. DOMAINS OF POSITIVITY OR SELF-DUAL CONVEX CONES

10.1 Definition

A subset Y of V is called *convex* if $x, y \in Y$ and $0 \leq \alpha \leq 1$ implies $\alpha x + (1-\alpha)y \in Y$. A subset Y of V is called a *cone* if $x \in Y$ and $0 < \alpha \in \mathbb{R}$ implies $\alpha x \in Y$.

By $a \to a^{<,>}$ $a^{<,>}(x) := \langle a,x\rangle$ we get an isomorphism of V onto its dual space V^*. The image Y^* of an open convex cone Y in V under this isomorphism

$$Y^* = \{\lambda \in V^* \,/\, \lambda(x) > 0 \quad \text{for all} \quad 0 \neq x \in \bar{Y}\} \qquad (10.1)$$

($\bar{Y}$ denotes the closure of Y in some given topology of V) is an open convex cone in V^*, called the *dual cone* of Y. The image in V of Y^* by the inverse isomorphism is

$$Y^{<,>} = \{x \in V \,/\, \langle x,y\rangle > 0 \quad \text{for all} \quad 0 \neq y \in \bar{Y}\} \qquad (10.2)$$

called the *$<,>$ - dual* cone of Y. Y is called *self-dual* if $Y = Y^{<,>}$ (not for every open convex cone there is a bilinear form with respect to which it is self-dual). An open convex cone Y with $Y^* \neq \emptyset$ is called a *domain of positivity* of $(V,<,>)$ if it is self-dual. We write $\mathrm{Pos}(V,<,>)$ in this case.

Theorem: An open non-empty subset Y of V is a domain of positivity with respect to $<,>$ iff

(a) $x, y \in Y$ imples $\langle x,y\rangle > 0$ and

(b) $\langle x,y\rangle > 0$ for all $0 \neq y \in \bar{y}$ implies $x \in Y$. (10.3)

Theorem: If Y is a domain of positivity then

(a) $x \in Y$ iff $\langle z,x\rangle > 0$ for all $0 \neq z \in \bar{Y}$ and

(b) $x \in \bar{Y}$ iff $\langle z,x\rangle \geqslant 0$ for all $z \in Y$. (10.4)

Domains of positivity are maximal in the sense that two for the same bilinear form necessarily are equal. However, there may be several bilinear forms leading to the same domain of positivity (see below).

10.2 Order and equivalence relations for open convex cones

In the following Y denotes an open convex cone in V with $Y^* \neq \emptyset$ (this condition is equivalent with the fact that $\bar{Y}$ contains no one-dimensional subspace of V, i.e. $x \in \bar{Y}$ and $-x \in \bar{Y}$ implies $x = 0$).

We introduce a partial order $\leq$ in V by

$$x \leq y \quad \text{iff} \quad y - x \in \bar{Y} \quad . \tag{10.5}$$

Especially, we have $y \in \bar{Y}$ iff $0 \leq y$. This relation is archimedian, compatible with the vector space structure and every $<,>$ on V is monotone. Besides (10.5) there is a transitive but not reflextive relation

$$x \leq y \quad \text{iff} \quad y - x \in Y \quad , \tag{10.6}$$

which again is archimedian, compatible and in which $<,>$ is monotone. In general, however, $x < y$ does not imply $x \leq y$ and $x \neq y$.

$\leq$ can be used to define an equivalence relation in $\bar{Y}$: Two points $x,y \in \bar{Y}$ are called equivalent if there are $0 < \alpha$ and $0 < \beta$ in $\mathbb{R}$ with $x \leq \beta y$ and $y \leq \alpha x$. The archimedian property implies that Y itself is a full equivalence class. The boundary decomposes in general in several classes.

10.3 The automorphism group of a domain of positivity

For an open convex cone Y in V we call $A \in Gl(V,\mathbb{R})$ an automorphism of Y if $AY = Y$. Example: $D_\lambda \in \text{Aut } Y$ for $\lambda > 0$. Aut Y is a closed subgroup of $Gl(V,\mathbb{R})$ and Aut $\bar{Y}$ = Aut Y. For $A \in gl(V,\mathbb{R})$ the adjoint transformation $A^{<,>}$ of A (with respect to $<,>$) is defined by $<A^{<,>}x,y> = <x,Ay>$. With the help of $(Y^{<,>})^{<,>} = Y$ one can prove that $\text{Aut}(Y^{<,>}) = (\text{Aut } Y)^{<,>}$. In a domain of positivity hence $A \in \text{Aut Pos}(V,<,>)$ implies $A^{<,>} \in \text{Aut Pos}(V,<,>)$.

Theorem: $\text{Pos}(V,<,>)$ is a domain of positivity with respect to the non-degenerate bilinear form τ iff there is an $A \in \text{Aut Pos}(V,<,>)$ with $A^{<,>} = A$ and $\tau(x,y) = <Ax,y>$. (10.7)

This shows that there can be several bilinear forms for which Y is a domain of positivity (an example is given below). Y is called homogeneous if Aut Y acts transitively on Y.

10.4 Examples

One class of examples is given by the set Y of all sefadjoint and positive definite endomorphisms in (V,τ) where τ is a positive definite symmetric bilinear form. Y is a domain of positivity with respect to the positive definite bilinear form trace (AB). Its closure is given by the set of selfadjoint, positive semidefinite transformations of V. It is homogeneous with respect to its automorphism group

$$\{\phi_A \ / \ \phi_A B := A^\tau B\ A \quad , B \in Y,\ A \in Gl(V, \mathbb{R})\} \quad . \tag{10.8}$$

$\phi : A \to \phi_A$ is an epimorphism $Gl(V,\mathbb{R}) \to \mathrm{Aut}\ Y$ with kernel $\pm\, id_V$. Hence Aut Y is covered twice by $Gl(V,\mathbb{R})$.

Another class of examples is given by the circular cones: Given $t \in V$ with $<t,t> = 1$, the bilinear form

$$\tau(x,y) := 2<t,x><t,y> - <x,y> = -<S_t x,y> \ , \tag{10.9}$$

c.f. (3.5), is symmetric and non-degenerate since $\tau(z,x) = 0$ for all $x \in V$ imples $2<t,z>t = z$ or $2<t,z> = <t,z>$ or $<t,z> = 0$ or $<z,x> = 0$ for all $x \in V$, hence $z = 0$. The subset

$$Y = \{y \in V \ / \ <t,y> > 0,\ <y,y> > 0\} \tag{10.10}$$

of V is open and not empty since it contains t. Choosing a basis in V in which the matrix of <,> is diagonal it is easy to see that

$$\mathrm{sign}<,> = (n_1,n_2) \ \text{ iff } \mathrm{sign}\ \tau = (n_2+1,n_1-1) \ . \tag{10.11}$$

Hence τ is positive definite iff <,> has the Lorentz signature. In this case Y is the interior of the forward light cone, example (d) in Section 3 and

Theorem: Y is a domain of positivity for <,> and τ. (10.12)

The boundary of Y is the forward light cone. Since $S_t t = -t$, the (with respect to <,> and τ) selfadjoint transformation $-S_t$ fulfills the condition for A in (10.7). Note that the interior of the null cone is not convex for <,> not positive definite.

Theorem: Y is homogeneous with respect to Aut Y, which is the direct product of the connected dilatation group with $\mathrm{Aut}_0(V,<,>) \cup S_t \mathrm{Aut}_0(V,<,>)$, where S_t is space inversion in a diagonal basis with $t = (1,0,\dots,0)$. (10.13)

Trivially Aut Y is exactly the group which preserves the order relation $\leq$. Consequently this is the result of Zeeman [Mi p. 101] [Si p.2] which expresses the idea of causality in Minkowski space.

11. JORDAN ALGEBRAS

A vector space V is called a (commutative) Jordan algebra if there is a bilinear composition $\top$ on V such that

(J1) $x \mathbin{\top} y = y \mathbin{\top} x$ (symmetry)
(J2) $x \mathbin{\top} ((x \mathbin{\top} x) \mathbin{\top} y) = (x \mathbin{\top} x) \mathbin{\top} (x \mathbin{\top} y)$ ("Jordan identity").

This composition is power associative, i.e $x^i \mathbin{\top} x^k = x^{i+k}$, where x^i is defined by recursion and $i, k \geqslant 1$.

Examples: (a) Every associative algebra $\mathcal{A}$ is a Jordan algebra with respect to the anticommutator $[x,y]_+ = \frac{1}{2}(xy + yx)$. This Jordan algebra is written $\mathcal{A}^+$.

(b) Given an associative algebra and an involutive antiautomorphism †, i.e. $(x^\dagger)^\dagger = x$ and $(xy)^\dagger = y^\dagger x^\dagger$, the set of symmetric elements $x^\dagger = x$ is a Jordan subalgebra of (a).

(c) A pseudo-orthogonal vector space $(V,<,>)$ can be given a Jordan algebra structure for any given $t \in V$ by

$$x \mathbin{\top} y = \langle x,t\rangle\, y + \langle y,t\rangle\, x \qquad x,y \in V. \tag{11.1}$$

Remark: Like for Lie algebras there are Lie triple systems, there are Jordan triple systems for Jordan algebras. The Jordan triple system associated to the Jordan algebra (c), after representation gives the Duffin Kemmer algebra, which was used in the theory of linear relativistic wave equations for the description of spin zero and spin one particles with non-vanishing mass.

The left multiplication in a Jordan algebra is defined by $L(x)y = x \mathbin{\top} y$. Contrary to Lie algebras it is no ("adjoint") representation since $x \to L(x)$ is no Jordan homomorphism into $gl(V,\mathbb{R})^+$. (J2) states that $L(x)$ and $L(x^2)$ commute for all x. x and y commute if $L(x)$ and $L(y)$ commute. If the Jordan algebra has a unit element e (by a standard procedure one can always adjoint a unit element if there is none) then $x \in V$ is said to be invertible if (a) there is an element $x^{-1} \in V$ with $x^{-1} \mathbin{\top} x = e$ and (b) x^{-1} commutes with x. For $P(x) := 2L^2(x) - L(x^2)$ the mapping $x \to P(x)$ is called the square mapping. One has $P(e^x) = e^{2L(x)}$ and the fundamental formula

$$P(P(x)y) = P(x)P(y)P(x) \quad , \tag{11.2}$$

which is easy to verify for example (a) where the square mapping reduces to $P(x)y = xyx$.

Theorem: x is invertible if $P(x)$ is invertible, i.e. iff $\det P(x) \neq 0$, and $x^{-1} = P(x)^{-1}x$, $P(x)^{-1} = P(x^{-1})$. (11.3)

The symmetric bilinear form $\tau(x,y) = \operatorname{trace} L(x \mathbin{\top} y)$ is associative, i.e. $\tau(x \mathbin{\top} y, z) = \tau(x, y \mathbin{\top} z)$. The Jordan algebra is called commutative if all elements commute with each other, central if the center, i.e. the commutative Jordan subalgebra of all elements which commute with all elements of the algebra, is $\mathbb{R}e$, semisimple if τ is non-degenerate, simple if it is not zero and has only the trivial ideals {0} and V, formal real (sometimes one says compact instead) if $x^2 + y^2 = 0$ implies $x + y + 0$ (equivalently if

τ is positive definite). The Jordan algebras (b) are formal real. If the Jordan algebra is semi-simple then it has a unit element, if it is formal real then it is semi-simple, if it is semi-simple (or formal real) then it is the direct sum of simple (or simple and formal real) ideals, if it is formal real and simple then it is central simple. A Jordan algebra is called exceptional if it is not isomorphic to a Jordan subalgebra of some $\mathcal{A}^+$, i.e. if it has no faithful finite-dimensional representation. The simple Jordan algebras are classified into seven classes and the formal real, simple ones are contained in this classification in the same way as the compact simple Lie algebras are contained in the simple ones.

From the axions (J1) and (J2) one can prove

$$[[L(x),L(y)]_-,L(z)]_- = L((y\tau z)\tau x - y\tau(z\tau x)) \ . \tag{11.4}$$

This relation has a number of consequences: (a) For every Jordan algebra (V,τ) the totality of left multiplications $L(V)$ is a Lie triple system. (b) The Lie algebra generated by the commutators of $L(V)$ is $L(V) \oplus [L(V),L(V)]_-$; it is called the structure Lie algebra of (V,τ). (c) The mappings $[L(x),L(y)]_-$ are derivations of (V,τ), called inner derivations. The Lie algebra of inner derivations is an ideal in the Lie algebra of all derivations. A semi-simple Jordan algebra has only inner derivations.

Both derivation algebras are Lie algebras of infinitesimal transformations of the corresponding automorphism groups of (V,τ), the normal subgroup of inner automorphisms being defined by this property.

12. THE RELATION BETWEEN DOMAINS OF POSITIVITY AND SYMMETRIC SPACES

12.1 Domains of positivity and Jordan algebras

In the following (V,τ) is a formal real Jordan algebra with unit element e and bilinear form τ.

Theorem: The connectivity component of e in the set of invertible elements of (V,τ) is a homogeneous domain of positivity with respect to τ. (12.1)

We write $\mathrm{Pos}(V,\tau)$ for this domain of positivity. Put $x^0 = e$.

Theorem: $\mathrm{Pos}(V,\tau) = \{e^x \ / \ x \in V\} = \{x^2 \ / \ x \in V \text{ invertible}\} = \{y \in V \ / \ L(y)$ positive definite with respect to $\tau\}$ = connectivity component of e of $\{y \in V \ / \ P(y)$ positive definite with respect to $\tau\}$;

$\overline{\mathrm{Pos}(V,\tau)} = \{x^2 \ / \ x \in V\} = \{ y \in V \ / \ L(y)$ positive semi-definite with respect to $\tau\}$. (12.2)

Obviously the boundary of $\mathrm{Pos}(V,\tau)$ is the set of x^2 where x runs through the set of non-invertible elements of (V,τ). Conversely to (12.1) M. Koecher has given a construction of a formal real Jordan algebra for every homogeneous domain of positivity. We need some elementary concepts for it: The directional derivative of a real valued function ϕ on V in the direction of $u \in V$ is

$$\Delta^u{}_x\phi(x) = \lim_{\mu\to 0} \frac{1}{\mu} (\phi(x + \mu u) - \phi(x))$$

for x in some open subspace of V. $u \to \Delta^u{}_x\phi(x)$ is a linear form on V. Hence there is a unique $\mathrm{grad}\phi(x) \in V$ such that $\langle \mathrm{grad}\phi(x),u\rangle = \Delta^u{}_x\phi(x)$. $\mathrm{grad}\phi(x)$ is called the gradient of ϕ. The real valued function $\omega(x)$, defined by

$$\omega(x) = \int_{\mathrm{Pos}(V,\langle,\rangle)} \exp(-\langle x,y\rangle)dy \quad ,$$

is called the norm or the invariant of $\mathrm{Pos}(V,\langle,\rangle)$, since $\omega(Ax) = \det(A)\omega(x)$ for all $A \in \mathrm{Aut\ Pos}(V,\langle,\rangle)$. Then $y \to y^\# := -\mathrm{grad}\log\omega(y)$ is an involutive mapping (of homogeneity degree -1) of $\mathrm{Pos}(V,\langle,\rangle)$ onto itself, which has exactly one fixed point, say e.

$$\lambda(u,v,w) = \Delta^u{}_y\,\Delta^v{}_y\,\Delta^w{}_y\,\omega(y)\big|_{y = e} \tag{12.3}$$

is a symmetric trilinear form on V. By

$$\lambda(u,v,w) = \langle u \tau v,w\rangle \tag{12.4}$$

we get a symmetric algebra composition τ on V and

Theorem: (V,τ) is a formal real Jordan algebra with unit element e and $\mathrm{Pos}(V,\langle,\rangle) = \mathrm{Pos}(V,\tau)$. In (V,τ) one has $y^\# = y^{-1}$ and $\omega(y) = (\det P(y))^{1/2}$. (12.5)

This establishes a functor from homogeneous domains of positivity onto formal real Jordan algebras. The exponential representation of $\mathrm{Pos}(V,\tau)$ in (12.2) shows the analogy with the functor of Lie groups onto Lie algebras. Clearly one can identify the tangent space in e of the manifold $\mathrm{Pos}(V,\tau)$ with the vector space V. This should be compared with the identification of the Lie algebra (i.e. the tangent space in id_V) of $\mathrm{Gl}(V,\mathbb{R})$ with $\mathrm{gl}(V,\mathbb{R})$, where $\mathrm{Gl}(V,\mathbb{R})$ is an open subspace of $\mathrm{gl}(V,\mathbb{R})$ as well. However, contrary to the relation between domains of positivity and Jordan algebras, there are cases for which such an embedding of the Lie group into the Lie algebra is not possible.

12.2 Domains of positivity and symmetric spaces

Theorem: The set of invertible elements in a Jordan algebra (V,τ) is a symmetric space with the multiplication

$$x \cdot y = P(x)y^{-1} \qquad x,y \in V.$$

Its Lie triple system in the tangent space of the base point e can be identified to V and

$$[xyz] = [[L(x),L(y)]_-,L(z)]_-(e) = x\tau(y\tau z) - y\tau(x\tau z). \qquad (12.6)$$

From (12.5) and (12.1) every homogeneous domain of positivity is the connectivity component of e in a formal real Jordan algebra. Hence it is a symmetric space with the multiplication and the Lie triple system given by (12.6).

Theorem: (a) $\mathrm{Aut}(V,\tau)$ is the isotropy group of e in $\mathrm{AutPos}(V,\tau)$.

(b) Polar decomposition: Every element of $\mathrm{AutPos}(V,\tau)$ can be written uniquely in the form $P(y)A$ with $y \in \mathrm{Pos}(V,\tau)$ and $A \in \mathrm{Aut}(V,\tau)$.

(c) $P(y) \in \mathrm{Aut\,Pos}(V,\tau)$ for $y \in \mathrm{Pos}(V,\tau)$ and the group generated by these $P(y)$ is the group of displacements of the symmetric space $\mathrm{Pos}(V,\tau)$ after restriction to $\mathrm{Pos}(V,\tau)$; hence it acts transitively.

(d) $y \in \mathrm{Pos}(V,\tau)$ implies $y^{-1} \in \mathrm{Pos}(V,\tau)$. (12.7)

The proof of (a) and (b) is rather involved. The first statement of (c) follows from (b); the second follows from $S(x)S(y)z = P(x)(P(y)z^{-1})^{-1} = P(x)P(y^{-1})z$, where we used (11.3) and (11.2). Since $P(e) = \mathrm{id}_V$ the group of displacements is generated by the $P(y)$ with $y \in \mathrm{Pos}(V,\tau)$, where $P(y)$ is to be restricted to $\mathrm{Pos}(V,\tau)$. Note that $P(x) = S(x)S(e) = Q(x)$ after restriction to $\mathrm{Pos}(V,\tau)$. (d) follows from $e \in \mathrm{Pos}(V,\tau)$ and $e \cdot y = P(e)y^{-1} = y^{-1}$.

Corollary: $\mathrm{AutPos}(V,\tau)$ when restricted to $\mathrm{Pos}(V,\tau)$ is a subgroup of the automorphism group of the symmetric space $\mathrm{Pos}(V,\tau)$. (12.8)

A verification using (11.2) and (11.3) shows that $P(y)$ is an automorphism of the symmetric space $\mathrm{Pos}(V,\tau)$, i.e. we have $P(y)(x \cdot z) = (P(y)x) \cdot (P(y)z)$ if $y \in \mathrm{Pos}(V,\tau)$. From (11.3) and $P(Ax) = A\,P(x)\,A^{-1}$ every $A \in \mathrm{Aut}(V,\tau)$ is such an automorphism too. The polar decomposition then gives the corollary.

(c) in (12.7) shows that the group of displacements of the symmetric space $\mathrm{Pos}(V,\tau)$ is the restriction of a linear transformation group on V. This is not true for the full automorphism group of the symmetric space $\mathrm{Pos}(V,\tau)$ since the symmetric automorphism $S(e)$ has not this property: $S(e)y = y^{-1}$ and $S(e)$ consequently has the degree of homogeneity -1.

13. THE JORDAN ALGEBRA OF MINKOWSKI SPACE

A pseudo-orthogonal vector space $(V,\langle\,,\rangle)$ can be given a Jordan algebra structure

$$x \tau y = \langle x,t\rangle y + \langle y,t\rangle x - \langle x,y\rangle t \qquad t,x,y \in V \tag{13.1}$$

for every t; compare (11.1) which defines a different Jordan algebra. In the following we assume $\langle t,t\rangle = 1$. Then t is the unit element of this Jordan algebra which usually is written $[V,\langle,\rangle,t]$.

Let us choose $V = \mathbb{R}\, t \oplus V_o$ such that the direct vector space sum is orthogonal with respect to $\langle,\rangle$. Writing $x = \xi\, t \oplus x_o$, $y = \eta\, t \oplus y_o$, $\langle x,y\rangle = \xi\eta + \langle x_o,y_o\rangle_o$, $\langle,\rangle_o$ being the restriction of $\langle,\rangle$ to V_o, we get

$$(\xi\, t \oplus x_o)\tau(\eta\, t \oplus y_o) = \langle -S_t x,y\rangle t \oplus \xi y_o + \eta x_o \quad . \tag{13.2}$$

Let us write as before $\tau(x,y) := -\langle S_t x,y\rangle = \xi\eta - \langle x_o,y_o\rangle_o$. Then it is easy to prove that

Lemma: (a) trace $L(x\tau y) = n\tau(x,y)$ is the bilinear form of the Jordan algebra $[V,\langle,\rangle,t]$;

(b) $x \in V$ is invertible iff $\langle x,x\rangle \neq 0$. In this case

$$x^{-1} = -S_t R(x) = -\frac{1}{\langle x,x\rangle} S_t x = \frac{1}{\langle x,x\rangle}(2\langle x,t\rangle t - x);$$

(c) The automorphism group of the Jordan algebra $[V,\langle,\rangle,t]$ is $\mathrm{Aut}(V_o,\langle,\rangle_o)$ and its Lie algebra of derivations is $\mathrm{der}(V_o,\langle,\rangle_o)$. (13.3)

From (a) follows that $[V,\langle,\rangle,t]$ is semi-simple and in addition formal real iff $\langle,\rangle$ has the Lorentz signature. In fact it is central-simple. From (b) we get the symmetric multiplication (3.3) on the set of invertible elements (which are the elements outside the null cone) by calculating $x\cdot y$ according to (12.6) and by noting that

$$P(x) = \langle x,x\rangle\, S_x S_t \,. \tag{13.4}$$

Iff $\langle,\rangle$ has the Lorentz signature, the interior of the forward light cone is the connectivity component of t in this space and hence the (homogeneous) domain of positivity of the Jordan algebra $[V,\langle,\rangle,t]$.

The Koecher construction of a (formal real) Jordan algebra for every domain of positivity in Section 12 can be applied in the same way to a generalization of the latter, the socalled ω-domains. The resulting Jordan algebra is semi-simple. Conversely the connectivity component of the unit element in the set of invertible elements of a semi-simple Jordan algebra is such a ω-domain. A ω-domain is a domain of positivity iff it is convex. Applying this construction to the connectivity component of some t in $(V,\langle,\rangle)$ with $\langle t,t\rangle = 1$ in the interior of the null cone, one arrives at the Jordan algebra (13.1). Conversely the connectivity component of t in the set of invertible elements of this Jordan algebra is an example of an ω-domain, which is convex iff $\langle,\rangle$ has the Lorentz signature. The norm of this ω-domain is $\omega(y) = \langle y,y\rangle^{n/2}$, which can be verified easily from (12.5) and (13.4), noting that the determinant of the reflection S_x is -1.

It is easy to prove that $L(V_o) \subset der(V,<,>)$. Using $[L(V),L(V)]_- = [L(V_o),L(V_o)]_- = der(V,\tau) = der(V_o,<,>_o) \subset der(V,<,>)$ since (V,τ) is simple and (c) in (13.3)), and the fact that $x_o \to L(x_o)$ is a vector space isomorphism $V_o \to L(V_o)$, a dimensional argument gives the

Theorem: The structure Lie algebra of $[V,<,>,t]$ is the direct sum of the dilatations $L(\lambda t) = D_\lambda$ and the pseudo-orthogonal Lie algebra $der(V,<,>)$. (13.5)

Hence it coincides with the Lie algebra of the group of displacements (3.10).

Let $e_1,\ldots,e_{n-1}$ be a basis of V_o in which the matrix I_o of $<,>_o$ is diagonal. Then (13.2) reads

$$e_i \tau e_k = -I_{ik}\, t \quad . \tag{13.6}$$

Suppose we have realized $[V,<,>,t]$ in an associative algebra $\mathcal{A}^+$. Then $2e_i\tau e_k = e_ie_k + e_ke_i$ and (13.6) becomes the defining relation of the Clifford algebra over the pseudo-orthogonal vector space $(V_o, -<,>_o)$. Since this algebra has a finite-dimensional faithful representation, the Jordan algebra $[V,<,>,t]$ is special.

Koecher [Koe 69 p. 132] has described a construction of a group of birational transformations in every Jordan algebra. From (b) in (13.3) one can show that the connectivity component of the identity in this group for the Jordan algebra $[V,<,>,t]$ is the connected conformal group of $(V,<,>)$. Conversely Meyberg [Koe 69 p.19] has shown that the subspace $\delta\tilde{K}_V$ and $\delta\tilde{T}_V$ in (7.4) of the conformal Lie algebra carry the composition (13.1) in a natural way [Ti 71b].

14. THE JORDAN ALGEBRA OF NON-RELATIVISTIC SPIN OBSERVABLES

Given the Jordan algebra $[V,<,>,e]$ with unit element e and $V = \mathbb{R}\, e \oplus V_o$, $<x,y> = \xi\eta - <x_o,y_o>_o$, where $<,>_o$ is the usual scalar product in the three-dimensional euclidean space. We write e instead of t since there will be no time interpretation of t although $(V,<,>)$ is of Minkowsky type.

$v \in V$ is called involutive if $v \tau v = e$. The set of involutive elements of $[V,<,>,e]$ is given by $\pm$ e and by the sphere

$$\{v_o \in V_o \,/\, <v_o,v_o>_o = 1\} \quad . \tag{14.1}$$

The set of involutive elements is a symmetric subspace of the space (12.6) because of $(v\cdot v')^{-1} = (P(v)v'^{-1})^{-1} = P(v^{-1})v' = v^{-1}\cdot v'^{-1}$. In every Jordan algebra the mapping $v \to \frac{1}{2}(e-v)$ is a bijection of the set of involutive elements onto the set of idempotent elements $m \tau m = m$; the inverse is given by $m \to e-2m$. e is mapped onto 0, -e onto e and (14.1) onto the sphere

$$\{m \in V \ / \ m = \frac{1}{2} e \oplus m_o \text{ with } \langle m_o, m_o\rangle_o = \frac{1}{4}\} \ . \qquad (14.2)$$

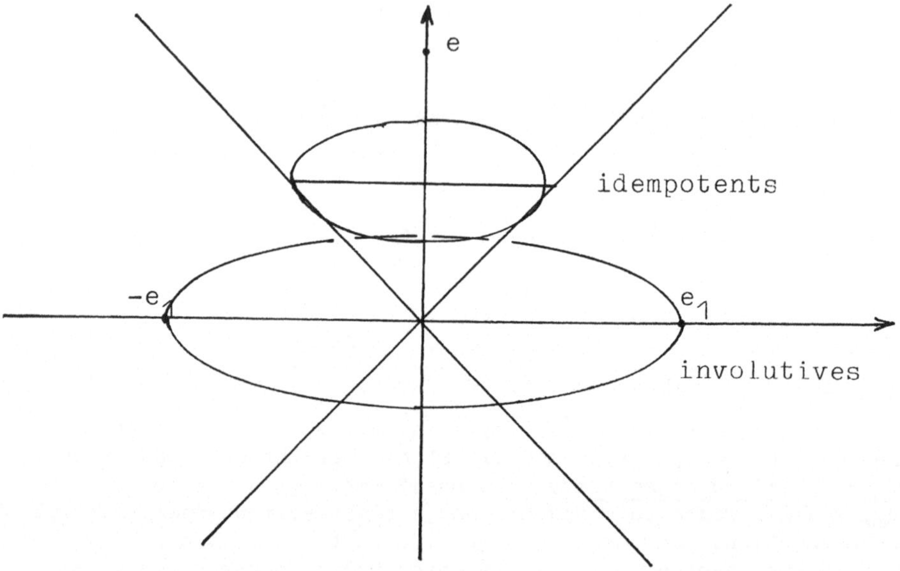

Since the only invertible idempotent is e the sphere (14.2) lies inside the null cone. Every element in V_o can be represented in polar coordinates in the form $x_o = \rho \cos\phi \sin\theta \, e_1 + \rho \sin\phi \sin\theta \, e_2 + \rho \cos\theta \, e_3$, where e_1, e_2, e_3 is the basis of V_o in which the matrix of $\langle,\rangle_o$ is id_3. The elements of (14.1) have the representation

$$v_o = \cos\phi \sin\theta \, e_1 + \sin\phi \sin\theta \, e_2 + \cos\theta \, e_3 \ . \qquad (14.3)$$

They can be identified with the observable spin in direction θ, ϕ; usually they are called polarization vectors. The pure statistical operators are defined by the set of (non-trivial) idempotents (14.2). The statistical operator of the state with the spin pointing in the direction θ,ϕ is given by $m_{\theta,\phi} = \frac{1}{2}(e + v_o)$.

Since n = 4 we have from (13.3)(a) for every pure state

$$\frac{1}{2} \text{ trace } L(m_{\theta,\phi}) = \frac{1}{2} \text{ trace } L(m_{\theta,\phi} \ \tau \ e) = 2 \ \tau(m_{\theta,\phi}, e) = 1 \quad (14.4)$$

Let us define for every observable $x \in V$ the expectation value of x in the state characterized by $m_{\theta,\phi}$ by

$$\langle\theta,\phi|x|\theta,\phi\rangle = \frac{1}{2} \text{ trace } L(m_{\theta,\phi} \tau x) = 2\tau(m_{\theta,\phi}, x) \ . \qquad (14.5)$$

Obviously $\langle\theta,\phi|v_o|\theta,\phi\rangle = 1$ as a special case of

$$<\theta',\phi'|v_o|\theta',\phi'> = \cos\theta\ \cos\theta' + \sin\theta\ \sin\theta'\ \cos(\phi-\phi'). \quad (14.6)$$

The mixed states are given by the interior of the sphere (14.2), the corresponding polarizations being given by the interior of the sphere (14.1).

In physics one uses a faithful representation of the Jordan algebra [V,<,>,e] by complex hermitian 2 x 2 matrices

$$v_o \rightarrow \sigma_{\theta,\phi} = \begin{pmatrix} \cos\theta & \sin\theta e^{-i\phi} \\ \sin\theta e^{i\phi} & -\cos\theta \end{pmatrix}, \quad (14.7)$$

i.e., e_1, e_2, e_3 are represented by the three Pauli matrices. e is the two-dimensional unit matrix and the Jordan product τ is represented by the anticommutator.

It is easy to prove that x and y in [V,<,>,e] commute (i.e. L(x) and L(y) commute) iff they commute as matrices. This makes it possible to define commensurability of observables in terms of commutativity in the Jordan algebra.

PART III / HALFSPACES OF JORDAN ALGEBRAS AND BOUNDED SYMMETRIC DOMAINS

A third type of possible applications of symmetric spaces is given in quantum field theory, where the symmetric spaces in question are generalized upper half planes and the corresponding bounded symmetric domains. Using the theory of Bergman kernels, those domains can be used for the construction of a class of useful Hilbert spaces in the theory of distributions. Especially one halfspace, the "tubular cone" of Minkowski space was used in this sense [Rü], see also this lecture notes and [SW Sections 2.3,2.4].

Again, the unifying point of view is given by the theory of Jordan algebras. The following presentation follows historical lines. Section 15 starts with the two-dimensional case, the complex plane. The second part of this section is devoted to C.L. Siegel's generalization, using the space of n-dimensional symmetric matrices instead of the real numbers and the symplectic instead of the special linear group. Section 16 gives a further generalization due to M. Koecher, to include the physical relevant case of the "tubular cone". Koecher's generalization defines a generalized upper half plane in the complexification of an arbitrary semi-simple Jordan algebra together with a certain bounded symmetric domain, which is mapped bijectively onto the upper half plane. The "tubular cone" is the bounded symmetric domain of the complexification of the Jordan algebra of Minkowski space. This space which is not covered by Siegel's generalization, is described in Section 17, where we give the relation of the group of biholomorphic bijections of the "tubular cone" in Minkowski space to the conformal group as well. This group theoretical result is due to U. Hirzebruch.

A. Weyler has used the bounded symmetric domain of Minkowski space and its group of biholomorphic bijections (which at least locally is isomorphic to the conformal group of Minkowski space) for an explicit computation of Sommerfeld's fine structure constant and the mass ratio of proton and electron [Wy 69, 71], see also [Gi], [Ro].

15. THE SIEGEL HALF SPACE

15.1 The two-dimensional case

We recall some facts of analysis in the complex plane. Given the upper half plane $H = \mathbb{R} + iY$ (where Y is the set of positive real numbers), the mapping π defined by

$$\pi(\zeta) = (\zeta - i)\ (\zeta + i)^{-1} \tag{15.1}$$

is a biholomorphic bijection from H onto the unit disk = interior of the unit circle E

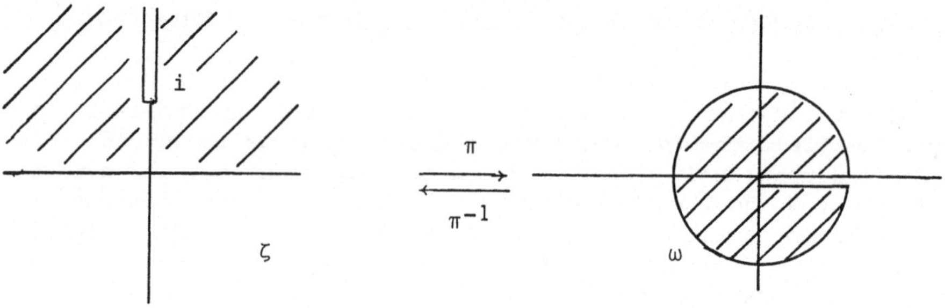

with the inverse mapping $\pi^{-1}(\omega) = -i(\omega+1)(\omega-1)^{-1}$. Given $A = \begin{smallmatrix}\alpha\beta\\ \gamma\delta\end{smallmatrix} \in SL(2,\mathbb{R})$, i.e. $\alpha\delta - \gamma\beta = 1$, the mapping

$$\zeta \to (\alpha\zeta + \beta)(\gamma\zeta + \delta)^{-1} = \tau_A(\zeta) \tag{15.2}$$

is a biholomorphic mapping of H onto H. More general, the group of biholomorphic mappings of H, Bihol H, is given by the set of transformations of type (15.2), and $\tau : A \to \tau_A$ is an epimorphism $\tau : SL(2,\mathbb{R}) \to$ Bihol H with kernel $\pm id_2$. Bihol H acts transitively on H since for every point in H there is a transformation of type (15.2) which maps it onto the point i. Bihol E is given by $\pi \cdot$Bihol $H \cdot \pi^{-1}$ and

$$\text{Bihol } E = \{\tau_A / \tau_A = \frac{\alpha\zeta+\beta^*}{\beta\zeta+\alpha^*}\ ,\quad \alpha\alpha^* - \beta\beta^* = 1\} \tag{15.3}$$

acts transitively on E. τ obviously is a group epimorphism τ: SU(1,1) $\to$ Bihol E. For $\gamma \neq 0$ the identity

$$\frac{\alpha\zeta+\beta}{\gamma\zeta+\delta} = \frac{\alpha(\gamma\zeta+\delta)+\beta\gamma-\alpha\delta}{\gamma(\gamma\zeta+\delta)} \quad \frac{\alpha}{\gamma} + \frac{1}{\gamma^2}(-(\zeta + \frac{\delta}{\gamma}))^{-1} \tag{15.4}$$

shows that the transformations

$$\zeta \to \zeta + \beta\,, \quad \zeta \to -\zeta^{-1}\,, \quad \zeta \to \frac{\gamma^{-1}\zeta}{\gamma} \tag{15.5}$$

generate Bihol H, i.e. every element of Bihol H is a product of such transformations. Moreover,

$$\begin{pmatrix} \gamma^{-1} & 0 \\ 0 & \gamma \end{pmatrix} = \begin{pmatrix} 1 & \gamma^{-1} \\ 0 & 1 \end{pmatrix}\begin{pmatrix} 0 & 1 \\ -1 & 0 \end{pmatrix}\begin{pmatrix} 1 & \gamma \\ 0 & 1 \end{pmatrix}\begin{pmatrix} 0 & 1 \\ -1 & 0 \end{pmatrix}\begin{pmatrix} 1 & \gamma^{-1} \\ 0 & 1 \end{pmatrix}\begin{pmatrix} 0 & 1 \\ -1 & 0 \end{pmatrix}$$

shows that already the first two (translations and inversions) generate Bihol H, since the matrices on the right hand side induce transformations of those types.

15.2 The n-dimensional case

The symplectic group $Sp(2n,\mathbb{R})$ is defined by

$$Sp(2n,\mathbb{R}) = \{B \in gl(2n,\mathbb{R}) \;/\; B^t\, I\, B = I\} \quad , \tag{15.6}$$

where I is the 2n x 2n matrix $\begin{pmatrix} 0 & id_n \\ -id_n & 0 \end{pmatrix}$. Note that $Sp(2,\mathbb{R})$ equals $SL(2,\mathbb{R})$ by definition. C.L. Siegel has given a generalization of the above results to the n-dimensional case: The generalized upper half plane

$$H_n = \{z = x + i\, y \in gl(n,C) \;/\; x \text{ symmetric real and } y \text{ symmetric real positive definite }\} \tag{15.7}$$

is the <u>Siegel half space</u> in n dimensions. The bounded set

$$E_n = \{z \in gl(n,C) \;/\; id_n - z^*z \text{ symmetric positive def.}\} \tag{15.8}$$

is the <u>generalized unit disk</u>. Its boundary is given by the unitary n x n matrices. The mapping

$$\pi : z \to (z - i\, id_n)(z + i\, id_n)^{-1} \tag{15.9}$$

is a biholomorphic bijection of H_n onto E_n with the inverse $\pi^{-1}(w) = i(id_n + w)(id_n - w)^{-1}$. For the 2n x 2n matrices $A = \begin{pmatrix} a & b \\ c & d \end{pmatrix}$ (a, b, c, d are n x n matrices) let us define as above

$$\tau_A z = (az + b)(cz + d)^{-1}, \quad \tau : A \to \tau_A$$

Theorem of Siegel: (a) τ_A with $A \in Sp(2n,\mathbb{R})$ is a well defined biholomorphic mapping of H_n onto itself;

(b) Bihol H_n acts transitively on H_n.

(c) $\tau: Sp(2n,\mathbb{R}) \to$ Bihol H_n is a group epimorphism with kernel $\pm id_{2n}$. (15.10)

Moreover Bihol H_n is generated by the translations $T_b(z) = z + b$ with b a symmetric and real n x n matrix and by the inversion $z \to -z^{-1}$. This implies that $Sp(2n,\mathbb{R})$ is generated by I and the matrices $\begin{pmatrix} id_n & b \\ 0 & id_n \end{pmatrix}$ (hence every symplectic matrix has the determinant +1). A dimensional argument shows that Bihol E_n no longer is the image of some unitary group in 2n dimensions. Obviously $\tau_A = \pi$ if $A = \begin{pmatrix} id_n & -i\, id_n \\ id_n & i\, id_n \end{pmatrix}$ and $\tau_A = \pi^{-1}$ if $A = \begin{pmatrix} i\, id_n & i\, id_n \\ -id_n & id_n \end{pmatrix}$.

15.3 The symmetric space realization

Given a unitary n x n matrix $u = a + ib$, it is easy to prove that

$$u = a + ib \to a \otimes id_2 + b \otimes \begin{pmatrix} 0 & 1 \\ -1 & 0 \end{pmatrix} = \begin{pmatrix} a & b \\ -b & a \end{pmatrix} \tag{15.11}$$

is an isomorphism of the Lie groups $U(n) \to Sp(2n,\mathbb{R}) \cap SO(2n,\mathbb{R})$. It is straightforward to prove an analogous result for the Lie algebras. Let us denote this real representation of U(n) by $U(n,\mathbb{R})$. The mapping $A \to I A I^{-1} =: \sigma(A)$ is an involutive automorphism of $Sp(2n,\mathbb{R})$ and its subgroup of fixed points $Sp(2n,\mathbb{R})_\sigma$ is exactly the connected group $U(n,\mathbb{R})$. Applying the results of Section 4, the mapping $q : A\,U(n,\mathbb{R}) \to A\sigma(A)^{-1} = AA^t$ (from the definition of the symplectic group) is an isomorphism of symmetric spaces

$$q = Sp(2n,\mathbb{R})/U(n,\mathbb{R}) \to Sp(2n,\mathbb{R})^\sigma \tag{15.12}$$

where $Sp(2n,\mathbb{R})^\sigma$ is the set of symmetric and positive definite, symplectic matrices. The symmetric multiplication becomes

$$AA^t \cdot BB^t = q(AU(n,\mathbb{R}))\cdot q(BU(n,\mathbb{R})) = q(A\sigma(A^{-1})\sigma(B)U(n,\mathbb{R}))$$

$$= q(AA^tB^{-1t}U(n,\mathbb{R})) = (AA^t)(BB^t)^{-1}(AA^t).$$

The mappings

$$AU(n,\mathbb{R}) \to \tau_A(i\, id_n), \qquad Sp(2n,\mathbb{R})/U(n,\mathbb{R}) \to H_n$$

$$AA^t \to \tau_A(i\, id_n), \qquad Sp(2n,\mathbb{R})^\sigma \to H_n$$

are well defined holomorphic diffeomorphisms onto H_n. Hence H_n

becomes a symmetric space with the symmetric multiplication

$$\tau_A(i\, id_n)\cdot\tau_B(i\, id_n) = \tau_{A A^{t}\bar{B}^{-1t}}(i\, id_n) \quad ,$$

where A, B $\in Sp(2n,\mathbb{R})$. By means of π, the symmetric multiplication can be transported to E_n as well. The Lie triple relations of the isomorphic symmetric spaces $E_n \cong H_n \cong Sp(2n,\mathbb{R})^\sigma \cong Sp(2n,\mathbb{R})/U(n,\mathbb{R})$ can be calculated easily from the eigenspace of eigenvalue - 1 of σ in the matrix Lie algebra $sp(2n,\mathbb{R}) = \{B \in gl(2n,\mathbb{R}) / B^t I + I B = 0\}$.

16. HALFSPACES AND BOUNDED SYMMETRIC DOMAINS

16.1 Halfspaces

In the following (V,τ) is a formal real Jordan algebra with unit element e and domain of positivity Y. The <u>halfspace</u> of (V,τ) is the subspace

$$H(V,\tau) = \{x + iy \;/\; x \in V,\; y \in Y\} = V + iY \tag{16.1}$$

of the complexification $V \oplus iV$ of (V,τ). The following theorem is due to U. Hirzebruch and M. Koecher:

Theorem: (a) Every $z \in H(V,\tau)$ is invertible in $V \oplus iV$;
(b) Bihol $H(V,\tau)$ is generated by the translations T_a with $a \in V$, the transformations $z \to Wz$, $W \in$ Aut Y and $Wz = Wx + i\,Wy$, and the involutive transformation $j_e : z \to -z^{-1}$;
(c) Bihol $H(V,\tau)$ acts transitively on $H(V,\tau)$. (16.2)

Like in (15.1) (b) can be sharpened by substituting for Aut Y its subgroup Aut(V,τ), c.f. (12.7)(a). $H(V,\tau)$ has not more connectivity components than Aut(V,τ). The mapping j_e is an involutive biholomorphic bijection of $H(V,\tau)$ onto itself with the only fixed point e. From (c) for all $z \in H(V,\tau)$ there is a transformation $g_z \in$ Bihol $H(V,\tau)$ such that $g_z(z) = e$. Hence $j_z = g_z^{-1} \circ j_e \circ g_z$ is an involutive biholomorphic bijection of $H(V,\tau)$ onto itself with the only fixed point z. Consequently $H(V,\tau)$ is a symmetric space with the symmetric multiplication

$$z\cdot z' = j_z(z') = (g_z^{-1} \circ j_e \circ g_z)(z') \quad . \tag{16.3}$$

Obviously the group of displacements of $H(V,\tau)$ is a subgroup of Bihol $H(V,\tau)$.

16.2 Bounded symmetric domains

The biholomorphic mapping

$$\pi : z \to (z - i\,e)\,\tau\,(z + i\,e)^{-1} \qquad (16.4)$$

is a bijection of $H(V,\tau)$ onto the bounded domain

$$E(V,\tau) = \{x + i\,y \in V \oplus i\,V \;/\; -e<x<e,\ -e<y<e\} \qquad (16.5)$$

with the inverse $\pi^{-1}(w) = i(e + w)\tau(e - w)^{-1}$. For this domain there are the equivalent realizations

$$\begin{aligned} E(V,\tau) &= \{w \in V \oplus i\,V \;/\; id_n - P(w)P(w^*) \text{ positive definite}\} \\ &= \{w \in V \oplus i\,V \;/\; w=P(u)r,\ u^{-1} = u^*, -e<r<e\}\ . \end{aligned} \qquad (16.6)$$

The group of biholomorphic mappings of $E(V,\tau)$ onto itself is $\pi \circ \text{Bihol}\ H(V,\tau) \circ \pi^{-1}$, which acts transitively. The biholomorphic mapping $\tilde{j}_o = \pi \circ j_e \circ \pi^{-1}$, $\tilde{j}_o : w \to -w$ is an involutive bijection of $E(V,\tau)$ onto itself with the only fixed point 0. Since Bihol $E(V,\tau)$ acts transitively, for all $w \in E(V,\tau)$ there is a transformation $\tilde{g}_w \in$ Bihol $E(V,\tau)$ with $\tilde{g}_w(w) = 0$. The biholomorphic involutions $\tilde{j}_w = \tilde{g}_w^{-1} \circ \tilde{j}_o \circ \tilde{g}_w$ of $E(V,\tau)$ leave only the point w fixed. Consequently, $E(V,\tau)$ is a symmetric space as well, hence a <u>bounded symmetric domain.</u>

16.3 The unitary elements

The compact subset

$$U(V,\tau) = \{u \in V \oplus i\,V \;/\; u^{-1} = u^*\} \qquad (16.7)$$

of $V \oplus i\,V$ is called the set of <u>unitary</u> elements of $V \oplus i\,V$. It is the socalled <u>Šilov boundary</u> of $E(V,\tau)$ and is contained in the 2n-1 dimensional boundary of $E(V,\tau)$ with dim $U(V,\tau) = n$. We have the equivalent realisation

$$U(V,\tau) = \{u \in V \oplus i\,V \;/\; u = e^{ix} \text{ with } x \in V\}\ .$$

Note that from the definition of invertibility in Jordan algebras, $u^{-1} = u^*$ is equivalent $u\tau u^* = e$ and $L(u)L(u^*) = L(u^*)L(u)$.

Remark: There is a subgroup of Bihol $U(V,\tau)$, called the <u>unitary group</u> of $V \oplus i\,V$, which acts transitively on $U(V,\tau)$.

Theorem: The image $\pi(V)$ of the "real axis" V in $V \oplus i\,V$ is given by those elements e^{ix} with $x \in V$, for which 1 is no eigenvalue of $L(e^{ix})$. The closure of $\pi(V)$ is $V \oplus i\,V$. (16.8)

For the proof see [Br p. 97]. In the special case i) of section 15, there is only one element on the unit circle which is not in $\pi(\mathbb{R})$, namely 1.

17. THE HALFSPACE OF MINKOWSKI SPACE

17.1 The tubular cone, its bounded symmetric domain and the space of unitary elements

In the following, $<,>$ has the Lorentz signature, i.e. the bilinear form of $[V,<,>,t]$, $\tau(x,y) = -<S_t x,y>$, is positive definite. The halfspace of $[V,<,>,t]$, sometimes called "tubular cone" is given by

$$H(V,<,>,t) = \{z = x + i\,y \in V \oplus i\,V \,/\, x \in V,\ y \in Y\}$$
$$= \{z \in V \oplus i\,V \,/\, <z-z^*,z-z^*> < 0,\ \frac{1}{2i} <z-z^*,t> > 0\}, \qquad (17.1)$$

where Y is the interior of the forward light cone. From (16.5) we see, that its bounded symmetric domain is

$$E(V,<,>,t) = \{u+iv \in V \oplus i\,v \,/\, t-u,\ t+u,\ t-v,\ t+v \in Y\} \qquad (17.2)$$

It is identical with the domain

$$E(V,<,>,t) = \{w \in V \oplus i\,V \,/\, \tau(w,w^*) < \frac{1}{2}(1+|\tau(w,w)|^2) < 1\} \qquad (17.3)$$

[Koe 69 p. 638], [Hi p. 416]. It is described in [Sie p. 158], [PC Section 8] and [Hua], where it is called "of type IV". Another characterisation is given in [Koe 69 p. 137 1.4 d] with the help of pairings. Such a pairing was calculated in [Ti 71b] for $[V,<,>, t]$. The result is

$$E(V,<,>,t) = \{w \in V \oplus iV/(2-<w,w^*>)\,id_V - o(w,w^*) > 0\} \qquad (17.4)$$

where $o(a,b)$ is the linear transformation $x \to <b,x>a - <a,x>b$ in the pseudo-orthogonal Lie algebra $der(V,<,>)$ and ">" means positive definite with respect to $\tau(z,w^*)$.

The boundary of $E(V,<,>,t)$ is shown in [PC p. 73] to be

$$\partial E(V,<,>,t) = \{w \in V \oplus iV/\tau(w,w^*) = \frac{1}{2}(1+|\tau(w,w)|^2) \leq 1\} \qquad (17.5)$$

It contains the Šilov boundary $U(V,<,>,t)$ for which $1 = \tau(t,t) = \tau(u\tau u^*,t) = \tau(u,u^*)$, $u \in U(V,<,>,t)$, since τ is associative. π maps the real axis V into (not onto from theorem (16.8)) $U(V,<,>,t)$. Hence we have $\pi(V) \subset U(V,<,>,t) \subset \partial E(V,<,>,t)$.

$$C(V,<,>,t) = \{e^{i\phi}x \,/\, x \in V,\ \tau(x,x) = 1,\ \phi \in \mathbb{R}\} \qquad (17.6)$$

is called the characteristic boundary of $E(V,<,>,t)$ [Hua]. It is easy to verify that $C(V,<,>,t)$ is a subset of $\partial E(V,<,>,t)$ but not of $U(V,<,>,t)$. We have $\dim\partial E(V,<,>,t) = 2n-1$, and $n = \dim U(V,<,>,t) = \dim C(V,<,>,t)$ [Lo II p. 174]. It remains to prove that π maps the boundary

$$\partial H(V,<,>,t) = \{x+iy \in V \oplus iV / x \in V \text{ and } <y,y> = 0\} \qquad (17.7)$$

onto $\partial E(V,<,>,t)$. Using $\pi(z) = t - 2i(z+it)^{-1}$ and (13.3b) we get

$$\pi(z) = (<z,z>-1+2i<z,t>)^{-1}(<z,z>t-2i<z,t>t+t+2iz) \ ,$$

from which one verifies $|<\pi(x),\pi(x)>|^2 = 1$ for real x.

17.2 The role of the conformal group

Following U. Hirzebruch, we give the relation of the conformal group to Bihol H(V,<,>,t). With the notation of Section 7, define the n+1-dimensional subset D of $C \oplus V \oplus i\,V \oplus C = \tilde{V}_C$ by

$$D = \{\tilde{z} \in \tilde{V}_C / \lessdot\tilde{z},\tilde{z}\gtrdot = 0,\ \lessdot\tilde{z}^*,\tilde{z}\gtrdot > 0,\ \operatorname{Im}\frac{\zeta_o}{\zeta_{n+1}} < 0\} \qquad (17.8)$$

In [Hi Section 12] it is shown that for $\tilde{z} = \zeta_o \oplus z \oplus \zeta_{n+1}$

$$\Gamma : D \to V \oplus i\,V\ , \qquad \Gamma : \tilde{z} \to \frac{z}{\zeta_{n+1}-\zeta_o} \qquad (17.9)$$

is a mapping onto H(V,<,>,t) and

$$\tilde{\Gamma} : V \oplus i\,V \to \tilde{V}_C\ ,\ \tilde{\Gamma}(z) = <z,z> -1 \oplus 2z \oplus <z,z> + 1 \qquad (17.10)$$

is a mapping from H(V,<,>,t) onto D, such that $\Gamma o \tilde{\Gamma}$ is the identity on H(V,<,>,t). Now instead of (7.3) define a mapping $\Gamma(\tilde{A})$ by

$$\Gamma(\tilde{A})(z) = \Gamma(\tilde{A}\tilde{\Gamma}(z)) \qquad (17.11)$$

for $\tilde{A} \in \operatorname{Aut}(\tilde{V},\lessdot,\gtrdot)$. $\Gamma(\tilde{A})$ transforms H(V,<,>,t) biholomorphically onto itself if $\tilde{A}$ is only in a certain subgroup of index two in $\operatorname{Aut}(\tilde{V},\lessdot,\gtrdot)$. Hence half of the connectivity components of $\operatorname{Aut}(\tilde{V},\lessdot,\gtrdot)$ are ruled out. Like in Section 7, $\Gamma : \tilde{A} \to \Gamma(\tilde{A})$ is an onto homomorphism of those connectivity components onto Bihol H(V,<,>,t) now, with the kernel $\pm id_{n+2}$. In [Hi] only the matrices (7.5) are used for the proof of "onto".

An analogous result is shown for Bihol E(V,<,>,t) by [Sie p. 158ff]. Let us write $\tilde{z} = \zeta_o \oplus \zeta_1 \oplus z_o \oplus \zeta_{n+1}$, hence z_o is a n-1 vector. The defining mapping is the projective transformation

$$\Gamma : \{\tilde{z} \in \tilde{V}_C / \lessdot\tilde{z},\tilde{z}\gtrdot = 0,\ \lessdot\tilde{z},\tilde{z}^*\gtrdot > 0,\ \operatorname{Im}\frac{\zeta_o}{\zeta_1} > 0\} \to E(V,<,>,t)$$

given by $\Gamma(\tilde{z}) = (\zeta_o+i\zeta_1)^{-1}(z_o \oplus \zeta_{n+1})$. Then the set of all $\Gamma(\tilde{A})$, $\Gamma(\tilde{A})$ being defined by $\Gamma(\tilde{A})\Gamma(\tilde{z}) = \Gamma(\tilde{A}\tilde{z})$, is Bihol E(V,<,>,t) if the $\tilde{A}$'s are in the same subgroup G of index two in $\operatorname{Aut}(\tilde{V},\lessdot,\gtrdot)$ as above.

17.3 The symmetric space realization

It is easy to see that this mapping Γ maps $\tilde{I}$ into the involutive mapping $w \to -w$ and that the subgroup G' of G, such that $\Gamma(G')$ leaves the point $0 \in E(V,<,>,t)$ fixed, $\Gamma(G') = \mathrm{Iso}(0)$, is isomorphic to $G \cap O(2,\mathbb{R}) \otimes O(n,\mathbb{R})$. $-\mathrm{id}_{n+2} \in G$ implies $-\mathrm{id}_{n+2} \in G'$. Summing, we have the isomorphisms

$$H(V,<,>,t) \cong \mathrm{Bihol}\ H(V,<,>,t)/\mathrm{Iso}(it) \cong E(V,<,>,t) \cong \mathrm{Bihol}\ E(V,<,>,t)/\mathrm{Iso}(0) \cong G/G \cap O(2,\mathbb{R}) \otimes O(n,\mathbb{R}) \tag{17.12}$$

of symmetric spaces. Note that $\pi(it) = 0$. The last space can be described with the results of Part I as follows (we restrict ourselves to the components of the identity only): $\sigma(\tilde{A}) = I_{2,n}\tilde{A}I_{2,n}$ is an involutive automorphism of $SO_0(2,n;\mathbb{R})$ with $SO_0(2,n;\mathbb{R})^\sigma = SO(2,\mathbb{R}) \otimes SO(n,\mathbb{R})$, compare (6.13).

REFERENCES

[BH] A.O. Barut and R.B. Haugen, Theory of the Conformally Invariant Mass, in these Proceedings, 1972.

[Br] H. Braun, Analytische Aspekte der Jordan-Algebren, Lecture Notes of the University of Hamburg, 1969.

[BK] H. Braun, M. Koecher, Jordan-Algebren, Springer, Berlin-Heidelberg-New York, 1966.

[Ch] C. Chevalley, Theory of Lie Groups I, Princeton University Press, 1964.

[Cl] A.J. Clark, The Representations of the Special Conformal Group in High Energy Physics, Princeton University Thesis, 1968.

[Gi] R. Gilmore, Scaling of Wyler's Expression for α, Phys. Rev. Lett. 28, 462-464 (1972).

[GKM] D. Gromoll, W. Klingenberg and W. Meyer, Riemannische Geometrie im Grossen, Springer Lecture Notes in Mathematics 55, Berlin-Heidelberg-New York, 1968.

[He] S. Helgason, Differential Geometry and Symmetric Spaces, Academic Press, New York-London, 1962.

[Hi] U. Hirzebruch, Halbräume und ihre Holomorphen Automorphismen, Math. Annalen 153, 395-417 (1964).

[Hua] L.K. Hua, Harmonic Analysis of Functions of Several Complex Variables in the Classical Domains, American Mathematical Society, Providence, Rhode Island, 1963.

[Hu] N.E. Hurt, Jordan Algebras and Quantizable Dynamical Systems, Lett. N. Cim. 1, 473-474 (1971).

[Ja] G. Janssen, Formal-Reelle Jordan-Algebren unendlicher Dimension und verallgemeinerte Positivitätsbereiche, Jour. R. Ang. Math. 249, 173-200 (1971).

[Jo] P. Jordan, Über eine Verallgemeinerungsmöglichkeit des Formalismus der Quantenmechanik, Nachr. d. Ges. d. Wiss. Göttingen 209-217 (1933).

[Ka] F.A. Kaempffer, Concepts of Quantum Mechanics, Academic Press, New York-London, 1965.

[Koe 62] M. Koecher, Jordan Algebras and their Applications, Lecture Notes of the University of Minneapolis, 1962.

[Koe 68] M. Koecher, Konvexe Kegel, Lecture Notes of the University of München, 1968.

[Koe 69] M. Koecher, Bounded Symmetric Domains, Rice University Press, Houston, Texas, 1969.

[Ku] R.S. Kulkarni, Curvature Structures and Conformal Transformations, Jour. Diff. Geom. 4, 425-451 (1970).

[Lo I] O. Loos, Symmetric Spaces I: General Theory, Benjamin, New York-Amsterdam, 1969.

[Lo II] O. Loos, Symmetric Spaces II: Compact Spaces and Classification, Benjamin, New York-Amsterdam, 1969.

[Mi] L. Michel, Applications of Group Theory to Quantum Physics. Algebraic Aspects, in Springer Lecture Notes in Physics 6, Berlin-Heidelberg-New York, 1970.

[OS] I. Oszváth and E.L. Schücking, The Finte Rotating Universe, Ann. of Phys. 55, 166-204 (1969).

[PC] I.I. Piatetsky-Chapiro, Géométrie des domaines classique et théorie des fonctions automorphes, Dunoc, Paris, 1966.

[Pi] G. Pinski, On the Embedding of Inhonogeneous Pseudo-Orthogonal Groups, N. Chim. 18, 1086-1107 (1967).

[Ro] B. Robertson, Wyler's Expression for the Fine-Structure Constant α, Phys. Rev. Lett. 27, 1545-1547 (1971).

[Rü] W. Rühl, Distributions on Minkowski Space and their Connection with Analytic Representations of the Conformal Group, Commun. Math. Phys. 27, 53-86 (1972).

[Sie] C.L. Siegel, Analytic Functions of Several Complex Variables, Princeton Lecture Notes (1948-1949).

[Si] D. Simms, Lie Groups and Quantum Mechanics, Springer Lecture Notes in Mathematics 52, Berlin-Heidelberg-New York, 1968.

[SW] R.F. Streater and A.S. Wightman, Die Prinzipien der Quantenfeldtheorie, B.I. Hochschultaschenbücher 435/435a, Mannheim-Zürich, 1969.

[Ti 71a] H. Tilgner, On Non-Linear Transformations in Vector Spaces. Colineation and Conformal Groups, to appear in J. Math. Phys. (1974).

[Ti 71b] H. Tilgner, Symmetric Lie Algebras of Non-Linear Transformations of Conformal Type in Quantum Mechanics, to appear in Int. J. Theor. Phys. (1973).

[Ti 72] H. Tilgner, Globally Symmetric Pseudo-Riemannian Spaces in Cosmology, to appear in Rep. Math. Phys. (1973).

[To] P. Tondeur, Introduction to Lie Groups and Transformation Groups, Springer Lecture Notes in Mathematics 7, Berlin-Heidelberg-New York, 1965.

[Tits] J. Tits, Liesche Gruppen und Algebren, Lecture Notes of the University of Bonn, 1964.

[Wy 69] A. Wyler, L'espace symmétrique du groupe des equations de Maxwell, C.R. Acad. Sci. Paris 269, 743-745 (1969).

[Wy 71] A. Wyler, Les groupes des potentials de Coulomb et de Yukawa, C.R. Acad. Sc. Paris 271, 186-188 (1971).

BOUNDARY VALUES OF HOLOMORPHIC FUNCTIONS THAT BELONG TO HILBERT SPACES CARRYING ANALYTIC REPRESENTATIONS OF SEMISIMPLE LIE GROUPS*

W. Rühl
Universität Trier-Kaiserslautern
Fachbereich Physik
Kaiserslautern

ABSTRACT. Unitary irreducible representations of the discrete series for the group SU(1,1) and SU(2,2) are realized in Hilbert spaces of holomorphic functions over certain domains in C_1, respectively C_4. The boundary values of these holomorphic functions are distributions whose local singularities can be characterized by the group invariants. In the case of half integral spin for SU(1,1) and even scale dimension for SU(2,2) the boundary distributions are represented as a differential operator applied to a square integrable function, the order of the differential operator being related with the spin respectively the scale dimension. In the case of SU(2,2) the holomorphic functions can be defined on the field theoretic tube domain. Thus we obtain Hilbert spaces of distributions over Minkowski space with a conformally invariant norm that carry unitary irreducible representations of the conformal group SU(2,2).

0. PRELIMINARIES

In these preliminaries we want to discuss some general notions of the theory of functions of several complex variables and of the theory of Hilbert spaces of such functions that occur repeatedly in the following lectures. The domains on which these functions are defined and holomorphic, respectively antiholomorphic, are homogeneous spaces for the Lie groups SU(1,1) and SU(2,2). For the sake of simplicity we restrict our investigation to these two groups. In the first case we have to deal with a single complex

---

* Lectures presented at the International Advanced Study Institute on Mathematical Physics in Istanbul, Turkey, 1972.

variable, a case to which physicists are well accustomed in general. In the case of SU(2,2), however, we have functions of four complex variables. But the generalization explicitly needed is so straightforward that our results can serve as an example to gain insights into the behaviour of holomorphic functions of several variables.

The homogeneous spaces of the groups SU(1,1) and SU(2,2) occur in two different realizations each: A "compact" realization which for the group SU(1,1) is the open unit circle, and a "generalized upper half plane" realization that is identical with the upper half plane for the group SU(1,1) and is the field theoretical tube domain for SU(2,2). Different realizations are analytic one-to-one maps of each other. The homogeneous spaces are obtained by dividing the groups by their maximal compact subgroups.

In general analytic representations are unitary representations realized in Hilbert spaces of holomorphic (or antiholomorphic) functions $f(z)$, $z = \{z_1, z_2, .. z_n\}$, defined on a certain domain D of the complex n-dimensional space C_n. D is a homogeneous space. Let us assume that D is the compact realization of the homogeneous space. Then we define a scalar product for any pair f_1, f_2 of holomorphic functions by

$$(f_1, f_2) = \int_D |dz| \, \overline{f_1(z)} f_2(z) \tag{0.1}$$

where

$$|dz| = \prod_{i=1}^{n} dx_i dy_i \quad , \quad z_i = x_i + iy_i \tag{0.2}$$

is the Bergman measure. In fact we shall have to use a different measure on D later on that makes the scalar product invariant under the actions of the group but this does not matter here. In this fashion we obtain a pre-Hilbert space $A^2(D)$ of all holomorphic functions $f(z)$ over D that are square integrable with respect to the Bergman measure (0.2). This space is obviously a subspace of the Hilbert space $\mathcal{L}^2(D)$ of all Bergman-square-integrable functions on D. Therefore any Cauchy sequence in $A^2(D)$ has a limit in $\mathcal{L}^2(D)$. Let

$$|z_i - z_i^0| < \delta \text{ for all } i,$$

define a polydisc $P_\delta(z^0)$ that lies completely in D (where "completely" means together with its closure). Then for any pair f_n, f_m of the Cauchy sequence we have

$$\|f_n - f_m\|^2 = \int_D |dz| \, |f_n(z) - f_m(z)|^2$$

$$\geq \int_{P_\delta(z_0)} |dz| \, |f_n(z) - f_m(z)|^2$$

$$= (2\pi)^n \sum_{k_i \geq 0} |a_{k_1 k_2 \ldots k_n}|^2 \prod_{i=1}^{n} \{\frac{\delta^{2k_i+2}}{2k_i+2} \cdot \frac{1}{(k_i!)^2}\}$$

$$\geqq (\pi\delta^2)^n |a_{00\cdots 0}|^2 \tag{0.3}$$

where the Taylor expansion

$$f_n(z)-f_m(z) = \sum_{k_i\geqq 0} a_{k_1k_2\ldots k_n} \prod_{i=1}^{n} \frac{(z_i-z_i^{0})^{k_i}}{k_i!} \tag{0.4}$$

that converges absolutely uniformly in the polydisc $P_\delta(z^0)$ has been inserted. In other terms, we have obtained the inequality

$$\|f_n-f_m\| \geqq (\pi\delta^2)^{\frac{n}{2}} |f_n(z^0)-f_m(z^0)| \tag{0.5}$$

It follows that the Cauchy sequence converges uniformly on each closed polydisc in D. Consequently the limit functions lies in $A^2(D)$ and not only in $\mathcal{L}^2(D)$. $A^2(D)$ is a Hilbert space.

One of the most important devices to be used by us is the Aronszajn-Bergman reproducing kernel. For the spaces of holomorphic functions with a Bergman measure it is denoted the Bergman kernel. Let the scalar product (f_1,f_2) (0.1) be connected with the norm $\|f\|$ in $A^2(D)$. The function value of f at a point z^0, $f(z^0)$ defines a linear continuous functional on $A^2(D)$, that is

$$|f(z^0)| \leqq C_{z^0} \|f\| \tag{0.6}$$

This assertion can be proved easily. In fact, due to the inequality (0.3), we have (replace f_n-f_m by f)

$$|f(z^0)| \leqq (\pi\delta^2)^{-\frac{n}{2}} \|f\| \tag{0.7}$$

where δ is the radius of any polidisc $P_\delta(z^0)$ lying in D. It therefore depends on z^0 as is permitted by (0.6). The estimate (0.6) has been proved this way. But due to Riesz' representation theorem [1] a linear continuous functional on a Hilbert space is generated by a unique vector of the Hilbert space through the scalar product, so that

$$f(z^0) = (K^B_{z^0}, f) \tag{0.8}$$

in our case. Therefore there exists a function $K_{z^0}(z)$ that lies in $A^2(D)$, i.e. that is holomorphic in z for fixed z^0 and is square integrable. This function is the Bergman kernel. We shall also use the notation

$$K^B_{z_1}(z_2) = K^B(z_1,z_2) \tag{0.9}$$

Applying Schwarz's inequality to (0.8) we get

$$|f(z^0)| \leqq \|K^B_{z^0}\| \, \|f\| \tag{0.10}$$

where the equality sign holds if and only if

$$f(z) = \alpha K^B_{z_0}(z), \ \alpha \text{ complex arbitrary} \tag{0.11}$$

This implies an extremal property of the Bergman kernel

$$\sup |f(z)| = \|K^B_z\| \qquad f \in A^2(D), \ \|f\| = 1 \tag{0.12}$$

that can in turn be used to define the Bergman kernel as we shall see in a moment.

For $f = K^B{}_{z_1}$ we obtain

$$K^B(z_1, z_2) = (K^B_{z_2}, K^B_{z_1}) \tag{0.13}$$

and

$$\|K^B_z\|^2 = K^B(z,z) \tag{0.14}$$

Hence from (0.13) we have hermiticity

$$K^B(z_1, z_2) = \overline{K^B(z_2, z_1)} \tag{0.15}$$

and from (0.14) positivity

$$K^B(z,z) > 0, \ z \in D \tag{0.16}$$

of the Bergman kernel. $K^B(z_1,z_2)$ is therefore antiholomorphic in z_1 for fixed z_2. Given $K^B(z,z)$ (say from (0.12) via (0.14)) one can reconstruct $K^B(z_1, z_2)$ by the replacements

$$\text{Re} z \to \frac{1}{2}(\bar{z}_1+z_2) \quad , \quad \text{Im } z \to \frac{1}{2i}(-\bar{z}_1+z_2)$$

in the Taylor expansion of $K^B(z,z)$ that can be shown to exist by general theorems.

If $\phi_n(z)$, $n = 0, 1, 2, \ldots$ is an orthonormal basis in $A^2(D)$, the Bergman kernel is obtained from the sum

$$K^B(z_1,z_2) = \sum_{n=0}^{\infty} \overline{\phi_n(z_1)}\, \phi_n(z_2) \tag{0.17}$$

If z_2 is fixed, this series converges uniformly for z_1 varying over compact subsets of D, and vice versa. To prove this we notice that

$$\overline{\phi_n(z_1)} = (\phi_n, K^B_{z_1}) \tag{0.18}$$

so that the series (0.17) becomes a Cauchy series for the element $K^B_{z_1}$ of $A^2(D)$. That such series (which are normconvergent a priori) are absolutely uniformly convergent on compact subsets was proved at the beginning.

The task of explicitly constructing the Bergman kernel can be solved by first finding an orthonormal basis and then summing the series (0.17). In the case of the unit circle D in one-dimensional complex space analytic polynomials in z (from now on simply: polynomials) form a dense subspace of $A^2(D)$. In the case of the domain D for the group SU(2,2) the situation looks more complicated. The domain is, however, of a very restricted type, namely it is a connected Reinhardt domain containing the origin. Reinhardt domains are defined to contain with every point $z = \{z_1, z_2, \dots z_n\}$ the point $e^{i\Theta}z = \{e^{i\Theta_1}z_1, e^{i\Theta_2}z_2, \dots e^{i\Theta_n}z_n\}$ for any real n-tupel $\Theta_1, \Theta_2, \dots \Theta_n$. In a connected Reinhardt domain containing the origin the Taylor expansion of a holomorphic function around the origin converges absolutely uniformly on every compact subset [2]. Therefore polynomials form a dense subspace in this case, too. It remains therefore to orthogonalize polynomials and sum the series (0.17).

Our homogeneous spaces are symmetric and hermitean spaces, where the latter notation means that an invariant Riemannian metric exists, [3]. This metric can be derived from the Bergman kernel, it is therefore denoted the Bergman metric. For the compact realizations of the symmetric hermitean spaces of the groups SO(n,2) a polynomial orthonormal basis of $A^2(D)$ can be given by group theoretic means [4]. It is a direct generalization of our construct for SO(4,2) (SU(2,2)). In fact the relation of the groups SU(1,1) and SU(2,2) with the pseudoorthogonal groups is as follows: Dividing by the respective central subgroups of Z_n of n elements we obtain

$$SU(1,1)/Z_2 \cong SO(1,2)$$

$$SU(2,2)/Z_4 \cong SO(4,2)/Z_2$$

Moreover are the pseudoorthogonal groups the automorphism groups of the domains D (up to an isomorphism).

Our domains D are geometrically convex and therefore domains of holomorphy. Their boundary is a 2n-1 dimensional manifold. An n-dimensional subset of the boundary is the Shilov boundary that is characterized by the property that any holomorphic function in D still continuous on the boundary assumes its maximum on the Shilov boundary. This implies in turn that given boundary values on the Shilov boundary determine the holomorphic function inside the domain uniquely. For further details and precise definitions we refer to the literature [5,6].

The Shilov boundaries admit a positive measure $d\rho$ that is quasiinvariant under the action of the automorphisms. Let us consider that dense subspace of $A^2(D)$ which contains functions f(z) being continuous still on the Shilov boundary. The space of polynomials is such space. In this subspace we can introduce a new norm

$$\|f\|_S^2 = \int_S |f(x)|^2 \, d\rho(x) \tag{0.19}$$

and a corresponding scalar product $(f_1,f_2)_S$. After completion we obtain a Hilbert space $\mathcal{H}(D)$ of holomorphic functions (this will be proved in the main text) which we call of "Hardy-Lebesgue type". This space again possesses an Aronszajn-Bergman kernel that we denote the "Szegö kernel". It has properties analogous to (0.8)-(0.18). The kernel $K_z^S(x)$ is the boundary value of a holomorphic function $K_z^S(z')$ in z' for fixed z if z' tends to the Shilov boundary. The function $K_z^S(z')$ is intimately related with the Bergman kernels in our case and has the same holomorphy properties. The formula

$$f(z) = (K_z^S, f)_S = \int_S \overline{K_z^S(x)}\, f(x)\, d\rho(x) \tag{0.20}$$

allows the extension of the boundary value $f(x)$ into the interior of the domain D and is therefore a special case of a Bergman-Weil formula generalizing the Cauchy integral formula to several complex variables and Shilov boundaries.

It turns out that the scalar product $(K_z^S, f)_S$ can be continuously extended to distributions f since K_z^S for fixed $z \in D$ can be considered as an element of a certain test function space, the functions and distributions being over the Shilov manifold. An appropriate way to analyze these distributions are Fourier series expansions in the case of the compact realizations and Fourier integrals in the case of the generalized-upper-half-plane realization. Correspondingly we choose our test functions spaces: $E(S)$ in the former case and $D_{L^2}(S)$ respectively $S(S)$ in the latter case (in Schwartz's notation, [7]).

1. THE DISCRETE SERIES OF SU(1,1)

1.1 Algebraic considerations, translations

We study the group SU(1,1) of matrices that can be written as

$$v = \begin{pmatrix} \alpha & \beta \\ \bar{\beta} & \bar{\alpha} \end{pmatrix}, \quad |\alpha|^2 - |\beta|^2 = 1 \tag{1.1}$$

These matrices satisfy

$$v^+ \sigma_3 = \sigma_3 v^{-1} \tag{1.2}$$

with σ_k, $k = 1,2,3$, the Pauli matrices. Let

$$z = x + iy \tag{1.3}$$

be a complex number. Then the subgroup U(1) of SU(1,1), consisting of matrices

$$u(\psi) = e^{\frac{i}{2}\psi\sigma_3}, \quad 0 \leq \psi < 4\pi \tag{1.4}$$

possesses cosets in SU(1,1) that can be characterized uniquely by the complex numbers z with $|z| < 1$, i.e. by the points of the open unit circle. In fact we introduce the notation

$$s(z) = \begin{pmatrix} N, & zN \\ \bar{z}N, & N \end{pmatrix}, \quad N = (1 - |z|^2)^{-\frac{1}{2}} \tag{1.5}$$

These are positive definite hermitean matrices of SU(1,1). Then any element $v \in SU(1,1)$ can be decomposed uniquely as

$$v = s(z)u(\psi) \tag{1.6}$$

This decomposition is the polar decomposition of v, that is: the decomposition of v into a positive definite hermitean matrix and a unitary matrix. The polar decomposition is known to be unique. The parameters are determined by

$$e^{i\frac{\psi}{2}} = \frac{\alpha}{|\alpha|}$$
$$z = \frac{\beta}{\bar{\alpha}} \tag{1.7}$$

The cosets of the unitary subgroup U(1) in SU(1,1) form a homogeneous space, that is: a manifold on which the group SU(1,1) acts transitively by right or left translations. If v obeys the representation (1.1) then its inverse is

$$v^{-1} = \begin{pmatrix} \bar{\alpha}, & -\beta \\ -\bar{\beta}, & \alpha \end{pmatrix} \tag{1.8}$$

Left translations are defined by

$$v^{-1}s(z) = s(z_v)u(\psi_v) \tag{1.9}$$

where due to (1.5) and (1.7)

$$e^{i\frac{\psi_v}{2}} = \frac{-\beta\bar{z} + \bar{\alpha}}{|-\beta\bar{z} + \bar{\alpha}|} \tag{1.10}$$

and

$$z_v = \frac{\bar{\alpha}z - \beta}{-\bar{\beta}z + \alpha} \tag{1.11}$$

These fractional linear transformations (1.11) form the automorphism group of the unit circle [8]. In order that the point $z = 0$ is a fixed point we must have $\beta = 0$ and consequently $|\alpha| = 1$, i.e. v must belong to the unitary subgroup U(1). In this case

$$z_v = \frac{\bar{\alpha}}{\alpha} z = (\bar{\alpha})^2 z \tag{1.12}$$

It follows that the subgroup Z_2 where Z_2 is the two-element centre of SU(1,1),

$$Z_2 = \{u(0), u(2\pi)\} \tag{1.13}$$

is the maximal subgroup leaving each point of the unit circle fixed (1.11). The automorphism group is the quotient group $SU(1,1)/Z_2$.

1.2 The discrete series of representations

We can obtain unitary representations of SU(1,1) by the method of induction, [9]. We consider a Hilbert space $\mathcal{L}^2(C)$ of square integrable functions on the unit circle $C = \{z \mid |z| < 1\}$.
We define

$$\mathcal{L}^2(C) = \{f(z) \mid f \text{ measurable}$$
$$\|f\|^2 = \int_C |dz| \, (1-|z|^2)^{-2} \, |f(z)|^2 < \infty\} \tag{1.14}$$

On this space we introduce the operator

$$T_v f(z) = e^{ik\psi_v} f(z_v) \tag{1.15}$$

where 2k is an arbitrary integer. The measure transforms as ($|dz|$ is the Bergman measure (0.2))

$$(1-|z|^2)^{-2}|dz| = (1-|z_v|^2)^{-2}|dz_v| \tag{1.16}$$

as follows from

$$\frac{dz_v}{dz} = (-\bar{\beta}z + \alpha)^{-2}, \quad |dz_v| = \left|\frac{dz_v}{dz}\right|^2 |dz| \tag{1.17}$$

and from (1.9)

$$(-\bar{\beta}z + \alpha)N(z) = N(z_v)e^{-i\frac{\psi_v}{2}} \tag{1.18}$$

Hence the measure appearing in (1.14) is invariant. The operator T_v is consequently isometric and we have $T_e = E$. If we prove that

$$T_{v_1} T_{v_2} = T_{v_1 v_2} \tag{1.19}$$

we obtain a representation by means of unitary operators.

We have to establish therefore that

$$e^{ik\psi_{v_1}(z)} \, e^{ik\psi_{v_2}(z_{v_1})} = e^{ik\psi_{v_1 v_2}(z)} \tag{1.20}$$

and

$$z_{v_2}\Big|_{z=z_{v_1}} = z_{v_1 v_2} \tag{1.21}$$

This is an elementary algebraic task. We start with the latter equation. In fact, in obvious notation

$$\frac{\bar{\alpha}_2 z_{v_1} - \beta_2}{-\bar{\beta}_2 z_{v_1} + \alpha_2} = \frac{(\bar{\alpha}_1\bar{\alpha}_2 + \bar{\beta}_1\beta_2)z - (\alpha_1\beta_2 + \beta_1\bar{\alpha}_2)}{-(\bar{\alpha}_1\bar{\beta}_2 + \bar{\beta}_1\alpha_2)z + (\alpha_1\alpha_2 + \beta_1\bar{\beta}_2)} = \frac{\bar{\alpha}_{12}z - \beta_{12}}{-\bar{\beta}_{12}z + \alpha_{12}} \tag{1.22}$$

which proves (1.21). Moreover we use (1.18) to express expi $\psi_v/2$ and insert it into (1.20). The equation reduces this way to

$$(-\bar{\beta}_1 z + \alpha_1)(-\bar{\beta}_2 z_{v_1} + \alpha_2) = (-\bar{\beta}_{12} z + \alpha_{12}) \tag{1.23}$$

This expression is proved analogously as (1.22).

The unitary representations obtained in this fashion are reducible in general. To see this we introduce the new functions

$$\hat{f}(z) = (1-|z|^2)^{-k} f(z) \tag{1.24}$$

The space $\mathcal{L}^2(C)$ of functions f(z) goes over into a space $\mathcal{L}^2_k(C)$ of functions f(z) by means of (1.24).
On this space T_v acts as

$$\begin{aligned} T_v\hat{f}(z) &= (1-|z|^2)^{-k} T_v f(z) \\ &= (1-|z|^2)^{-k} \left[\frac{(1-|z|^2)}{(1-|z_v|^2)}\right]^k (-\bar{\beta}z+\alpha)^{-2k} f(z_v) \\ &= (-\bar{\beta}z+\alpha)^{-2k} \hat{f}(z_v) \end{aligned} \tag{1.25}$$

The multiplier on the right hand side is holomorphic in C, so is z_v as a function of z. Therefore holomorphic functions on C form an invariant subspace $A^2_k(C)$ (this notation takes the k-dependence of the measure into account) of $\mathcal{L}^2_k(C)$, provided holomorphy of $\hat{f}$ and square integrability of f are compatible. Otherwise the subspace $A^2_k(C)$ is void.

The norm in the Hilbert space

$A^2_k(C)$ is

$$\|\hat{f}\|^2_k = c \int_C |dz| (1-|z|^2)^{2k-2} |\hat{f}(z)|^2, \quad c > 0 \tag{1.26}$$

with some constant c to be specified in a moment. The scalar product is defined correspondingly. The norm is well defined for $k \geqq 1$ at least.
It can, however, be extended to k = 1/2 by arguments into which we shall go later, if we choose the free constant c in (1.26) as

$$c = \frac{2k-1}{\pi} \tag{1.27}$$

This factor is adjusted such that for f(z) = 1 the norm is one.

In a similar fashion we may introduce new functions

$$\hat{\hat{f}}(z) = (1-|z|^2)^k f(z) \tag{1.28}$$

Taking the complex conjugate and inverse of expi $\psi_V/2$ in (1.18) we find

$$\begin{aligned} T_V\hat{\hat{f}}(z) &= (1-|z|^2)^k \, T_V f(z) \\ &= (1-|z|^2)^k \left[\frac{1-|z|^2}{1-|z_V|^2} \right]^{-k} (-\beta\bar{z}+\alpha)^{2k} f(z_V) \\ &= (-\beta\bar{z}+\alpha)^{2k} \, \hat{\hat{f}}(z_V) \end{aligned} \tag{1.29}$$

with the invariant norm

$$\|\hat{\hat{f}}\|_k^2 = c \int_C |dz| \; (1-|z|^2)^{-2k-2} |\hat{\hat{f}}(z)|^2 \tag{1.30}$$

Therefore for $k \leqq -1$ and by extension for $k = -1/2$ we have invariant Hilbert subspaces $A^{*2}_{|k|}(C)$ of antiholomorphic functions in $\mathcal{L}^2_k(C)$.

The spaces $A^2_k(C)$ and $A^{*2}_k(C)$, $k \geqq \frac{1}{2}$, 2k integral, carry the irreducible unitary representations of the discrete series of SU(1,1) [10].

1.3 The Bergman kernel

From now on we discuss only spaces of holomorphic functions, the antiholomorphic case can be treated analogously. The elements of $A^2_k(C)$ are denoted $f(z)$ instead of $\hat{f}(z)$ for the sake of simplicity. For the scalar product we use the form

$$(f_1,f_2)_k = \frac{2k-1}{\pi} \int |dz| \; (1-|z|^2)^{2k-2} \; \overline{f_1(z)} f_2(z) \tag{1.31}$$

The normalized powers $f_m(z) = N_m z^m$, m = 0, 1, 2, ... form an orthogonal basis in $A^2_k(C)$ according to the arguments presented in the "Preliminaries".

In order to find the Bergman kernel we have first to compute the normalization factors N_m

$$\begin{aligned} N_m^{-2} &= (z^m, z^m)_k \\ &= \frac{2k-1}{\pi} 2\pi \int_0^1 dr \; r^{2m+1} \; (1-r^2)^{2k-2} \\ &= \frac{m!(2k-1)!}{(2k+m-1)!} = \binom{2k+m-1}{m}^{-1} = (-1)^m \binom{-2k}{m}^{-1} \end{aligned} \tag{1.32}$$

The sum

$$K^B(z_1,z_2) = \sum_{m=0}^{\infty} N_m^2 \, \bar{z}_1^{\,m} \, z_2^{\,m} \tag{1.33}$$

can then be performed by elementary means and yields the Bergman kernel.

$$K^B(z_1,z_2) = (1-\bar{z}_1 z_2)^{-2k} \tag{1.34}$$

This series converges absolutely uniformly for

$$|z_1 z_2| \leqq 1-\varepsilon, \quad \varepsilon > 0 \tag{1.35}$$

whenever $k \geqq 1/2$ (and not only for $k \geqq 1$).

So far the spaces $A_k^2(C)$ were defined only for $k \geqq 1$. We want to show how the definition can be extended to $k = 1/2$. We consider the l^2-summable sequences $\{a_m\}$ with m running over nonnegative integers. They form a Hilbert space. In order to give this space a more than formal meaning we consider the series

$$\sum_{m=0}^{\infty} a_m z^m$$

We take account of the fact that for $k = 1/2$ all normalization factors N_m (1.32) are equal to one. Therefore we have for $|z| < 1$

$$\left|\sum_{m=0}^{M} a_m z^m\right|^2 \leqq \left(\sum_{m=0}^{\infty} |a_m|^2\right)\left(\sum_{m=0}^{\infty} |z|^{2m}\right) \leqq K^B_{k=\frac{1}{2}}(z,z) \sum_{m=0}^{\infty} |a_m|^2 \tag{1.36}$$

Therefore the series

$$f(z) = \sum_{m=0}^{\infty} a_m z^m = \sum_{m=0}^{\infty} a_m f_m(z)$$

defines a holomorphic function f(z) in C. Hence the elements of $A_{1/2}^2(C)$ can be identified with these holomorphic functions if one talks about function properties and with the l^2-summable sequences of the Taylor coefficients $\{a_m\}$ if one computes scalar products etc. Nevertheless, the Bergman kernal maintains its meaning under this extension, though the scalar product in

$$f(z) = (K_z^B, f)_k, z \in C \tag{1.38}$$

reduces to a (trivial) summation over the Taylor coefficients of f and K_z^B in the case $k = 1/2$. In particular Schwarz's inequality

$$|f(z)| \leqq \|f\|_k \,(1-|z|^2)^{-k} \tag{1.39}$$

that limits the increase of f(z) at the boundary of the unit circle holds true also for the case $k = 1/2$.

1.4 Fourier series on the boundary

The boundary S of C is the unit circle $|z| = 1$. We consider the space $\mathcal{L}^2(S)$ of functions on S

$$\mathcal{L}^2(S) = \{g(\xi)\,|\,g \text{ measurable and periodic, } g(\xi+2\pi) = g(\xi)$$
$$\|g\|_S^2 = \frac{1}{2\pi}\int_0^{2\pi} d\xi\,|g(\xi)|^2 < \infty\} \tag{1.40}$$

In this space Fourier expansions

$$g(\xi) = \sum_{m=-\infty}^{+\infty} b_m e^{im\xi} \tag{1.41}$$

converge in the L^2-norm sense. By means of Parseval's formula the norm can be expressed as

$$\|g\|_S^2 = \sum_{m=-\infty}^{+\infty} |b_m|^2 \tag{1.42}$$

We can split each element of $\mathcal{L}^2(S)$ uniquely into two (overlapping) parts

$$g(\xi) = g_+(\xi) + g_-(\xi) - b_0 \tag{1.43}$$

$$g_\pm(\xi) = \sum_{m=0}^{\pm\infty} b_m e^{im\xi} \tag{1.44}$$

The part $g_+(\xi)$ can be considered as the boundary value of the function $f_+(z)$ defined by

$$f_+(z) = \sum_{m=0}^{\infty} b_m z^m \tag{1.45}$$

whereas $g_-(\xi)$ as the boundary value of an antiholomorphic function $f_-(z)$

$$f_-(z) = \sum_{m=0}^{\infty} b_{-m}\,\bar{z}^m \tag{1.46}$$

In fact, we concentrate on $f_+(z)$ and study the convergence of the series (1.45). We can rewrite it as

$$f_+(z) = \sum_{m=0}^{\infty} b_m N_m^{-1} f_m(z) \tag{1.47}$$

The coefficients of $f_+(z)$ in the basis $f_m(z)$ can be estimated by

$$\sum_{m=0}^{\infty} |b_m|^2 N_m^{-2} \leqq \sum_{m=0}^{\infty} |b_m|^2 < \infty \tag{1.48}$$

independently of k ($k \geqq 1/2$). To arrive at this estimate we make use of the fact that

$$\max_m N_m^{-2} = \max_m \frac{m!\,(2k-1)!}{(2k+m-1)!} = 1 \tag{1.49}$$

for all k. This assertion follows from the monotonic decrease of these numbers (constancy for $k = 1/2$)

$$N_{m+1}^{-2}/N_m^{-2} \leqq 1 \tag{1.50}$$

Therefore $f_+(z)$ is a holomorphic function on C that lies simultaneously in <u>all</u> Hilbert spaces $A_k^2(C)$.

We would next like to know what the solution of the inverse question is: Find the boundary value of a given element $f(z)$ of a Hilbert space $A_k^2(C)$. Simple arguments show that we should expect a distribution as boundary value. These arguments are based on the at-most-polynomial increase at the boundary as expressed by Schwarz's inequality (1.39). We have therefore to say a few words on distributions.

We use a space $E(C)$ of infinitely differentiable periodic functions on S, these are our test functions. The topology (we need it!) may be defined by the infinite set of norms

$$\|g\|_{\sup,m} = \sup_{l \leq m} \sup_{\xi} \left|\frac{d^l}{d\xi^l} g(\xi)\right| \tag{1.51}$$

Since the manifold on which the test functions are defined is compact, there are many equivalent definitions of the topology. Bounded linear functions on $E(C)$ are the distributions that form the dual space $E'(C)$. We denote them $\phi(\xi)$.

Such distributions (or generalized functions or singular functions) have "singularities" that cannot be expressed by giving the values of the distribution at points ξ. A correct way is for example to write

$$\phi(\xi) = \left(1 + \frac{d}{d\xi}\right)^l g(\xi), \quad g \in \mathcal{L}^2(S) \tag{1.52}$$

where the derivative of $g(\xi)$ has to be taken in the weak sense, that is: integrating by parts formally we perform the differentiations on the test functions. Each distribution can in fact be given in this elegant form. Another and more detailed description of the singularities is possible by means of the Fourier series. That is one of the reasons we introduced them, indeed. If we give a series

$$\phi(\xi) = \sum_{m=-\infty}^{+\infty} b_m e^{im\xi} \tag{1.53}$$

this series converges in the distribution sense (that means: integrate term by term with a test function and sum afterwards) if and only if a real number ρ exists such that

$$|b_m| \leq C(1+|m|)^{\rho} \tag{1.54}$$

for all m. The limit of the series is a distribution, and in turn can every distribution be expanded into a series (1.53) with coefficients satisfying (1.54). The label ρ runs over the whole real axis instead of only the nonnegative integers. This makes the description by Fourier series more powerful.

The parameters ρ and l are not independent. Indeed, let ϕ be a distribution with the expansion (1.53) and $\rho_0 \geq \rho$ be an integer. Then the series

$$g(\xi) = \sum_{m=-\infty}^{+\infty} \frac{b_m}{(1+im)^{\rho_0+1}} e^{im\xi} \tag{1.55}$$

converges in the sense of the L^2-norm so that $g \in \mathcal{L}^2(S)$. This follows from the estimate

$$\sum_{m=-\infty}^{+\infty} \left|\frac{b_m}{(1+im)^{\rho_0+1}}\right|^2 \leqq \sum_{m=-\infty}^{+\infty} C^2 \frac{(1+|m|)^{2\rho}}{(1+m^2)^{\rho_0+1}} < \infty \tag{1.56}$$

Therefore we end up with

$$\phi(\xi) = (1+\frac{d}{d\xi})^{\rho_0+1} g(\xi) \tag{1.57}$$

We can therefore always find an integer l in (1.52) with

$$l \geqq \rho+1 \tag{1.58}$$

The condition (1.58) is not necessary, however, but only sufficient in general.

In turn, if the differential representation (1.52) holds true, then the Fourier coefficients b_m of ϕ can be estimated by the coefficients a_m of g by

$$|b_m| = |a_m| \ (1+m^2)^{\frac{l}{2}} \leqq (\sup_m |a_m|) \ (1+|m|)^l \tag{1.59}$$

The condition (1.54) is satisfied therefore for all real $\rho \geqq l$. We formulate the result of the problem posed at the beginning in the following theorem:

Theorem A_1. Given any $f(z) \in A_k^2(C)$. Then f(z) tends towards a distribution boundary value $\phi(\xi)$ in the limit

$$\lim_{t\to 1-} f(te^{i\xi}) = \phi(\xi) \tag{1.60}$$

This limit is assumed in the distribution topology sense. The distribution ϕ has singularities described either by $\rho = k-1/2$ or by $l \geqq k-1/2$ (in particular $l = k-1/2$ if 2k is odd).

For the proof we start from the expansion of f(z) in terms of the basis $f_m(z)$

$$f(z) = \sum_{m=0}^{\infty} a_m \left[\frac{(2k+m-1)!}{m!(2k-1)!}\right]^{\frac{1}{2}} z^m \tag{1.61}$$

Therefore we put

$$b_m = a_m \left[\frac{(2k+m-1)!}{m!(2k-1)!}\right]^{\frac{1}{2}} \tag{1.62}$$

The coefficients b_m can easily be estimated by (see (1.68), (1.69))

$$|b_m| \leqq C(1+|m|)^{k-\frac{1}{2}} \tag{1.63}$$

with, say

$$C = \sup_m |a_m| \tag{1.64}$$

This proves the first assertion $\rho = k-1/2$. Now to the second assertion $l \geqq k-1/2$. We have

$$1+iz\,\frac{d}{dz} = 1 + \frac{d}{d\xi} \tag{1.65}$$

on the boundary $z = e^{i\xi}$. This allows us to rewrite (1.61) as

$$f(z) = \left(1+iz\frac{d}{dz}\right)^{l} \sum_{m=0}^{\infty} a_m \left[\frac{(2k+m-1)!}{m!\,(2k-1)!}\right]^{\frac{1}{2}} \frac{z^m}{(1+im)^{l}} \tag{1.66}$$

The coefficients in this series can easily be estimated (see below)

$$\sum_{m=0}^{\infty} |a_m[\ldots]^{\frac{1}{2}}|^2 \frac{1}{(1+m^2)^{l}} = \sum_{m=0}^{\infty} |a_m|^2 \frac{(2k+m-1)!}{m!\,(2k-1)!\,(1+m^2)^{l}}$$

$$\leqq 2^{l} \sum_{m=0}^{\infty} |a_m|^2 < \infty\,, \quad l \geqq k - \frac{1}{2} \tag{1.67}$$

So that the second assertion is also proved.

Both in (1.64) and in (1.67) the inequality

$$A_m = \frac{(2k+m-1)!}{m!\,(2k-1)!\,(1+m)^{2k-1}} \leqq 1 \tag{1.68}$$

for all $m \geqq 0$ has been made use of. This inequality is established first by proving the monotonous non-increase

$$\frac{A_{m+1}}{A_m} = \frac{(2k+m)\,(m+1)^{2k-1}}{(m+1)\,(m+2)^{2k-1}} = \frac{1+\dfrac{2k-1}{m+1}}{\left(1+\dfrac{1}{m+1}\right)^{2k-1}} \leqq 1 \tag{1.69}$$

and afterwards taking the maximum at $m = 0$.

As a corollary we have obtained that any $f(z) \in A_k^2(C)$ can be presented in the form

$$f(z) = \left(1+iz\frac{d}{dz}\right)^{l} f_0(z) \tag{1.70}$$

with $l \geqq k-1/2$ and $f_0(z) \in A_{1/2}^2(C)$. If $2k$ is odd, we can again take $l = k-1/2$.

A distribution $\phi(\xi)$ can be split into parts $\phi_\pm(\xi)$ by means of its Fourier expansion just as a function $g(\xi)$ of $\mathcal{L}^2(S)$. We assume that $\phi(\xi)$ is equal to its positive part $\phi_+(\xi)$ in the next theorem.

Theorem B_1. Let $\phi(\xi)$ be equal to its positive part $\phi_+(\xi)$ and such that

$$\phi(\xi) = \left(1+\frac{d}{d\xi}\right)^{l} g(\xi)\,,\quad g \in \mathcal{L}^2(S) \tag{1.71}$$

Then $\phi(\xi)$ possesses a holomorphic extension f(z) into C that lies in all $A_k^2(C)$ with $k \geqq l+1/2$.

The proof can be established as follows. The extenion f(z) is obtained from the extension $f_l(z)$ of $g(\xi)$ by

$$f(z) = (1+iz\frac{d}{dz})^l f_l(z) \tag{1.72}$$

The expansion coefficients a_m of f(z) in terms of the basis functions $f_m(z)$ and b_m of $g(\xi)$ in terms of powers of $e^{i\xi}$ are related by

$$a_m = (1+im)^l \left[\frac{m!(2k-1)!}{(2k+m-1)!}\right]^{\frac{1}{2}} b_m \tag{1.73}$$

Therefore we can estimate

$$\sum_{m=0}^{\infty} |a_m|^2 = \sum_{m=0}^{\infty} (1+m^2)^l \frac{m!(2k-1)!}{(2k+m-1)!} |b_m|^2$$

$$\leqq \sum_{m=0}^{\infty} A_m^{-1} |b_m|^2 \leqq (2k-1)! \sum_{m=0}^{\infty} |b_m|^2 \tag{1.74}$$

whenever $2l \leqq 2k-1$. For the final estimate we made use of the fact that

$$A_m \geqq \lim_{m'\to\infty} A_{m'} = [(2k-1)!]^{-1} \tag{1.75}$$

for all m, see (1.68) and the arguments following it. This completes the proof.

We notice that the spaces with k half odd integral play a special role. If we put $l = k-1/2$ in Theorem A_1 and $k = l+1/2$ in Theorem B_1, we have a simple one-to-one relation between the set of distributions $\phi(\xi) = \phi_+(\xi)$ of fixed degree l of the singularities and the space $A_k^2(C)$ of holomorphic functions in C that are their extensions. In other words: The distributions

$$\phi(\xi) = 1+\frac{d}{d\xi})^l g(\xi), \quad g(\xi) = g_+(\xi) \in \mathcal{L}^2(S) \tag{1.76}$$

make up a Hilbert space that carries the unitary irreducible representation with label $k = l+1/2$ of the discrete series of SU(1,1).

1.5 The Cauchy integral formula and the Hardy-Lebesgue space

For any given distribution $\phi(\xi) \in E'(S)$ the extension $f_+(z)$ of $\phi_+(\xi)$ into the unit circle can obviously also be obtained by means of the Cauchy integral formula

$$f_+(z) = \frac{1}{2\pi i} \int_S \frac{\phi(\xi)}{e^{i\xi}-z} d(e^{i\xi}) \quad z \in C \tag{1.77}$$

This formula can easily be rewritten if we introduce the Szegö kernel

$$K^S(z_1,z_2) = (1-\bar{z}_1 z_2)^{-1} = K^S_{z_1}(\xi),\ z_2 = e^{i\xi} \tag{1.78}$$

We extend the scalar product of $\mathcal{L}^2(S)$ such that in

$$(g_1,g_2)_S = \frac{1}{2\pi}\int_0^{2\pi} d\xi\ \overline{g_1(\xi)} g_2(\xi) \tag{1.79}$$

g_2 is allowed to become a distribution of $E'(S)$ whereas g_1 is restricted to test functions of $E(S)$. Inserting the Szegö kernel into the Cauchy formula (1.77) we get

$$f_+(z) = (K^S_z, \phi)_S,\ z \in C \tag{1.80}$$

This is the most compact form for the holomorphic extension of any given distribution ϕ. If we finally expand the Szegö kernel in powers of $\bar{z}_1 z_2$ and take z_2 on the boundary, the equivalence of the extension by means of formulae (1.77), (1.80) and the Fourier series method (Theorem B_1) turns out. We notice that the limit $z \to e^{i\xi}$ gives us back ϕ_+ even if we started from ϕ and if $\phi \neq \phi_+$. The negative part $(\phi_- - b_0)$ is projected out by taking the scalar product with the Szegö kernel. We shall see later that the projection of a distribution $\phi(\xi)$ onto its positive part is connected with the problem of cutting a distribution on the real axis into two parts with supports on the positive and negative real axis, respectively. The Szegö kernel provides us with a "canonical" solution of this problem that is unique.

It is obvious from (1.34) that the Szegö kernel (that is independent of k) coincides with the Bergman kernel for k = 1/2. For fixed $z_1 \in \bar{C}$ ($\bar{C}$ denotes the closure of C) $K^S_{z_1}(z_2)$ is holomorphic in $z_2 \in C$ and for fixed $z_1 \in C$ it lies in $A^2_{1/2}(C)$ and consequently in all $A^2_k(C)$. Moreover we know already that any square integrable boundary value $g(\xi) = g_+(\xi)$ possesses a holomorphic extension $f_+(z)$ in $A^2_{1/2}(C)$. In turn let f(z) be continuous in $\bar{C}$ and holomorphic in C. We define the Hardy-Lebesgue norm of f(z) and a corresponding scalar product by

$$\|f\|_S = \|g\|_S \tag{1.81}$$

where $g(\xi)$ is the boundary value of f(z). In this fashion we obtain a pre-Hilbert space of holomorphic functions. It contains all powers z^m, m = 0, 1, 2, ..., that are orthonormal without any further normalization factor. For k = 1/2 the basis vectors $f_m(z)$ reduce to the same powers. It follows that the completion of the pre-Hilbert space leads to the Hilbert space $A^2_{1/2}(C)$, in other words: this space is the Hardy-Lebesgue Hilbert space and its Aronszajn-Bergman kernel is the Szegö kernel. We may also consider the measure

$$\frac{2k-1}{\pi}(1-|z|^2)^{2k-2}\,|dz|$$

on C as a function of k. For k = 1/2 this measure concentrates on the boundary of C and yields there the measure $(2\pi)^{-1}d\xi$.

Finally we consider the kernel

$$(K^S_{z_1}, K^S_{z_2})_k = M_k(z_1,z_2),\ z_{1,2}\in C \qquad (1.82)$$
$$\hat{M}_k(\xi) = M_k(e^{i\xi},1)$$

that exists due to the arguments just presented, whenever the two arguments do not ccincide on the boundary of the unit circle. Its meaning becomes obvious through the following discussion. Let

$$f_{1,+}(z) = (K^S_z,\phi_1)_S$$
$$f_{2,+}(z) = (K^S_z,\phi_2)_S \quad , f_{1,+}\ ,\ f_{2,+}\in A^2_k(C) \qquad (1.83)$$

for any fixed k. Then

$$(f_{1,+}, f_{2,+})_k = \frac{2k-1}{\pi}\int_C |dz|\,(1-|z|^2)^{2k-2}\ \times$$
$$\times\ \{\frac{1}{2\pi}\int_0^{2\pi} d\xi_1\, K^S_z(\xi_1)\,\overline{\phi_1(\xi_1)}\ \frac{1}{2\pi}\int_0^{2\pi} d\xi_2\,\overline{K^S_z(\xi_2)}\phi_2(\xi_2)\} \qquad (1.84)$$
$$= (2\pi)^{-2}\int_0^{2\pi} d\xi_1\,\overline{\phi_1(\xi_1)}\int_0^{2\pi} d\xi_2\,\hat{M}_k\,(\xi_1-\xi_2)\,\phi_2(\xi_2)$$

It defines the SU(1,1) invariant scalar product in the Hilbert space of distributions on the boundary.
The power expansion of $K^S(z_1, z_2)$ leads to

$$M_k(z_1, z_2) = \sum_{m=0}^{\infty}(z_1\bar{z}_2)^m\frac{m!\,(2k-1)!}{(2k+m-1)!} \qquad (1.85)$$
$$= {}_2F_1(1,\ 1;\ 2k;\ z_1\bar{z}_2)$$

The function $\hat{M}_k(\xi)$ is singular at $\xi = 0$ with a pole of first order for k = 1/2, with a logarithmic singularity for k = 1, and is continuously differentiable k - 3/2 times there for k = 3/2. The regularization of the integral (1.84) that may be necessary has to be achieved by holomorphic extension into the interior of the circle.

1.6 The upper half plane

By the conformal mapping

$$w(z) = i\frac{1-z}{1+z}\quad ,\qquad z(w) = \frac{1+iw}{1-iw} \qquad (1.86)$$

the unit circle goes over into the upper half plane H_+. The automorphism group of the upper half plane H_+ [8] is the group of fractional linear transformations

$$w_a = \frac{a_{11}w+a_{21}}{a_{12}w+a_{22}} \tag{1.87}$$

where the matrices

$$a = \begin{pmatrix} a_{11} & a_{12} \\ a_{21} & a_{22} \end{pmatrix} \tag{1.88}$$

with real elements a_{ij} and det a = 1, form the group SL(2,R).

Inserting the function w = w(z) into (1.87) defines the function $w_a(z)$. Inserting the unknown parameter z_v into w(z) (1.86) gives us the function $w(z_v)$.
We put

$$w_a(z) = w(z_v) \tag{1.89}$$

We solve for $z_v = z_v(z)$ and obtain

$$z_v = \frac{\bar{\alpha}z-\beta}{-\bar{\beta}z+\alpha} \tag{1.90}$$

If we use the abbreviations

$$\begin{aligned} \alpha &= \tfrac{1}{2}\,[a_{11}+a_{22}+i(a_{12}-a_{21})] \\ \beta &= \tfrac{1}{2}\,[a_{11}-a_{22}-i(a_{12}+a_{21})] \end{aligned} \tag{1.91}$$

Moreover from (1.91) we find

$$|\alpha|^2 - |\beta|^2 = \det a = 1$$

Since (1.91) is identical with (1.11) we have thus established a global isomorphism between SU(1,1) and SL(2,R) that can be put into the form

$$a = u\,\begin{pmatrix} \alpha & \beta \\ \bar{\beta} & \bar{\alpha} \end{pmatrix} u^{-1}\ , \quad u = e^{i\frac{\pi}{4}\sigma_1} e^{-i\frac{\pi}{4}\sigma_3} \tag{1.92}$$

Next we set

$$F(w) = m_q(w) f(z(w)) \tag{1.93}$$

where f(z) is holomorphic in C and the multiplier is chosen as

$$m_q(w) = c(1-iw)^{-2k+q}\ , \quad c = \text{const.} \tag{1.94}$$

i.e. holomorphic in H_+ with a certain polynomial increase at infinity. This has been introduced in order to treat tempered distribution boundary values. q is arbitrary, though integral. The functions F(w) are holomorphic in H_+ and suffer a transformation if a fractional linear transformation is exerted on H_+. In fact, we want to end up with a formula of the type

$$T_a F(w) = \mu(a,w) F(w_a) \tag{1.95}$$

where the multiplier $\mu(a,w)$ is holomorphic on H_+.

In order to achieve this we define

$$T_a F(w) = m_q(w) T_v f(z(w)) = m_q(w)(-\bar{\beta} z(w)+\alpha)^{-2k} f(z_v(w)) \tag{1.96}$$

where (1.25) has been inserted. We set similarly as in (1.89)

$$z_v(w) = z(w_a) \tag{1.97}$$

Using finally

$$-\bar{\beta} z(w)+\alpha = (1-iw)^{-1}(a_{12}w+a_{22})(1-iw_a) \tag{1.98}$$

we end up with

$$T_a F(w) = (1-iw)^q (1-iw_a)^{-q} (a_{12}w+a_{22})^{-2k} F(w_a) \tag{1.99}$$

We see that $q = 0$ is a particularly simple, say "natural" case. However, if we want to deal with general tempered distributions, we cannot restrict ourselves to a discussion of this special case. If we would rather deal with Schwartz's distributions we must change the ansatz (1.94) by setting $q = 0$ and multiplying with a function $M(w)$ that increases more than polynomially at infinity. The method of constructing $M(w)$ for any given Schwartz distribution as boundary value, is contained in the mathematical literature [11].

We fix the constant c in (1.94) such that the norms of F and f are equal and the Bergman measure on H_+ assumes a simple form. We notice first that

$$|dz| = 4|1-iw|^{-4}|dw| \tag{1.100}$$

$$(1-|z|^2) = 4v|1-iw|^{-2}, \quad w = u+iv \tag{1.101}$$

Inserting this into the scalar product (1.31) we have

$$\begin{aligned}(f_1,f_2)_k = \frac{2k-1}{\pi} \int_{H_+} |dw| v^{2k-2} \; |1-iw|^{-2q} \\ \times \overline{F_1(w)} F_2(w) = (F_1,F_2)_{kq}\end{aligned} \tag{1.102}$$

if

$$c = 2^{2k-1} \tag{1.103}$$

This establishes a natural isomorphism of the Hilbert spaces $A_k^2(C)$

and $A^2_{k,q}(H_+)$, the latter being defined by the scalar product (1.102). The Bergman kernel is correspondingly

$$\begin{aligned} K^B(w_1,w_2) &= \overline{m_q(w_1)}m_q(w_2)K^B(z(w_1),z(w_2)) \\ &= \frac{1}{4}(1-iw_1)^q(1+i\bar{w}_2)^q\left[-\frac{i}{2}(w_1-\bar{w}_2)\right]^{-2k} \end{aligned} \tag{1.104}$$

Next we map $\mathcal{L}^2(S)$ on $\mathcal{L}^2(R)$ as follows. We set

$$G(u) = (1-iu)^{-1}g(\xi(u)) \tag{1.105}$$

where in correspondence with (1.86)

$$\begin{aligned} e^{i\xi(u)} &= \frac{1+iu}{1-iu}, \quad u = \mathrm{tg}\,\frac{1}{2}\xi, \quad -\pi \leqq \xi \leqq \mu \\ \xi(u) &= 2\,\mathrm{arctg}\,u, \quad \xi(0) = 0 \end{aligned} \tag{1.106}$$

If $g(\xi)$ is in $E(S)$ then G is infinitely differentiable, falls off at infinity as u^{-1} and possesses a simultaneous asymptotic expansion at $u = \pm\infty$

$$G(u) \cong \sum_{n=1}^{\infty} c_n u^{-n} \tag{1.107}$$

In fact, the function $g(\xi)$ is infinitely differentiable at $\xi = \pi$ from both sides and we have the asymptotic series (Taylor expansion)

$$g(\xi) \cong g(\pi) + (\xi-\pi)g'(\pi)+\frac{1}{2}(\xi-\pi)^2g''(\pi)+\ldots \tag{1.108}$$

Further, on the correct branches of arctg

$$\frac{1}{2}(\xi-\pi) \underset{|u|\to\infty}{\cong} -\frac{1}{u}+\frac{1}{3}\frac{1}{u^3}-\frac{1}{5}\frac{1}{u^5}+\ldots \tag{1.109}$$

Inserting (1.109) into (1.108) and both into (1.105) yield the coefficients c_n in (1.107). Finally we remark that the asymptotic expansion (1.107) can be differentiated term by term.

Inserting (1.105) into the formula (1.40) for the norm of g we obtain

$$\begin{aligned} \|g\|_S^2 &= \frac{1}{2\pi}\int_0^{2\pi} d\xi\,|g(\xi)|^2 \\ &= \frac{1}{\pi}\int_{-\infty}^{+\infty} du\,|G(u)|^2 = \|G\|_R^2 \end{aligned} \tag{1.110}$$

The functions G are square integrable with respect to the Lebesgue measure $\pi^{-1}du$ on R. The same measure enters the definition of the space $D_{L^2}(R)$ of test functions with the properties:

1. It consists of all infinitely differentiable functions that are square integrable together with all their derivatives;
2. an infinite set of norms is defined by

$$\|G\|^2_{m,2} = \sum_{l \leq m} \frac{1}{\pi} \int_R |G^{(l)}(u)|^2 du \tag{1.111}$$

for m = 0, 1, 2,...

The relation (1.105) defines an injection of $E(S)$ into $D_{L^2}(R)$ that is continuous. We prove the last assertion in the following fashion.

A null sequence in $E(S)$, i.e. a sequence of functions $g_n(\xi)$ that goes to zero with respect to each one of the infinite sequence of norms (1.51), defines a null sequence in $D_{L^2}(R)$ with respect to the norms (1.111). That is what we have to prove, the proof is established by some estimates. First

$$|G^{(l)}(u)| \leq c_l (1+u^2)^{-\frac{1}{2}} \sup_{n \leq l} \sup_{\xi} |g^{(n)}(\xi)| \tag{1.112}$$

with a certain numerical constant c_l that depends on l but not on g or G. Therefore

$$\frac{1}{\pi} \int_R |G^{(l)}(u)|^2 du \leq c_l^2 \{\sup_{n \leq l} \sup_{\xi} |g^{(n)}(\xi)|\}^2 \tag{1.113}$$

or finally

$$\|G\|^2_{m,2} \leq (m+1)(\sum_{l=0}^{m} c_l^2)\{\sup_{l \leq m} \sup_{\xi} |g^{(l)}(\xi)|\}^2 \tag{1.114}$$

This completes the proof.

Moreover the test function space $S(R)$ maps into $E(S)$ by the same mapping (1.105). One can show this by similar estimates. Consequently

$g \in E(S)$ implies $G \in D_{L^2}(R)$,

$G \in S(R)$ implies $g \in E(S)$

via (1.105). Since both injections are continuous it follows in turn that the mapping of distributions formally analogous to (1.105)

$$\psi(u) = (1-iu)^{-1} \phi(\xi(u)) \tag{1.115}$$

is such that

$\phi \in E'(S)$ implies $\psi \in S'(R)$

$\psi \in D'_{L^2}(R)$ implies $\phi \in E'(S)$

The argument goes as follows. For any given $\phi \in E'(S)$ we define ψ by

$$(G,\psi)_R = (g,\phi)_S \tag{1.116}$$

Since ϕ is continuous, a null sequence g_n in $E(S)$ implies $\lim (g_n, \phi)_S = 0$. Therefore also $\lim (G_n, \psi)_R = 0$. For any null sequence G_n in $S(R)$ g_n is a null sequence in $E(S)$. Consequently ψ is continuous on $S(R)$ and is a tempered distribution. In turn, let ψ be a distribution of $D_L{}'_2(R)$. Then for G_n a null sequence in $D_{L2}(R)$ $\lim\,(G_n,\psi)_R = 0$. The null sequences $g_n \in E(S)$ map on particular null sequences $G_n \in D_{L2}(R)$. Therefore for all null sequences g_n $\lim\,(g_n, \phi)_S = 0$, $\phi \in E'(S)$.

We emphasize once more that switching from the image of $E(S)$ under (1.105) that could also serve as a test function space to the spaces $D_{L2}(R)$ and $S(R)$, is motivated by the desire to deal with spaces of test functions with simple and known behaviour under Fourier and Hilbert transformations. Doing this we give up the goal of a complete characterization of the boundary distributions. Such characterization could anyway be obtained from Theorems A_1 and B_1 by some elementary algebra.

1.7 Extension of boundary values into the upper half plane, Hilbert transforms

For distributions on the real axis the extension problem coincides with the problem of Hilbert transformations. We start by introducing the Szegö kernel. We define it by

$$K^S_{w_1}(w_2) = K^S(w_1,w_2) = [2i(\bar{w}_1,-w_2)]^{-1} \tag{1.117}$$

Then for any $\psi \in D_{L2}(R)$

$$\begin{aligned} F_+(w) &= (K^S_w, \psi)_R = (1-iw)^{-1} f_+(z(w)) \\ &= (1-iw)^{-1} (K^S_{z(w)}, \phi)_S \end{aligned} \tag{1.118}$$

This relation between f_+ and F_+ is not the same as (1.93). In (1.118) ϕ and ψ are related as in (1.115). $F_+(w)$ is holomorphic in the upper half plane. Sinve for w going to the real axis, z tends to the unit circle and $f_+(z)$ to $\phi_+(\xi)$ in the $E'(S)$ topology, $F_+(w)$ tends to a distribution $\psi_+(u)$ in the topology of $S'(R)$ such that

$$\psi_+(u) = (1-iu)^{-1}\ \phi_+(\xi(u)) \tag{1.119}$$

We call $\psi_+(u)$ the positive part or the Hilbert transform of $\psi(u)$. It is uniquely determined for any $\psi \in D_L{}'_2(R)$.

The equation

$$F_+(w) = (K^S_w, \psi)_R \tag{1.120}$$

can also be written in the (formal) integral form

$$F_+(w) = \frac{1}{2\pi i} \int_{-\infty}^{+\infty} \frac{\psi(u')}{u'-w}\, du', \quad w \in H_+ \tag{1.121}$$

By a Fourier transformation

$$\psi(u) = \int_{-\infty}^{+\infty} e^{itu}\, \hat{\psi}(t)dt \tag{1.122}$$

we obtain a tempered distribution $\hat{\psi}(t)$ that can be represented by a locally square integrable function. We can cut it therefore uniquely into the two pieces $\hat{\psi}_{\pm}(t)$,

$$\hat{\psi}_{\pm}(t) = \Theta(\pm t)\ \hat{\psi}(t) \tag{1.123}$$

where $\Theta(t)$ is the characteristic function of the positive real axis. The Fourier integral

$$\int_{-\infty}^{+\infty} e^{itu}\ \hat{\psi}_{+}(t)\ dt \tag{1.124}$$

can be obtained by folding $\psi(u)$ with the Fourier transform of the step function $\Theta(t)$

$$\frac{i}{u+i0} = \int_{-\infty}^{+\infty} e^{itu}\ \Theta(t)dt \tag{1.125}$$

in this fashion the Fourier transform (1.124) turns out to be

$$\psi_{+}(u) = \frac{1}{2\pi i} \int_{-\infty}^{+\infty} \frac{\psi(u')}{u'-u-i0} du' \tag{1.126}$$

This is in agreement with (1.121) and by an extension of u in (1.124) into the upper half plane leads to a representation of $F_{+}(w)$ as a properly convergent Laplace transform of $\hat{\psi}_{+}(t)$. Summarizing we can say that taking the Hilbert transform of $\psi \in \mathcal{D}'_{L^2}(R)$ is equivalent with cutting its Fourier transform $\hat{\psi}$ into two pieces

$$\hat{\psi} = \hat{\psi}_{+} - \hat{\psi}_{-} \tag{1.127}$$

with support on the positive, respectively negative real axis. This cutting does not lead to ambiguities.

If we start from tempered distributions ψ the situation is slightly different. In general such distribution can be represented as

$$\psi(u) = \left(\frac{d}{du}\right)^{m} (1-iu)^{l} G(u),\ G \in \mathcal{L}^2(R) \tag{1.128}$$

Then the holomorphic extension is

$$F_{+}(w) = \left(\frac{d}{dw}\right)^{m} (1-iw)^{l} (K_{w}^{S}, G)_{R} \tag{1.129}$$

A priori the non-uniqueness of the representation (1.128) may entail a non-uniqueness of the holomorphic extension. The boundary value ψ_{+} of $F_{+}(w)$ is again assumed in the tempered distribution topology and we can call ψ_{+} the Hilbert transform of ψ just as in the preceding case.

Again we may try a Fourier transformation (1.122) on $\psi(u)$ that gives us the tempered distribution $\hat{\psi}(t)$, and cut $\hat{\psi}(t)$ into two parts as in (1.123)

$$\hat{\psi}(t) = \hat{\psi}_+(t) + \hat{\psi}_-(t) \tag{1.130}$$

This cutting of general tempered distributions is, however, by no means unique. In fact, we write

$$\hat{\psi}(t) = (\frac{d}{dt})^{\hat{m}}(1-it)^{\hat{l}}\hat{G}(t), \; \hat{G} \in \mathcal{L}^2(R) \tag{1.131}$$

Then we may add to $(1-it)^{\hat{l}}\hat{G}(t)$ any polynomial of maximal degree $\hat{m} - 1$ without changing $\hat{\psi}$, and by an appropriate choice of $\hat{l}$, maintaining the square integrability of $\hat{G}$. If we cut $\hat{G}$ into $\hat{G}_+$ and $\hat{G}_-$, then

$$\hat{\psi}_+(t) = (\frac{d}{dt})^{\hat{m}}(1-it)^{\hat{l}}\hat{G}_+(t) \tag{1.132}$$

is determined only up to

$$(\frac{d}{dt})^{\hat{m}} \Theta(t) \text{ times a polynomial of degree } \hat{m} - 1$$
$$= \sum_{n=0}^{\hat{m}-1} a_n \delta^{(n)}(t) \tag{1.133}$$

This entails that $\psi_+(u)$ is only determined up to a polynomial, too. After this digression we return to Equation (1.129).

The question still unanswered is whether $F_+(w)$ (1.129) is unique or only determined up to a polynomial in w. If we replace G(u) in (1.129) by

$$G(u) = (1-iu)^{-l}P_n(u) \tag{1.134}$$

where $P_n(u)$ is a polynomial of maximal degree n and

$$n + 1 \leqq \min (l,m)$$

the corresponding holomorphic function $F_+(w)$ vanishes identically. This can be most easily shown by a fractional decomposition of g. Therefore $F_+(w)$ (1.129) is unique indeed. It follows that using a formula of the type (1.129) amounts to a unique cutting procedure for the Fourier transform $\hat{\psi}$ of ψ.

1.8 Holomorphic extensions as elements of $A^2_{k,q}(H_+)$

In Section 1.4 we formulated two theorems by which we related the Hilbert spaces $A^2_k(C)$ containing the holomorphic extensions with the degree of the singularities of the boundary values. Our proofs were based on Fourier series expansions. On the real axis Fourier series expansions are not a very natural recipe. Therefore we develop a new technique that gives, however, less far reaching results. The main tool of our argument will be the Szegö kernel.

We notice first that for any $\psi \in \mathcal{D}'_{L^2}(R)$ we can find the representation

$$\psi(u) = \sum_{n \leqq l} \frac{d^n}{du^n} G_n(u) \ , \quad G_n \in \mathcal{L}^2(R) \tag{1.135}$$

for some l. Therefore from (1.118)

$$F_+(w) = \frac{1}{\pi} \int_R du \sum_{n=0}^{l} (-1)^n \left(\frac{d^n}{du^n} \overline{K_w^S(u)}\right) G_n(u) \tag{1.136}$$

The explicit form of the Szegö kernel (1.117) shows that this formula can be transformed into

$$F_+(w) = \sum_{n=0}^{l} \frac{d^n}{dw^n} (K_w^S, G_n)_R \tag{1.137}$$

which by Schwarz's inequality yields

$$|F_+(w)| \leqq \sum_n \left[\frac{d^n}{dw^n} \frac{d^n}{d\bar{w}^n} K^S(w,w)\right]^{\frac{1}{2}} \|G_n\|_R \tag{1.138}$$

We make use of

$$\frac{d^n}{dw^n} \frac{d^n}{d\bar{w}^n} K^S(w,w) = \frac{(2n)!}{2^{2n+2}} v^{-2n-1} \tag{1.139}$$

$$w = u + iv$$

and have finally

$$|F_+(w)| \leqq \sum_{n=0}^{l} \frac{[(2n)!]^{\frac{1}{2}}}{2^{n+1}} \|G_n\|_R \, v^{-n-\frac{1}{2}} \tag{1.140}$$

Now let $\psi(u)$ be a tempered distribution. Then we have in a similar fashion from (1.129)

$$|F_+(w)| \leqq \|G\|_R \left\{\frac{d^m}{dw^m} \frac{d^m}{d\bar{w}^m} (1-iw)^l (1+i\bar{w})^l \, K^S(w,w)\right\}^{\frac{1}{2}} \tag{1.141}$$

The curly bracket can be estimated now

$$\leqq C \, \|G\|_R \, (1+|w|^2)^{\frac{1}{2}l} \, v^{-m-\frac{1}{2}} \tag{1.142}$$

so that

$$|F_+(w)| \leqq M(|w|) \, v^{-m-\frac{1}{2}} \tag{1.143}$$

$$M(|w|) = M_0 \, (1+|w|^2)^{\frac{1}{2}l} \tag{1.144}$$

with e.g.

$$M_0 = C \, \|G\|_R \tag{1.145}$$

Analogous expressions were first obtained by Tillmann [11] from the same premises.

With these bounds on the holomorphic functions at the real axis and at infinity it is easy to obtain a criterion for whether $F_+(w)$ lies in $A^2_{k,q}(H_+)$ or not. We simply estimate whether the integral

$$\|F_+\|^2_{k,q} = \frac{2k-1}{\pi} \int_{H_+} |dw|\; v^{2k-2}\; |1-iw|^{-2q}\; |F_+(w)|^2 \tag{1.146}$$

converges or not. The increase $v^{-m-1/2}$ in (1.142), (1.143) if w tends towards the real axis, that creates the distribution singularities on the boundary, can be cancelled by the factor v^{2k-2} in (1.146), whereas the factor $M(|w|)$ in (1.143) can be compensated by an appropriately chosen factor $|1-iw|^{-2q}$ in the integral (1.146). Let us assume that q is nonnegative. Then

$$|1-iw|^{-2q} \leqq (1+|w|^2)^{-q}, \quad \text{Im } w \geqq 0 \tag{1.147}$$

An integral of the kind

$$\int_{H_+} |dw|\; (1+|w|^2)^{-q+1}\; v^{2k-2m-3} \tag{1.148}$$

converges if and only if

$$\begin{aligned} &2k - 2m - 3 \geqq 0 \\ &2k - 2m - 3 - 2q + 2l \leqq -3 \end{aligned} \tag{1.149}$$

as can be seen immediately after introducing polar coordinates. Simplifying these conditions we have

$$m + q - l \geqq k \geqq m + \frac{3}{2} \tag{1.150}$$

In order that any solution k exists in (1.150), we must have

$$q \geqq l + \frac{3}{2} \tag{1.151}$$

We see that our constraint on k is stronger than in Theorem B_1 ($k \geqq m + 1/2$). An analogue to (1.151) did not occur earlier because we had not to deal with tempered distributions.

The inverse problem (analogous to the content of Theorem A_1) can be solved by the now standard estimate

$$|F(w)| \leqq \|F\|_{k,q}\; K^B(w,w)^{\frac{1}{2}} \tag{1.152}$$

By means of (1.104) we deduce from this formula

$$|F(w)| \leqq \tfrac{1}{2}\; \|F\|_{k,q}\; |1-iw|^q v^{-k} \tag{1.153}$$

We can insert this estimate into Tillmann's formulas and obtain with their help a derivative representation for the tempered distribution that appears as the boundary value. For the details we refer to the literature [11].

Of course, the boundary values $\psi(u)$ of functions $F(w)$ $A^2_{k,q}$ form a Hilbert space themselves since the relation between boundary value and holomorphic extension is unique. These Hilbert spaces of distribution whose scalar product can be expressed by means of a certain convolution kernel integral, carry the representations "k, holomorphic" of the discrete series of SU(1,1). We shall not discuss the problem of classifying explicitly the distributions that belong to one such Hilbert space, but instead we will be content with the estimates given above. In any case it can be expected that for the case 2k odd, there is an elementary description for the distribution space.

2. THE DISCRETE SERIES OF SU(2,2)

Before we start our discussion we want to point out that to a far extent the situation is analogous to the case of the group SU(1,1) and that we shall skip therefore over many details. On the other hand the pecularities of the higher dimensional group, in particular those connected with holomorphic functions of several variables, will be emphasized. More details can moreover be found in the original article [12]. The results presented are in general less complete than in the case of the group SU(1,1). More work can therefore be done on these problems.

2.1 Algebraic considerations

The group SU(2,2) consists of complex four-by-four matrices that are grouped into two-by-two submatrices

$$M = \begin{pmatrix} A & B \\ C & D \end{pmatrix} \tag{2.1}$$

The matrix M is assumed to satisfy the constraint

$$M^+H = HM^{-1} \tag{2.2}$$

as well as

$$\det M = 1 \tag{2.3}$$

The matrix H is given by

$$H = \begin{pmatrix} -E & O \\ O & E \end{pmatrix} \tag{2.4}$$

where E is the 2 x 2 unit matrix. The constraint (2.2) is equivalent with the set of three relations for the submatrices

$$A^+A - C^+C = E$$
$$D^+D - B^+B = E \tag{2.5}$$
$$A^+B - C^+D = 0$$

that together with (2.3) can also be used to define the group SU(2,2). The submatrices A and D possess inverses as follows from (2.5).

The maximal compact subgroup of SU(2,2) consists of the matrices

$$A = K_1, \quad D = K_2, \quad B = C = 0 \tag{2.6}$$

where $K_{1,2}$ are both unitary. It possesses cosets in SU(2,2) that can be uniquely characterized by complex 2 x 2 matrices Z,

$$Z = \begin{pmatrix} z_{11} & z_{12} \\ z_{21} & z_{22} \end{pmatrix} \tag{2.7}$$

satisfying the constraint

$$E - Z^+Z > 0 \quad \text{(this denotes that the matrix is positive definite)} \tag{2.8}$$

The constraint (2.8) defines the domain D in the space C_4 of the four variables z_{ij}, i, j = 1,2, in (2.7). Before we prove the characterization of the cosets by the matrices Z we want to show that D possesses a compact closure, is geometrically convex, and a Reinhardt domain.

The first assertion follows from the fact, that the sum of the eigenvalues of $E - Z^+Z$

$$\mathrm{Tr}(E-Z^+Z) = 2 - \sum_{ij} |z_{ij}|^2 \tag{2.9}$$

must be positive, implying that D is a subdomain of the sphere

$$\sum_{ij} |z_{ij}|^2 < 2 \tag{2.10}$$

If Z_1 and Z_2 are in D then

$$x \neq 0 \ , \ (x, (E-Z^+Z)x) = (x,x) - (Zx, Zx) > 0 \tag{2.11}$$

for both Z_1 and Z_2 and any complex 2-vector x. This condition (2.11) is in turn also sufficient for the positivity of $E - Z^+Z$. Inserting

$$Z = tZ_1 + (1-t)Z_2 \ , \ 0 \leqq t \leqq 1 \tag{2.12}$$

into (2.11) yields a polynomial of second order in t with the coefficient

$$-(x, (Z_1^+ - Z_2^+)(Z_1 - Z_2)\ x) \tag{2.13}$$

of the quadratic term that is nonpositive. Since the polynomial assumes two positive values as $t = 0$ and $t = 1$, it must be positive in the whole interval $0 \leqq t \leqq 1$. This proves the second assertion (convexity). Moreover with $Z \in D$, $K_1 Z K_2$ for any unitary matrices K_1 and K_2 lies in D. This is true in particular for diagonal matrices $K_{1,2}$ that can be chosen such that under multiplication of Z with K_1 and K_2 the elements z_{ij} of Z go over into $e^{i\Theta_{ij}} . z_{ij}$ with arbitrary phases Θ_{ij}. Therefore D is a Reinhardt domain (see the "Preliminaries").

Next we return to the problem of characterizing the cosets of the maximal compact subgroup by the matrices Z. In fact, we can uniquely decompose M as

$$\begin{pmatrix} A & B \\ C & D \end{pmatrix} = \begin{pmatrix} N_1, & ZN_2 \\ Z^+N_1, & N_2 \end{pmatrix} \begin{pmatrix} K_1 & 0 \\ 0 & K_2 \end{pmatrix} \tag{2.14}$$

where the rightmost factor is in the maximal compact subgroup. N_1 and N_2 are positive definite hermitean and given by the polar decomposition of A and D

$$A = N_1 K_1, \quad D = N_2 K_2 \tag{2.15}$$

whereas Z follows from

$$Z = BD^{-1} = (CA^{-1})^+ \tag{2.16}$$

The constraints (2.5) imply moreover

$$\begin{aligned} N_1^{-2} &= (E - ZZ^+) \\ N_2^{-2} &= (E - Z^+Z) \end{aligned} \tag{2.17}$$

Remembering that the positivity of $E - ZZ^+$ implies the positivity of $E - Z^+Z$ and vice versa, our assertion is proved completely. We can formulate it: D is a homogeneous space for the group SU(2,2).

On this homogeneous space we define left translations in the familiar fashion. Since it is impossible to give a simple explicit form for M^{-1}, we use the notation

$$M^{-1} = \begin{pmatrix} A & B \\ C & D \end{pmatrix} \tag{2.18}$$

instead of (2.1) in this context. From

$$\begin{pmatrix} A & B \\ C & D \end{pmatrix} \begin{pmatrix} N_1 & ZN_2 \\ Z^+N_2 & N_2 \end{pmatrix} = \begin{pmatrix} N_1' & Z'N_2' \\ Z'^+N_1' & N_2' \end{pmatrix} \begin{pmatrix} K_1' & 0 \\ 0 & K_2' \end{pmatrix} \tag{2.19}$$

we find that the left translations have the form

$$Z' = (AZ+B)(CZ+D)^{-1} \tag{2.20}$$

The factors of this expression cannot be commuted in general.

We introduce the Lebesgue measure $|dZ|$ on D

$$|dZ| = \prod_{i,j=1,2} d(\mathrm{Re}Z_{ij})\, d(\mathrm{Im}Z_{ij}) \tag{2.21}$$

By arguments as those leading to (1.25), (1.26) we can show that the operators

$$T_M f(Z) = [\det(CZ+D)]^{-n} f(Z') \tag{2.22}$$

and the norm

$$\|f\|_n^2 = c \int_D |dZ|\ [\det(E-Z^+Z)]^{n-4}\ |f(Z)|^2 \tag{2.23}$$

define a unitary representation of SU(2,2) in the Hilbert space $A_n^2(D)$ of all holomorphic and square integrable (with respect to the norm (2.23)) functions on D where n is allowed to assume integral values not smaller than four. We choose c again such that the norm of the function $f(z) = 1$ is one. This allows us to extend the definition of the representation to $n = 2$ and $n = 3$ as we shall see in a moment.

The representations just constructed belong to the holomorphic branch of the discrete series d_0 [13,14] of SU(2,2), they are irreducible in particular. Each member of this series d_0 is characterized by three labels n, j_1, j_2, where $2j_1$ and $2j_2$ are also nonnegative integers. They characterize the SL(2,C) content (j_1, j_2) of the little group representation. We shall stick to the case $j_1 = j_2 = 0$ in these lectures. The representations obtained in this special case are those defined by (2.22), (2.23). The antihomomorphic branch can be treated analogously.

2.2 The polynomial basis in $A_n^2(D)$ and the Bergman kernel

Since D is a connected Reinhardt domain containing the origin $Z = 0$ each $f \in A_n^2(D)$ can be expanded in a Taylor series around the point $z_{ij} = 0$ that converges absolutely uniformly in any compact subset of D. Therefore all polynomials in the z_{ij} form a dense subspace in $A_n^2(D)$.

We choose homogeneous polynomials, namely

$$\Delta^{jm}_{q_1q_2}(Z) = (N^{jm})^{-1} (\det Z)^m\, D^j_{q_1q_2}(Z)$$

$$m = 0, 1, 2, \ldots, \quad 2j = 0, 1, 2, \ldots \tag{2.24}$$

$$-j \leq q_{1,2} \leq +j$$

The polynomials $D^j_{q_1q_2}(Z)$ are known from the representations of the group SU(2), or more precisely: They define a contravariant spinor representation of GL(2,C) with spin j. We choose the convention

$$D^{j}_{q_1q_2}(Z) = \left[\frac{(j+q_1)!(j-q_1)!}{(j+q_2)!(j-q_2)!}\right]^{\frac{1}{2}}$$
$$\times \sum_{s} \binom{j+q_2}{s}\binom{j-q_2}{s-q_1-q_2} z_{11}^{s} z_{12}^{j+q_1-s} z_{21}^{j+q_2-s} z_{22}^{s-q_1-q_2} \tag{2.25}$$

The polynomials (2.24) are homogeneous in the variables z_{ij} of degree

$$N = 2j + 2m \tag{2.26}$$

By elementary combinatorics one can show that for a fixed degree N there are S_N

$$S_N = \frac{1}{6}(N+1)(N+2)(N+3) \tag{2.27}$$

of the polynomials (2.24), these are moreover linearly independent. The easiest way to establish the proof of the last assertion is to restrict the polynomials on U(2) and show that they are orthogonal with respect to the Haar measure of U(2). On the other hand there are also precisely S_N linearly independent polynomials of the type

$$z_{11}^{n_{11}} z_{12}^{n_{12}} z_{21}^{n_{21}} z_{22}^{n_{22}}$$

for a fixed degree N of homogeneity

$$N = \sum_{ij} n_{ij} \tag{2.28}$$

This proves that the polynomials (2.24) form a complete set.

We have still to show that these polynomials (2.24) are orthogonal and normalized, in other words: we must compute the normalizing factor N^{jm} still. This is best done in the following fashion. We split each matrix Z "canonically", that is to say as

$$Z = u_1\begin{pmatrix}\lambda_1 & 0\\ 0 & \lambda_2\end{pmatrix}u_2 \tag{2.29}$$

with complex numbers $\lambda_{1,2}$ and unitary matrices $u_{1,2}$ of the special form

$$u_1 = e^{i\phi_1\sigma_3}e^{i\vartheta_1\sigma_2}$$
$$u_2 = e^{i\vartheta_2\sigma_2}e^{i\phi_2\sigma_3} \tag{2.30}$$

This decomposition (2.29) can be made unique by restricting the phases appropriately. The matrices $u_{1,2}$ represent cosets of SU(2)/U(1) in SU(2). The homogeneous space of these cosets carries a measure (for either subscript 1 and 2)

$$d'\mu(u_1) = \frac{1}{2\pi} d\phi_1 dt_1$$

$$d'\mu(u_2) = \frac{1}{2\pi} d\phi_2 dt_2 \tag{2.31}$$

$$0 \leqq \phi \leqq 2\pi, \quad t = \cos^2 \frac{\vartheta}{2}, \; 0 \leqq t \leqq 1$$

With $|d\lambda| = d(\mathrm{Re}\lambda)d(\mathrm{Im}\lambda)$ we can then write

$$|dZ| = J\, d'\mu(u_1) d'\mu(u_2) |d\lambda_1| |d\lambda_2| \tag{2.32}$$

After some algebra we find

$$J = \frac{1}{2} \pi^2 \, (|\lambda_1|^2 - |\lambda_2|^2)^2 \tag{2.33}$$

In the parameters u_1, u_2, λ_1, λ_2 the domain D is defined by $|\lambda_1| < 1$, $|\lambda_2| < 1$.

These expressions are inserted into the scalar product for elements of the basis (2.24). The orthogonality of the basis elements follows immediately from the orthogonality of $D^j_{q_1q_2}(u)$ on SU(2) and from the orthogonality of the functions $(\lambda/|\lambda|)^m$ on the unit circle. It remains to calculate the norms.
After some elementary algebra it remains to do the integral

$$\begin{aligned}(N^{jm})^2 = {} & \frac{c(2\pi)^4}{8(2j+1)^2} \int_0^1 |\lambda_1| d|\lambda_1| \int_0^1 |\lambda_2| d|\lambda_2| \\ & \times (|\lambda_1|^2 - |\lambda_2|^2)^2 |\lambda_1\lambda_2|^{2m} (1-|\lambda_1|^2)^{n-4} (1-|\lambda_2|^2)^{n-4} \\ & \times \frac{|\lambda_1|^{4j+2} - |\lambda_2|^{4j+2}}{|\lambda_1|^2 - |\lambda_2|^2}\end{aligned} \tag{2.34}$$

This integral can be evaluated by elementary means and yields

$$(N^{jm})^2 = c\pi^4 \, \frac{(n-3)!\,(n-4)!\,m!\,(m+2j+1)!}{(2j+1)\,(m+n-2)!\,(m+2j+n-1)!} \tag{2.35}$$

In order that $N^{00} = 1$ we put

$$c = \pi^{-4}(n-1)(n-2)^2(n-3) \tag{2.36}$$

With this normalization we extend the definition (2.24) to n = 2 and n = 3.

The Bergman kernel is defined by the series

$$\begin{aligned}K^B(Z_1, Z_2) &= \sum_{jmq_1q_2} \overline{\Delta^{jm}_{q_1q_2}(Z_1)}\, \Delta^{jm}_{q_1q_2}(Z_2) \\ &= \sum_{jm} (N^{jm})^{-2} \, [\det(Z_1^+ Z_2)]^m \Big(\sum_q D^j_{qq}(Z_1^+ Z_2)\Big)\end{aligned} \tag{2.37}$$

This series converges absolutely uniformly for compact subsets of D and yields (also for n = 2 and n = 3)

$$K^B(Z_1, Z_2) = [\det(E - Z_1^+ Z_2)]^{-n} \tag{2.38}$$

As usual we define l^2-sequences $a^{jm}_{q_1q_2}$ of coefficients belonging to holomorphic functions f(Z)

$$f(Z) = \sum_{jm} \sum_{q_1q_2} a^{jm}_{q_1q_2} \Delta^{jm}_{q_1q_2}(Z) \tag{2.39}$$

for the cases n = 2 and n = 3. These holomorphic functions form the Hilbert spaces $A^2_{2,3}(D)$. Scalar products must be evaluated, however, by the sequences of coefficients, and not by the fomula (2.23) that does not make sense in these cases.

2.3 Fourier expansions on the Shilov boundary

Points on the boundary of D are obtained if either λ_1 or λ_2 in the canonical decomposition of Z (2.29) are of modulus one. If both are of modulus one, Z itself is unitary and lies on the Shilov boundary of D. The Shilov boundary is a four-dimensional manifold, whereas the boundary ∂D is a seven-dimensional manifold.

We denote the points of the Shilov boundary by X. The Shilov boundary S can then be parametrized as

$$\begin{aligned} X &= e^{i\frac{\phi}{2}} u, \quad u \in SU(2) \\ e^{i\phi} &= \det X, \quad 0 \leq \phi \leq 2\pi \end{aligned} \tag{2.40}$$

By means of the normalized Haar measure $d\mu(u)$ on SU(2) we can therefore introduce the measure

$$d\mu(X) = \frac{1}{2\pi} d\phi d\mu(u) \tag{2.41}$$

on S that is in fact identical with the Haar measure on U(2).

Functions on S that are measurable and square integrable with respect to the measure (2.41) form the Hilbert space $\mathcal{L}^2(S)$. By means of the Peter-Weyl theorem for the group U(2) [15,16] we can expand any element g(X) of $\mathcal{L}^2(S)$ into a series

$$g(X) = \sum_{m=-\infty}^{+\infty} \sum_{j=0}^{\infty} \sum_{q_1,q_2=-j}^{+j} b^{jm}_{q_1q_2} (2j+1)^{\frac{1}{2}} e^{i(m+j)\phi} D^j_{q_1q_2}(u) \tag{2.42}$$

that converges in the sense of the norm of $\mathcal{L}^2(S)$.
Parseval's formula for this expansion is

$$\|g\|^2_S = \sum_{m=-\infty}^{+\infty} \sum_{j=0}^{\infty} \sum_{q_1q_2=-j}^{+j} |b^{jm}_{q_1q_2}|^2 \tag{2.43}$$

We denote scalar products and norms in $\mathcal{L}^2(S)$ by the subscript S.

Equations (2.42) and (2.43) can be interpreted as defining an isomorphism between $\mathcal{L}^2(S)$ and l^2-summable sequences $b^{jm}_{q_1q_2}$.

Contrary to the case of SU(1,1) we can split each $g(X) \in \mathcal{L}^2(S)$ into <u>three</u> parts

$$\begin{aligned} g_+(X) &= \sum_{j=0}^{\infty} \sum_{m=0}^{\infty} \sum_{q_1q_2=-j}^{+j} \cdots \\ g_-(X) &= \sum_{j=0}^{\infty} \sum_{m=-\infty}^{-2j} \sum_{q_1q_2=-j}^{+j} \cdots \\ g_0(X) &= \sum_{j=0}^{\infty} \sum_{m=-2j+1}^{-1} \sum_{q_1q_2=-j}^{+j} \cdots \end{aligned} \tag{2.44}$$

The positive part g_+ can be holomorphically extended into D with respect to <u>all</u> variables, the negative part g_- possesses an antiholomorphic extension into D with <u>all</u> variables, whereas the remainder g_0 does not possess any such extension. We shall recognize later that this decomposition of g corresponds in a sense to cutting the Minkowski space into a positive timelike, a negative timelike, and a spacelike subdomain, respectively. All three parts of g lie in $\mathcal{L}^2(S)$. They add up to

$$g = g_+ + g_- + g_0 - b^{00}_{00} \tag{2.45}$$

We treat the positive part first. We make use of

$$e^{i(m+j)\phi} D^j_{q_1q_2}(u) = (\det X)^m D^j_{q_1q_2}(X) \tag{2.46}$$

The right hand side is obviously the boundary value of $(\det Z)^m D^j_{q_1q_2}(Z)$. Therefore we attempt to define the holomorphic extension of $g_+(X)$ by

$$f_+(Z) = \sum_{m=0}^{\infty} \sum_{j=0}^{\infty} \sum_{q_1q_2=-j}^{+j} b^{jm}_{q_1q_2} (2j+1)^{\frac{1}{2}} N^{jm} \Delta^{jm}_{q_1q_2}(Z) \tag{2.47}$$

In fact it is easy to show that the factors $(2j+1)^{1/2}N^{jm}$ are bounded for all $n \geq 2$, consequently (2.47) defines a holomorphic function $f_+(z)$ that lies in all $A^2_n(D)$.

In order to find the antiholomorphic extension of the negative part we start from the identity

$$\begin{aligned} e^{i(m+j)\phi} D^j_{q_1q_2}(u) &= (-1)^{q_1-q_2} e^{i(m+j)\phi} D^j_{-q_2,-q_1}(u_1^+) \\ &= (-1)^{q_1-q_2} (\det X^+)^{-m-2j} D^j_{-q_2,-q_1}(X^+) \end{aligned} \tag{2.48}$$

that can be proved from the representation (2.25). With the shorthands

$$\begin{aligned} m' &= -m-2j \\ \tilde{b}^{jm'}_{q_1q_2} &= (-1)^{q_1-q_2} b^{jm}_{-q_2,-q_1} \end{aligned} \tag{2.49}$$

we can write the antiholomorphic extension as

$$f_-(Z) = \sum_{m'=0}^{\infty} \sum_{j=0}^{\infty} \sum_{q_1 q_2=-j}^{+j} \tilde{b}^{jm'}_{q_1 q_2} (2j+1)^{\frac{1}{2}} N^{jm'} \Delta^{jm'}_{q_1 q_2}(Z^+) \qquad (2.50)$$

The positive parts $g_+(X)$ define the subspace $\mathcal{L}^2_+(S)$. The holomorphic extensions of all elements of $\mathcal{L}^2_+(S)$ form the Hardy-Lebesgue space of holomorphic functions on D. Due to the unique connection between boundary value and extension the two Hilbert spaces are isomorphic in a natural fashion. The Hardy-Lebesgue space contains at least all those holomorphic functions that are still continuous on S. These functions form a dense subspace of any $A^2_n(D)$. The Hardy-Lebesgue norm $\|f\|_S$ of such function f whose boundary value is g is identical with $\|g\|_S$ by definition,

$$f(Z) = \sum_{m=0}^{\infty} \sum_{j=0}^{\infty} \sum_{q_1 q_2=-j}^{+j} a^{jm}_{q_1 q_2} \Delta^{jm}_{q_1 q_2}(Z)$$

$$g(X) = \sum_{m=0}^{\infty} \sum_{j=0}^{\infty} \sum_{q_1 q_2=-j}^{+j} b^{jm}_{q_1 q_2} (2j+1)^{\frac{1}{2}} e^{i(m+j)\phi} D^j_{q_1 q_2}(u) \qquad (2.51)$$

$$\|g\|^2_S = \sum_{jmq_1q_2} |b^{jm}_{q_1 q_2}|^2 = \|f\|^2_S$$

Now for n = 2 we have

$$N^{jm} = (2j+1)^{-\frac{1}{2}} \qquad (2.52)$$

so that $a^{jm}_{q_1 q_2} = b^{jm}_{q_1 q_2}$ and consequently

$$\|f\|^2_S = \sum_{jmq_1q_2} |a^{jm}_{q_1 q_2}|^2 = \|f\|^2_2 \qquad (2.53)$$

Hence the Hardy-Lebesgue space coincides with $A^2_2(D)$. Its Aronszajn-Bergman kernel is the Szegö kernel

$$K^S(Z_1, Z_2) = [\det(E - Z_1^+ Z_2)]^{-2} = K^S_{Z_1}(Z_2) \qquad (2.54)$$

We note finally that the holomorphic extension $f_+(Z)$ of $g_+(X)$ (2.47) can be obtained from

$$f_+(Z) = (K^S_Z, g)_S = (K^S_Z, g_+)_S \qquad (2.55)$$

where Z_2 in (2.54) has been put on the Shilov boundary S.

2.4 Distributions on the Shilov boundary

We use the method of Fourier series to treat the distributions on the Shilov boundary S. These distributions make up the space $E'(S)$. Any $\phi(X) \in E'(S)$ can be expanded in the series

$$\phi(X) = \sum_{m=-\infty}^{\infty} \sum_{j=0}^{\infty} \sum_{q_1 q_2=-j}^{+j} b^{jm}_{q_1 q_2} (2j+1)^{\frac{1}{2}} (\det X)^m D^j_{q_1 q_2}(X) \qquad (2.56)$$

such that with the notation

$$\sigma^{jm} = \sum_{q_1 q_2=-j}^{+j} |b^{jm}_{q_1 q_2}|^2 \tag{2.57}$$

the quantity σ^{jm} is bounded by some polynomial in j and $|m|$,

$$\sigma^{jm} \leqq P(j,|m|) \tag{2.58}$$

In turn any formal series (2.56) satisfying (2.58) converges in the distribution topology to an element

$$\phi \in E'(S).$$

Derivative representations of distributions play an important role in the present case as in the case of SU(1,1). Because of the several variables we have to deal with, there are differential operators of many types. Two of them are of particular importance. The first one is Euler's differential operator

$$O_1(Z) = \sum_{i,j=1,2} z_{ij} \frac{\partial}{\partial z_{ij}} \tag{2.59}$$

that possesses homogeneous functions as eigenfunctions. Applied to the homogeneous polynomials of our basis (2.24) we find

$$O_1(Z)\Delta^{jm}_{q_1 q_2}(Z) = N\Delta^{jm}_{q_1 q_2}(Z) \tag{2.60}$$

with the homogeneity N (2.26). The other operator is of second order

$$O_2(Z) = (\det Z)\left(\frac{\partial^2}{\partial z_{11}\partial z_{22}} - \frac{\partial^2}{\partial z_{12}\partial z_{21}}\right) \tag{2.61}$$

The basis elements (2.24) are also eigenfunctions with respect to $O_2(Z)$

$$O_2(Z)\Delta^{jm}_{q_1 q_2}(Z) = m(m+2j+1)\Delta^{jm}_{q_1 q_2}(Z) \tag{2.62}$$

The proof of the last equation is not quite trivial, it can be established either by a straightforward algebraic computation or by the group theoretic arguments that are presented in the Appendix.

Both operators $O_1(Z)$ and $O_2(Z)$ are connected with the Casimir operators of the group U(2) that acts on the Shilov boundary by left translations. If we map U(2) on the unit sphere in four dimensional space, the basis elements (2.24) go over into spherical harmonics that are eigenfunctions of the operators $O_{1,2}(Z)$.

We show next that any distribution of $E'(S)$ can be given in either derivative form

$$\phi(X) = [1+O_1(X)]^{l_1} g_1(X)$$
$$\phi(X) = O(X,s)^{l_2} g_2(X) \tag{2.64}$$

where

$$O(X,s) = O_2(X) + sO_1(X) + s(s+1) \tag{2.65}$$

with nonnegative integers l_1, l_2, $s-1$ and $g_{1,2}(X) \in \mathcal{L}^2(S)$. If ϕ is equal to its positive part ϕ_+, then $g_{1,2}$ can be chosen from $\mathcal{L}^2_+(S)$. We assume that this is done in the sequel.

Let the degree of the polynomial (2.58) be τ. Then we can estimate σ^{jm} also by

$$\sigma^{jm} \leqq C(1+2m+2j)^{\tau} \tag{2.66}$$

with some constant C. The expansion (2.56) can then be rewritten as

$$\begin{aligned}\phi(X) = [1+O_1(X)]^{l_1} \sum_{m=0}^{\infty} \sum_{j=0}^{\infty} \sum_{q_1q_2=-j}^{+j} \\ \times\, b^{jm}_{q_1q_2} (1+N)^{-l_1} (2j+1)^{\frac{1}{2}} (\det X)^m D^j_{q_1q_2}(X)\end{aligned} \tag{2.67}$$

In this series the coefficients can be estimated by

$$\sum_{jm} \sigma^{jm} = \sum_{N=0}^{\infty} (\sum_{j,N \text{ fixed}} \sigma^{jm}) \leqq \tfrac{1}{2}C \sum_{N=0}^{\infty} (1+N)^{\tau-2G}(2+N) < \infty \tag{2.68}$$

whenever $\tau-2l_1 < -2$. Therefore we may identify the sum in (2.67) with $g_1(X) \in \mathcal{L}^2_+(S)$ with $l_1 > \frac{1}{2}\tau + 1$.

The second equation (2.64) is a bit more difficult to prove. The eigenvalue of $O(X,s)$ on $\Delta^{jm}_{q_1q_2}(Z)$ is $(m+s)(m+2j+s+1)$. It is again obvious that from (2.58) one can find an integer $\hat{\tau}$ such that

$$\sigma^{jm} \leqq C'[(m+s)(m+2j+s+1)]^{\tau} \tag{2.69}$$

for any fixed $s \geqq 1$.
But then (2.56) becomes

$$\begin{aligned}\phi(X) = O(X,s)^{l_2} \sum_{m=0}^{\infty} \sum_{j=0}^{\infty} \sum_{q_1q_2=-j}^{+j} b^{jm}_{q_1q_2} \\ \times\, [(m+s)(m+2j+s+1)]^{-l_2} (2j+1)^{\frac{1}{2}} (\det X)^m D^j_{q_1q_2}(X)\end{aligned} \tag{2.70}$$

and for the coefficients of this series we have

$$\sum_{j,m} \sigma^{jm} \leqq C' \sum_{m=0}^{\infty} \sum_{j=0}^{\infty} (m+s)^{\tau-2l_2} (m+2j+s+1)^{\tau-2l_2} < \infty \tag{2.71}$$

if $\tau-2l_2 < -1$.

After these preliminaries we turn directly to the propositions asserting a certain connection between the holomorphic functions of $A^2_n(D)$ and the degree of singularities of their boundary values.

Theorem A_2. Let $f(Z)$ be an element of $A^2_n(D)$. Then $f(Z)$ approaches a distribution $\phi(X) \in E'(S)$ in the sense of the topology of $E'(S)$ if Z tends to S. ϕ can be represented in both derivative forms

$$\phi(X) = 0(X,s)^{l_2} g_2(X) \tag{2.72}$$

or

$$\phi(X) = (1+0_1(X))^{l_1} g_1(X) \tag{2.73}$$

for $g_{1,2}(X) \in \mathcal{L}^2(S)$, any $s \geqq 1$ and

$$l_1 \geqq n-2, \quad l_2 \geqq \tfrac{1}{2}n-1 \tag{2.74}$$

Proof: We expand $f(Z)$ in the basis (2.24) with expansion coefficients $a^{jm}_{q_1q_2}$. Then $f(Z)$ can be represented as

$$\begin{aligned} f(Z) = (1+0_1(Z))^{n-2} \sum_{m=0}^{\infty} \sum_{j,q_1q_2} b^{jm}_{q_1q_2} \\ \times (2j+1)^{\frac{1}{2}} (\det Z)^m D^j_{q_1q_2}(Z) \end{aligned} \tag{2.75}$$

with

$$b^{jm}_{q_1q_2} = (2j+1)^{-\frac{1}{2}} (N^{jm})^{-1} (2j+2m+1)^{-n+2} a^{jm}_{q_1q_2} \tag{2.76}$$

One can show that the factor of $a^{jm}_{q_1q_2}$ assumes its maximum at $m = j = 0$ and that this maximum is one,

$$|b^{jm}_{q_1q_2}| \leqq |a^{jm}_{q_1q_2}| \tag{2.77}$$

Therefore the series in (2.75) converges in the $A^2_2(D)$ norm and further

$$f(Z) = (1+0_1(Z))^{n-2} f_1(Z), f_1 \in A^2_2(D) \tag{2.78}$$

The assertion (2.73) follows if we let Z tend to S.

The assertion (2.72) is more of the kind of Theorem A_1 (n corresponds to 2k !). The proof is similar to that of (2.72) only the estimate differs. Instead of (2.76) we have

$$b^{jm}_{q_1q_2} = (2j+1)^{-\frac{1}{2}} (N^{jm})^{-1} [(m+s)(m+2j+s+1)]^{-l_2} a^{jm}_{q_1q_2} \tag{2.79}$$

With

$$C_1(n,s)^{-1} = \inf_m \{[(n-1)!(n-2)!]^{\frac{1}{2}} \frac{m!(m+s)^{n-2}}{(m+n-2)!}\} > 0 \tag{2.80}$$

we obtain

$$|b^{jm}_{q_1q_2}| \leqq C_1(n,s) \; |a^{jm}_{q_1q_2}| \tag{2.81}$$

Again we have as a corollary

$$f(Z) = 0(Z,s)^{l_2} f_2(Z), \; f_2 \in A^2_2(D), \; l_2 \geqq \tfrac{1}{2}n-1, \; s \geqq 1. \tag{2.82}$$

This completes the proof.

Theorem B_2. Let $\phi(\ddot{X})$ be a distribution of $E'(S)$ that can be written as

$$\phi(X) = O(X,s)^{1}g(X), \quad g \in \mathcal{L}^2_+(S) \tag{2.83}$$

for any $s \geqq 1$. Then ϕ possesses a holomorphic extension $f(Z)$ into D that lies in $A^2_n(D)$ with $n \geqq 2l+2$.

Proof: In fact, the holomorphic extension is obtained from

$$\begin{aligned} f(Z) &= O(Z,s)^{1}f_1(Z) \\ f_1(Z) &= (K^S_Z,g)_S \end{aligned} \tag{2.84}$$

It remains to show only that if $f_1 \in A^2_2(D)$ then $O(Z,s)^{1}f_1 \in A^2_n(D)$ for any $n \geqq 2l+2$. In order to establish this assertion we have to verify the square summability of $a^{jm}_{q_1q_2}$

$$a^{jm}_{q_1q_2} = (2j+1)^{\frac{1}{2}}N^{jm}[(m+s)(m+2j+s+1)]^{1}\, b^{jm}_{q_1q_2} \tag{2.85}$$

where $b^{jm}_{q_1q_2}$ are the expansion coefficients of $f_1(Z)$ in the basis (2.24). With the definition

$$C_2(n,s) = \sup_m\{[(n-1)!(n-2)!]^{\frac{1}{2}} \frac{m!(m+s)^{n-2}}{(m+n-2)!}\} < \infty \tag{2.86}$$

We can estimate

$$|a^{jm}_{q_1q_2}| \leqq C_2(n,s)|b^{jm}_{q_1q_2}| \tag{2.87}$$

This completes the proof.

As in the case of SU(1,1) for half of the representations, namely <u>even</u> n the boundary values of the holomorphic functions of $A^2_n(D)$ are characterized by the derivative representation

$$\phi(X) = O(X,s)^{\frac{1}{2}n-1}g(X), \quad g \in \mathcal{L}^2_+(S) \tag{2.88}$$

that is: distributions ϕ are boundary values for $A^2_n(D)$ if and only if they allow the representation (2.88).

Finally we mention that another version of Theorem B_2 can be given that uses the operator $O_1(Z)$ but cannot be used to characterize boundary distributions in a unique fashion. It says: Given a distribution

$$\phi(X) = (1+O_1(X))^{1}g(X), \quad g \in \mathcal{L}^2_+(S) \tag{2.89}$$

then ϕ possesses a holomorphic extension $f(Z)$ that lies in $A^2_n(D)$ with $n \geqq 2l+2$. We leave the proof to the interested reader (see also [12]).

2.5 The tube domain

The domain D can be mapped analytically and one-to-one on a tube domain T (or a "generalized upper half plane") in C_4 by a mapping analogous to (1.86), i.e. a Cayley transformation. It is explicitly given by

$$W = i(E-Z)(E+Z)^{-1}$$
$$Z = (E-iW)^{-1}(E+iW) \qquad (2.90)$$

where W is a complex 2 x 2 matrix as is Z. If Z is unitary, that is $Z \in S$, then W is hermitean. We put

$$W = w_0 E + \sum_{k=1}^{3} w_k \sigma_k \qquad (2.91)$$

so that the Shilov boundary S of D is mapped on real vectors w, $w = \{w_\mu,\ \mu = 0,1,2,3\}$. The tube domain T is characterized by

$$w = u+iv\ , \quad u, v \text{ real}$$
$$v_0 > \left(\sum_{k=1}^{3} v_k^2\right)^{\frac{1}{2}} \qquad (2.92)$$

that is: the imaginary part of w lies in the forward (open) light cone L_+. Here and in the sequel we skip over the elementary algebraic proofs.

The automorphisms (2.20) of D induce automorphisms of T of the fractional linear form

$$W' = (RW+S)(TW+Q)^{-1} \qquad (2.93)$$

where in analogy with (1.91) we can express the 2 x 2 matrices R, S, T, Q, by A, B, C, D as

$$\begin{pmatrix} R, & iS \\ -iT, & Q \end{pmatrix} = \frac{1}{2}\begin{pmatrix} E, & -E \\ +E, & E \end{pmatrix}\begin{pmatrix} A & B \\ C & D \end{pmatrix}\begin{pmatrix} E, & +E \\ -E, & E \end{pmatrix} \qquad (2.94)$$

where the matrices R, S, T, Q are subject to the restriction

$$R^+T = H_1$$
$$R^+Q = E+H_2-iH_3$$
$$S^+T = H_2+iH_3 \qquad (2.95)$$
$$S^+Q = H_4$$

and the unimodularity constraint

$$\det\begin{pmatrix} R, & iS \\ -iT, & Q \end{pmatrix} = 1 \qquad (2.96)$$

The constraints (2.95) require H_k, k = 1,2,3,4 to be arbitrary hermitean matrices and thus guarantee the pseudounitarity of M.

The action (2.93) of the elements of SU(2,2) is precisely the customary action of the conformal group on four-vectors in Minkowski space. The inhomogeneous SL(2,C) subgroup of SU(2,2) is obtained if T = 0, det R = 1. For the transformations of the subgroup SL(2,C) the bilinear form $w^2 = ww = w_0{}^2 - \sum_k w_k{}^2$ remains invariant. The Shilov boundary of T that consists of all real fourvectors w can therefore be identified with the (real) Minkowski space M_4. The dilations belong to T = S = 0, R = λE, λ > 0, and the special conformal transformations to S = 0, Q = E.
For more details see [12].

We consider next the space $A^2_{n,q}(T)$ of holomorphic functions that are obtained from $A^2_n(D)$ by the mapping (compare (1.93), (1.94)).

$$F(w) = m_q(w)\, f(Z(w)) \tag{2.97}$$

and

$$m_q(w) = 2^{2n-2}[\det(E-iW)]^{-n+q} \tag{2.98}$$

with integral q. The scalar product in $A^2_{n,q}(T)$ is by definition

$$\begin{aligned}(F_1,F_2)_{n,q} = c \int_T |dw|\ \overline{F_1(w)}F_2(w)\\ \times\ [(\mathrm{Im}w)^2]^{n-4}|[\det(E-iW)]|^{-2q}\end{aligned} \tag{2.99}$$

with c as in (2.36). By (2.97) and (2.99) we have a natural isomorphism between $A^2_n(D)$ and $A^2_{n,q}(T)$, for arbitrary q. The operators T_M in $A^2_{n,q}(T)$, i.e. the unitary representation of SU(2,2), are given by

$$T_M F(w) = \mu_q(M,w)F(w') \tag{2.100}$$

with

$$\mu_q(M,w) = \{\det(E-iW)(E-iW')^{-1}\}^q[\det(TW+Q)]^{-n} \tag{2.101}$$

The Bergman kernel is

$$\begin{aligned}K^B(w_1,w_2) = 2^{-4}[\det(E+iW_1^+)(E-iW_2]^q\\ \times\ \{\det[+\tfrac{i}{2}(W_1{}^+-W_2)]\}^{-n}\end{aligned} \tag{2.102}$$

We map the space $\mathcal{L}^2(S)$ isomorphically on the space $\mathcal{L}^2(M_4)$ by the definitions

$$\begin{aligned}&G(u) = [\det(E-iU)]^{-2}\ \ g(X(u))\\ &g \in \mathcal{L}^2(S),\quad G \in \mathcal{L}^2(M_4)\end{aligned} \tag{2.103}$$

with the scalar product in $\mathcal{L}^2(M_4)$ defined by

$$(G_1,G_2)_{M_4} = \left(\frac{2}{\pi}\right)^3 \int_{M_4} d^4u\, \overline{G_1(u)}G_2(u) = (g_1,g_2)_S \tag{2.104}$$

Finally the Szegö kernel is given by

$$K^S(w_1,w_2) = 2^{-4}[(\bar{w}_1-w_2)^2]^{-2} \tag{2.105}$$

This kernel can be used for holomorphic extensions into T by (see (1.118))

$$F_+(w) = (K^S_w,G)_{M_4} = [\det(E-iW)]^{-2}(K^S_{Z(w)},g)_S \tag{2.106}$$

where G is in $\mathcal{L}^2(M_4)$.

With these definitions we can go on discussing the holomorphic extensions on distributions of $D'_{L^2}(M_4)$ and $S'(M_4)$ in complete analogy to the case of SU(1,1) where the space M_4 replaces the real axis R. There is one point of general interest which we want to discuss here. Namely, the "generalized Hilbert transform" $F_+(w)$ (2.106) for a distribution $\psi(u) \in D'_{L^2}(M_4)$ can also be obtained by a Laplace transform from a distribution $\hat{\psi}_+(t)$, viz. by the integral

$$F_+(w) = \int_{M_4} e^{iwt}\, \hat{\psi}_+(t)d^4(t) \tag{2.107}$$

that converges properly.

This Laplace transform representation of $F_+(w)$ is obtained by first transforming $\psi(u) \in D'_{L^2}(M_4)$ (as a formal integral)

$$\psi(u) = \int_{M_4} e^{iut}\, \hat{\psi}(t)d^4t \tag{2.108}$$

and then cutting the locally square integrable tempered distribution $\hat{\psi}(t)$ by the characteristic function $\xi(t)$ of the forward light cone

$$\xi(t) = \begin{cases} 1 & t \in L_+ \\ 0 & t \notin L_+ \end{cases} \qquad \hat{\psi}_+(t) = \xi(t)\hat{\psi}(t) \tag{2.109}$$

We notice that (2.107) is certainly equivalent with folding the distribution $\psi(u)$ with the Fourier (Laplace-) transform of the characteristic function $\xi(t)$ of the forward light cone. This Fourier transform of $\xi(t)$ yields the Szegö kernel

$$\tau(w) = (2\pi)^{-4} \int e^{iwt}\xi(t)d^4t = \left(\frac{2}{\pi}\right)^3 K^S(0,w) \tag{2.110}$$

This shows that (2.106) and (2.107) are equivalent forms for the holomorphic extension.

Tempered distributions of $S'(M_4)$ are represented as

$$\psi(u) = D_u^m [\det(E-iU)]^k G(u) \quad G \in \mathcal{L}^2(M_4) \tag{2.111}$$

Their holomorphic extensions can be obtained from

$$F_+(w) = D_w^m [\det(E-iW)]^k (K_w^S, G)_{M_4} \tag{2.112}$$

By means of the Szegö kernel and Schwarz's inequality one can again estimate the increase of $F_+(w)$ at the boundary of T. This estimate can be used to prove that the holomorphic extension lies in all Hilbert spaces $A^2_{n,q}(T)$ satisfying

$$q+|m|-2k+2 \geqq n \geqq 2|m|+6 \tag{2.113}$$

On the other hand one obtains in this fashion estimates for the Laplace transforms of tempered distributions with support in the forward light cone that sharpens the assertion of a classic theorem [17].

APPENDIX

The operator $O_2(Z)$

We define the differential operators H_k, k= 1,2,3, that act on holomorphic functions over D by

$$\lim_{\varepsilon\to 0} \frac{i}{\varepsilon} [f(e^{\frac{i}{2}\varepsilon\sigma_k}z)-f(z)] = H_k f(z) \tag{A-1}$$

and find (note $[H_1, H_2]= i H_3$)

$$H_1 = -\frac{1}{2} [\sum_{i=1,2} (z_{2i}\frac{\partial}{\partial z_{1i}} + z_{1i}\frac{\partial}{\partial z_{2i}})]$$

$$H_2 = \frac{i}{2} [\sum_{i=1,2} (z_{2i}\frac{\partial}{\partial z_{1i}} - z_{1i}\frac{\partial}{\partial z_{2i}})] \tag{A-2}$$

$$H_3 = -\frac{1}{2} [\sum_{i=1,2} (z_{1i}\frac{\partial}{\partial z_{1i}} - z_{2i}\frac{\partial}{\partial z_{2i}})]$$

Then we define the Casimir operator

$$O_3(Z) = \sum_{k=1}^{3} H_k^2 \tag{A-3}$$

that acts on the basis elements (2.24) like

$$O_3(Z)\Delta^{jm}_{q_1q_2}(Z) = j(j+1)\Delta^{jm}_{q_1q_2}(Z) \tag{A-4}$$

Explicitely we find by direct computation

$$O_3(Z) = \frac{1}{4} O_1^2(Z) + \frac{1}{2} O_1(Z) - O_2(Z) \tag{A-5}$$

The eigenvalues of the eigenfunctions (2.24) of $O_2(Z)$ are therefore

$$\frac{1}{4} N^2 + \frac{1}{2} N - j(j+1) = m(m+2j+1) \ . \tag{A-6}$$

REFERENCES

1. K. Yosida, Functional Analysis, Springer Verlag, Berlin-Heidelberg-New York, 1968.
2. L. Hörmander, An Introduction to Complex Analysis in Several Variables, D. Van Nostrand, Princeton, 1966, Theorem 2.4.5.
3. S. Helgason, Differential Geometry and Symmetric Spaces, Academic Press, New York and London, 1962.
4. L.K. Hua, Am. Math. Soc. (2), **32**, 195.
5. A.S. Wightman, Analytic Functions of Several Complex Variables, in: Relations de dispersion et particules élémentaires, ed. by C. De Witt and R. Omnès, Hermann, Paris, 1960.
6. M.A. Neumark, Normierte Algebren, Deutscher Verlag der Wissenschaften, Berlin, 1959, Section 13.
7. L. Schwartz, Théorie des distributions, Hermann, Paris, 1966.
8. H. Behnke and F. Sommer, Theorie der analytischen Funktionen einer komplexen Veränderlichen, Springer-Verlag, Berlin-Göttingen-Heidelberg, 1955, p. 324, Satz 25.
9. W. Rühl, The Lorentz Group and Harmonic Analysis, Benjamin, New York, 1970, Section 5-8.
10. V. Bargmann, Ann. Math. **48**, 568 (1947).
11. H.G. Tillmann, Math. Z. **59**, 61 (1953); **76**, 5 (1961); **77**, 125 (1961).
12. W. Rühl, Comm. Math. Phys. **27**, 53 (1972).
13. M.L. Graev, Dokl. Akad. Nauk SSSR **98**, 517 (1954).
14. A. Esteve and P.G. Sona, Nuovo Cimento **32**, 473 (1964).
15. F. Peter und H. Weyl, Math. Ann. **97**, 737 (1927).
16. Ref. 2, Section 32, Theorem 5.
17. R.F. Streater and A.S. Wightman, PCT, Spin and Statistics, and all that, New York, 1964, Theorem 2-10.

THE SEMISIMPLE SUBALGEBRAS OF THE ALGEBRA B_3 (SO(7)) AND THEIR INCLUSION RELATIONS

Bruno Gruber
Physics Department, Southern Illinois University
Carbondale, Ill. 62901

1. INTRODUCTION

In this article the semisimple subalgebras of the simple Lie algebra B_3 are determined explicitly. This classification of the semisimple subalgebras of B_3 has a twofold purpose.

Firstly, B_3 is a physically interesting algebra, used in the shell theory of atomic spectroscopy. It contains the exceptional algebra G_2 as a subalgebra, which was found to be of utmost importance for the success of theoretical calculations for $\ell = 3$ electrons (ℓ is the orbital angular momentum) in rare earth spectroscopy. Moreover, B_3 contains A_3 as a subalgebra. The algebra A_3 is again of physical importance. It corresponds to Wigner's SU(4), which in turn has two SU(2) x SU(2) subgroups ($A_1 + A_1$ subalgebras), one of which is the Spin-Isotopic Spin subalgebra. B_3 also contains A_2 as a subalgebra, used both in particle physics (unitary spin) and nuclear physics (Elliott's SU(3)). The algebra A_2, finally, contains two subalgebras of type A_1, namely A_1^1 and A_1^4, whereby A_1^1 is the Isotopic Spin subalgebra of particle physics while A_1^4 is the orbital angular momentum subalgebra of the Elliott model. Thus the algebra B_3 contains a considerable number of subalgebras which are of significance to physics. It is of interest to know how these subalgebras are embedded in the larger algebras containing them (in particular in B_3). Moreover, it is of interest to study the interrelationships of these algebras as subalgebras of B_3 as well as their relationship to subalgebras of B_3 which are isomorphic to them but of no physical significance.

Secondly, the explicit classification of the semisimple subalgebras of B_3 serves as an example for the general method of the classification of the semisimple subalgebras of a simple algebra as developed by Dynkin [1,2] and Malcev [3] in three rather voluminous articles. Apart from the general theory Dynkin also gave an explicit classification of the semisimple subalgebras of the exceptional Lie algebras. M. Lorente and the author determined all semisimple subalgebras of the classical Lie algebras up to rank 6, extending Dynkin's definition of defining vector to that of defining matrix [4]. The determination of the semisimple subalgebras of the algebra B_3 is based upon the rules for such a classification as compounded and formulated in ref. [4].

2. CLASSIFICATION SCHEME

We distinguish between three types of subalgebras of G: the regular subalgebras (r-subalgebras), the S-subalgebras, and the R-subalgebras. The latter two types of subalgebras are called non-regular subalgebras (non-r-subalgebras) of G.

The definition of these subalgebras is given as follows. Let Σ denote the root system of the algebra G and $\tilde{\Sigma}$ the root system of a subalgebra $\tilde{G}$ of G. Then we say $\tilde{G}$ is a

r-subalgebra of G, if $\tilde{\Sigma} \subset \Sigma$,

and a

non-r-subalgebra of G, if $\tilde{\Sigma} \not\subset \Sigma$.

Let $\tilde{G}'$ with root system $\tilde{\Sigma}'$ denote an arbitrary r-subalgebra of G. A non-r-subalgebra $\tilde{G}$ of G, having a root system $\tilde{\Sigma}$, is called

S-subalgebra of G, if there exists no proper r-subalgebra $\tilde{G}'$ of G containing $\tilde{G}$ as a subalgebra,

R-subalgebra of G, if there exists a proper r-subalgebra $\tilde{G}'$ of G containing $\tilde{G}$ as a subalgebra.

A R-subalgebra of G, is in turn, an R-subalgebra or an S-subalgebra of the r-subalgebra $\tilde{G}'$.

We have therefore, the following classification scheme for the subalgebras of a simple Lie algebra G:

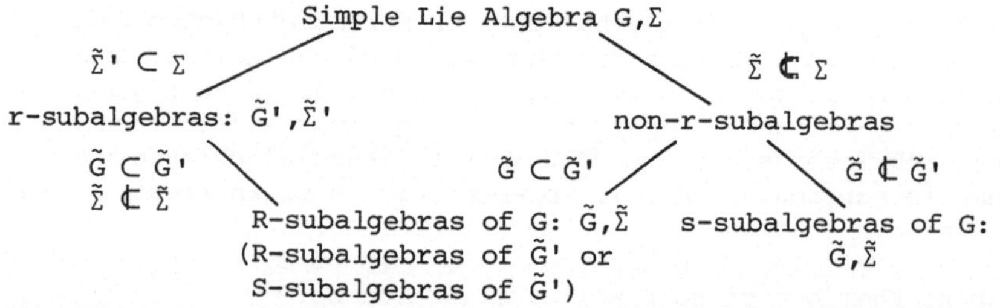

A subalgebra $\tilde{G}$ of an algebra G is called maximal if no proper subalgebra G' of G exists such that $G \supset G' \supset \tilde{G}$.

Each subalgebra $\tilde{G}$ of G can further be classified as maximal, non-maximal, simple and non-simple. Thus, we have the diagram

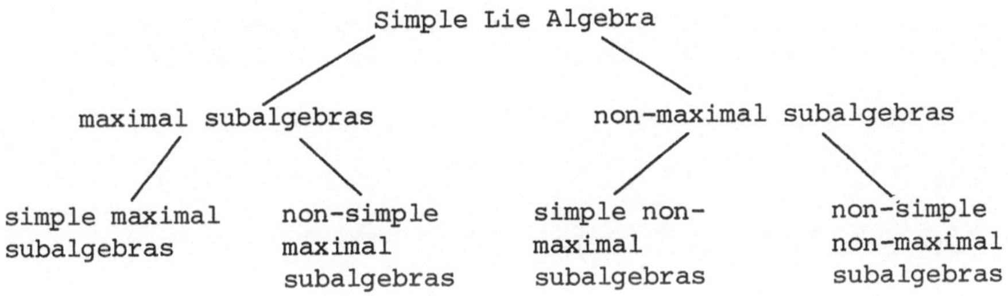

3. ACTUAL CLASSIFICATION

From the above discussion of subalgebras of a simple algebra, it can easily be deduced that a practical method for the classification consists in classifying successively maximal subalgebras. That is, the maximal subalgebras, r-subalgebras and S-subalgebras, of the simple algebra G are determined. For the maximal subalgebras obtained, again the maximal subalgebras are determined. This process is continued until no new maximal subalgebras are obtained.

Two difficulties arise if this classification is attempted. The first one is how to recognize two subalgebras, obtained in different steps, to be identical. This problem of identification of subalgebras will be dealt with below and is solved, partially, through the index of the embedding of a subalgebra in an algebra. The second problem is that in the course of the classification non-simple subalgebras will occur, and it thus becomes necessary to determine maximal subalgebras of non-simple algebras.

The maximal subalgebras of a simple algebra G will be determined in three steps:

(I) simple and non-simple r-algebras
(II) simple S-algebras
(III) non-simple S-algebras

The maximal subalgebras of a non-simple algebra G will be determined through the same three steps. Thus

(IV) simple and non-simple r-algebras
(V) simple S-algebras
(VI) non-simple S-algebras

Among the subalgebras of the algebra B_3 examples for all six cases (I)-(VI) can be found.

4. INDEX OF EMBEDDING; DEFINING MATRIX

A faithful embedding f of a subalgebra $\tilde{G}$ in an algebra G is an isomorphic mapping of the elements $\tilde{X} \varepsilon \tilde{G}$ onto elements $f(\tilde{X}) \varepsilon G$,

$$\tilde{X} \to f(\tilde{X}) \ ,$$

such that

$$[f(\tilde{X}_i), f(\tilde{X}_j)] = f([\tilde{X}_i, \tilde{X}_j]).$$

Two embeddings of a subalgebra $\tilde{G}$ in an algebra G are called equivalent if the two subalgebras are conjugate in G, i.e. related to each other by an inner automorphism. If U denotes an inner automorphism of G, then it holds for two equivalent embeddings f_1 and f_2 that

$$U f_1(\tilde{X}) = f_2(\tilde{X}), \text{ for all } \tilde{X} \varepsilon \tilde{G} \ .$$

In order to distinguish classes of equivalent embeddings, the <u>index</u> j_f of an embedding f is introduced.

A scalar product can be defined in G as

$$(X,Y) \equiv \mathrm{Tr}\ (\mathrm{ad}(X) \ \ \mathrm{ad}(Y)), \ X,Y \ \varepsilon \ G,$$

where ad(X) denotes the adjoint representation of the algebra G. This scalar product is invariant under inner automorphisms. The index j_f of an embedding f of an algebra $\tilde{G}$ in an algebra G is then defined by

$$(f(\tilde{X}), f(\tilde{Y})) = j_f(\tilde{X},\tilde{Y}) \ , \quad \tilde{X},\tilde{Y} \ \varepsilon \ \tilde{G} \ .$$

The scalar factor j_f is independent of $\tilde{X}$ and $\tilde{Y}$ and is the same for all equivalent embeddings.

If ϕ denotes a representation of the algebra G, then ϕf induces a representation for the subalgebra $\tilde{G}$. If ϕ is irreducible, ϕf may or may not be an irreducible representation of the subalgebra $\tilde{G}$.

Similarly to the index j_f of an embedding f, an index ℓ_ϕ can be defined for the representation ϕ. Thus, if we define a scalar product for the representation ϕ,

$$(\phi(X), \phi(Y)) = \mathrm{Tr}(\phi(X) \ \phi(Y)) \ , \ X,Y \ \varepsilon \ G \ ,$$

the index ℓ_ϕ of the representation ϕ is given by

$$(\phi(X), \phi(Y)) = \ell_\phi(X,Y) \ .$$

It can be shown that

$$j_f = \frac{\ell_{\phi f}}{\ell_\phi} \ .$$

Thus, the properties of linear representations of the algebra G and the subalgebra $\tilde{G}$ determine the index of the embedding f of $\tilde{G}$ in G. For given algebra G and given representation ϕ of G, the index j_f depends on the properties of the representation ϕf. Inequivalent embeddings of the same algebra $\tilde{G}$ are distinguished through different branchings of the representation ϕ of G under the restriction of ϕ to the subalgebra $\tilde{G}$.

The index j_f is not unique, i.e., the same index may characterize inequivalent classes. This is though not very frequent, but must nevertheless be kept in mind.

A unique characterization is provided through the defining matrix. The defining matrix defines, up to equivalence, the embedding of the Cartan subalgebra $\tilde{H}$ of the subalgebra $\tilde{G}$ in the Cartan subalgebra H of the algebra G. Choosing an orthormal base for the Cartan subspaces, the defining matrix is given as

$$f(\tilde{H}_i) = \sum_k f_{ik} H_k , \quad i = 1,2,\dots,n' \quad k = 1,2,\dots,n \quad n' \leq n,$$

where n' and n are the ranks of $\tilde{G}$ and G. It holds

$$\sum_{k=1}^{n} f_{ik} f_{mk} = j_f \delta_{im} .$$

All defining matrices related by an inner automorphism, i.e. related through the Weyl group, are equivalent. Due to different conventions for the length of the longest root of the simple Lie algebras (C_n), and due to the embedding of the Cartan subspace of the algebras A_n, G_2, E_6, E_7 and E_8 in a space with one more dimension, the following relations hold for the index j_f of an embedding:

$$\sum_{k=1}^{n} f_{ik} f_{jk} = \alpha_{ij} j_f ,$$

where for

(a) $G = A_n, B_n, D_n, G_2, F_4, E_6, E_7, E_8$

$\tilde{G} = B_{n'}, D_{n'}, F_4, (1 \leq n' \leq n)$, and for

$G = C_n, \tilde{G} = C_{n'}, (1 \leq n' \leq n)$,

$\alpha_{ij} = \delta_{ij}$

(b) $G = A_n, B_n, D_n, G_2, F_4, E_6, E_7, E_8$

$\tilde{G} = C_{n'}, (1 \leq n' \leq n)$,

$\alpha_{ij} = 2\delta_{ij}$

(c) $G = C_n$

$\tilde{G} = B_{n'}, D_{n'}, F_4, (1 \leq n' \leq n)$

$\alpha_{ij} = 1/2\ \delta_{ij}$

(d) $G = A_n, B_n, D_n, G_2, F_4, E_6, E_7, E_8$

$\tilde{G} = A_{n'}, G_2, E_6, E_7, E_8, (1 \leq n' \leq n),$

$$\alpha_{ij} = \frac{n'}{n'+1}, \text{ for } i = j,$$

$$\alpha_{ij} = \frac{-1}{n'+1}, \text{ for } i \neq j.$$

(e) $G = C_n$

$\tilde{G} = A_{n'}, G_2, E_6, E_7, E_8, (1 \leq n' \leq n),$

$$\alpha_{ij} = \frac{n'}{2(n'+1)}, \text{ for } i = j,$$

$$\alpha_{ij} = \frac{-1}{2(n'+1)}, \text{ for } i \neq j.$$

The defining matrix satisfies the following relations: For

(a) $G = B_n, C_n, D_n, F_4$

$\tilde{G} = A_{n'}, G_2, E_6, E_7, E_8, (1 \leq n' \leq n),$

$$\sum_{i=1}^{n'+1} f_{ik} = 0$$

(b) $G = A_n, G_2, E_6, E_7, E_8$

$\tilde{G} = B_{n'}, C_{n'}, D_{n'}, F_4$

$$\sum_{k=1}^{n+1} f_{ik} = 0$$

(c) $G = A_n, G_2, E_6, E_7, E_8$

$\tilde{G} = A_{n'}, G_2, E_6, E_7, E_8, (1 \leq n' \leq n),$

$$\sum_{i=1}^{n'+1} f_{ik} = c, \qquad \sum_{k=1}^{n+1} f_{ik} = c\frac{n+1}{n'+1},$$

where the constant c has been set equal to zero.

The defining matrix f_{ik} also describes the map $f^*(m) = m'$ of a weight m of G onto a weight m' of $\tilde{G}$, as well as the embedding of the roots of the subalgebra $\tilde{G}$ in the weight space of the algebra G. We have

$$[f^*(m)]_i = m'_i = \sum_{k=1}^{n} f_{ik}\, m_k \quad ,$$

$$[f(\alpha')]_k = \sum_{i=1}^{n'} \alpha'_i\, f_{ik} \; .$$

Some properties of the index j_f and of the defining matrix f_{ik} are listed in the following. These properties can be nicely observed in the classification of the semisimple subalgebras of B_3 given in the next section.

(i) j_f is a positive integer

(ii) If $G_1 \supset G_2 \supset G_3$ are simple algebras and if j_{f_1} is the index of G_2 in G_1 and j_{f_2} is the index of G_3 in G_2, then the index j_f of G_3 in G_1 is

$$j_f = j_{f_1} \cdot j_{f_2}$$

If $f^1{}_{kt}$ is the defining matrix for G_2 in G_1 and if f^2_{ik} is the defining matrix for G_3 in G_2, then the defining matrix f_{it} of G_3 in G_1 is given as

$$f_{it} = \sum_{k=1}^{n'} f^2{}_{ik}\, f^1{}_{kt} \; , \qquad \begin{array}{l} t = 1,2,\ldots,n \\ i = 1,2,\ldots,n'' \\ n \geq n' \geq n'' \end{array}$$

where n, n' and n" are the ranks of the algebras G_1, G_2 and G_3 respectively.

(iii) If $f^1, f^2, \ldots, f^s$ are embeddings of a simple algebra $\tilde{G}$ in a simple algebra G and if

$$[f^i(\tilde{x}), f^j(\tilde{y})] = 0 \quad \text{for } i \neq j \text{ and } \tilde{x}, \tilde{y} \in \tilde{G},$$

then

$$f = f^1 + f^2 + \ldots + f^s$$

is again an embedding of the subalgebra $\tilde{G}$ in G and the index j_f is given as

$$j_f = j_{f^1} + j_{f^2} + \ldots + j_{f^s}$$

Before proceeding to the classification of the subalgebras of B_3, an important theorem, obtained by Dynkin, is listed. This theorem is of importance for an understanding of the embedding $f(\tilde{E}_\alpha)$ of the generators $\tilde{E}_\alpha$ of a semisimple subalgebra in an algebra G. This theorem states (in part):

If $\Gamma_{\alpha'}$ is the set of roots of G which project onto a root α'

of $\tilde{G}$, the embedding of the element $\tilde{E}_{\alpha'}$ of $\tilde{G}$ in G is given as

$$f(\tilde{E}_{\alpha'}) = \sum_{\alpha\varepsilon\Gamma_{\alpha'}} c_{\alpha'\alpha} E_\alpha \ , \quad c_{\alpha'\alpha} \ \varepsilon \ C, \ E_\alpha \ \varepsilon \ G$$

with

$$f(\alpha') = \sum_{\alpha\varepsilon\Gamma_{\alpha'}} |c_{\alpha'\alpha}|^2 \cdot \alpha \ ,$$

$$j_f = \sum_{\alpha\varepsilon\Gamma_{\alpha'}} |c_{\alpha\alpha'}|^2 \ ,$$

$$\bar{c}_{\alpha'\alpha} = c_{-\alpha', -\alpha} \qquad \text{(complex conjugation)}.$$

5. CLASSIFICATION OF SEMISIMPLE SUBALGEBRAS OF B_3

In this section all semisimple subalgebras of B_3 are listed together with their index and defining matrix. For the case of the S-subalgebras the embedding $f(\tilde{E}_{\alpha'})$ is given explicitely and thus the complete embeddings of all subalgebras $\tilde{G}$ of B_3 in B_3 is given. These results are obtained following the rules as compiled in ref. [4].

The system of simple (positive) roots of B_3 is

$$\pi = \{e_1 - e_2, \ e_2 - e_3, \ e_3\} \ .$$

The lowest root is $(-e_1 - e_2)$. In the first step the maximal subalgebras of B_3 are listed according to the first three steps as explained in the preceding section.

Simple algebra B_3:

(I) All r-subalgebras and their defining matrices f_{ik} are:

A_3^1: o ———— o ———— o $\quad f(\tilde{H}_i) = H_i, \ i = 1,2,3$

$e_1 - e_2 \qquad e_2 \ e_3 \qquad -e_1 - e_2$ $(e_3 - e_4)$

$$f(\tilde{E}_{e_1-e_2}) = E_{e_1-e_2}$$

$$f(\tilde{E}_{e_2-e_3}) = E_{e_2-e_3}$$

$$f(\tilde{E}_{-e_1-e_2}) = E_{-e_1-e_2}$$

However, it is customary to embed the Cartan subspace of the algebra A_n in a space with one more dimension, such that for the generators H_i of the Carton subalgebra holds $\sum_i H_i = 0$. If this is done we have the map

$$e_1 - e_2 \rightarrow e_1 - e_2$$

$$e_2 - e_3 \to e_2 - e_3$$

$$-e_1 - e_2 \to e_3 - e_4$$

and we obtain for the defining matrix of A_3 in B_3

$$f_{ik} = \frac{1}{2}\begin{pmatrix} 1 & -1 & -1 \\ -1 & 1 & -1 \\ -1 & -1 & 1 \\ 1 & 1 & 1 \end{pmatrix} \qquad \begin{array}{l} f(\tilde{E}_{(1,-1,0,0)}) = E_{(1,-1,0)} \\ f(\tilde{E}_{(0,1,-1,0)}) = E_{(0,1,-1)} \\ f(\tilde{E}_{(0,0,1,-1)}) = E_{(-1,-1,0)} \end{array}$$

$$\sum_k f_{ik} f_{ik} = \frac{3}{4} j_f = \frac{3}{4} \qquad \begin{array}{l} f(1,-1,0,0) = (1,-1,0) \\ f(0,1,-1,0) = (0,1,-1) \\ f(0,0,1,-1) = (-1,-1,0) \end{array}$$

$$j_f = 1$$

$A_1^1 + A_1^1 + A_1^2$: o ($e_1 - e_2$) o ($-e_1-e_2$) • (e_3)

We want to treat all three A_1 subalgebras equally and embed the Cartan subspace in a 6-dimensional space (of which 3 dimensions are redundant). Thus we have the map

$$e_1 - e_2 \to e_1 - e_2$$

$$-e_1 - e_2 \to e_3 - e_4$$

$$e_3 \to e_5 - e_6$$

Thus we obtain for the defining matrix of $A_1^1 + A_1^1 + A_1^2$ in B_3

$$f_{ik} = \frac{1}{2}\left(\begin{array}{ccc} 1 & -1 & 0 \\ -1 & 1 & 0 \\ \hline -1 & -1 & 0 \\ 1 & 1 & 0 \\ \hline 0 & 0 & 2 \\ 0 & 0 & 2 \end{array}\right); \qquad \begin{array}{l} f(1,-1) = (1,-1,0), \\ f(\tilde{E}_{(1,-1)}) = E_{(1,-1,0)} \\ f(1,-1) = (-1,-1,0), \\ f(\tilde{E}_{(1,-1)}) = E_{(-1,-1,0)} \\ f(1,-1) = 2(0,0,1), \\ f(\tilde{E}_{(1,-1)}) = \sqrt{2}\, E_{(0,0,1)} \end{array}$$

$$\sum_k f_{ik}\, f_{ik} = \frac{1}{2} \cdot j_f \cdot \begin{cases} 1/2 \\ 1/2 \\ 1 \end{cases}$$

Setting $f(\tilde{E}_{(1,-1)}) = f(\tilde{E}_+)$, $f(\tilde{E}_{(-1,1)}) = f(\tilde{E}_-)$ we obtain

$$[f(\tilde{E}_+),\ f(\tilde{E}_-)] = f(\tilde{H})$$

$$[f(\tilde{H}),\ f(\tilde{E}_\pm)] = \pm\ f(\tilde{E}_\pm) \quad ,$$

i.e. just the commutation relations for the algebra A_1, where $f(\tilde{H})$ is for each of the three cases obtained from the defining matrix as H_1-H_2, $-H_1-H_2$ and $2H_3$ respectively.

B^1_2 :

e_2-e_3 e_3

$$f = \begin{matrix} 0 & 1 & 0 \\ 0 & 0 & 1 \end{matrix} \quad ; \quad \begin{aligned} f(\tilde{E}_{(1,-1)}) &= E_{(0,1,-1)} \\ f(\tilde{E}_{(0,1)}) &= E_{(0,0,1)} \\ j_f &= 1 \end{aligned}$$

$A^1_1 + A^1_1$: o o

e_1-e_2 $-e_1-e_2$

$$f = \frac{1}{2}\begin{pmatrix} 1 & -1 & 0 \\ -1 & 1 & 0 \\ \hline -1 & -1 & 0 \\ 1 & 1 & 0 \end{pmatrix} ; \quad \begin{aligned} f(\tilde{E}_+) &= E_{(1,-1,0)} \\ f(\tilde{E}_+) &= E_{(-1,-1,0)} \end{aligned}$$

The defining matrix f for $A^1_1 + A^1_1$ in B_3 is obtained from the defining matrix of the subalgebra $2A^1_1 + A^2_1$ by simply taking that part relating to the subalgebra $2A^1_1$ of $2A^1_1 + A^2_1$.

$A^1_1 + A^2_1$: o ●

e_1-e_2 e_3

$$f = \frac{1}{2}\begin{pmatrix} 1 & -1 & 0 \\ -1 & 1 & 0 \\ \hline 0 & 0 & 2 \\ 0 & 0 & -2 \end{pmatrix}$$

A_2^1 :

e_1-e_2 $\qquad$ e_2-e_3

$$f(\tilde{H}_i) = H_i - \frac{1}{3}\sum_{k=1}^{3} H_k \quad , \qquad i = 1,2,3$$

The sum over $\frac{1}{3}$ of the H_i's has been added in order to satisfy the condition $\Sigma H_i = 0$.

$$f_{ik} = \frac{1}{3}\begin{pmatrix} 2 & -1 & -1 \\ -1 & 2 & -1 \\ -1 & -1 & 2 \end{pmatrix}$$

$$f(1,-1,0) = (1,-1,0)$$
$$f(0,1,-1) = (0,1,-1)$$
$$f(\tilde{E}_{(1,-1,0)}) = E_{(1,-1,0)}$$
$$f(\tilde{E}_{(0,1,-1)}) = E_{(0,1,-1)}$$

$$\sum_k f_{ik} f_{ik} = \frac{2}{3} j_f = \frac{6}{9} \quad , \; j_f = 1$$

A_1^1 : ○ e_1-e_2

$$f = \frac{1}{2}\begin{pmatrix} 1 & -1 & 0 \\ -1 & 1 & 0 \end{pmatrix}, \quad f(1,-1) = (1,-1,0)$$
$$f(\tilde{E}_+) = E_{(1,-1,0)}$$

A_1^2 : ● e_3

$$f = \begin{pmatrix} 0 & 0 & 1 \\ 0 & 0 & -1 \end{pmatrix} \qquad f(1,-1) = (0,0,2)$$
$$f(\tilde{E}_+) = \sqrt{2}\, E_{(0,0,1)}$$

The <u>maximal</u> r-subalgebras of B_3 are A_3^1, $2A_1^1 + A_1^2$.

(II) Simple maximal S-subalgebras of B_3

If for the algebra B_3 the simple maximal S-subalgebras are determined according to Dynkin's rules, one of the rare exceptions is met. The S-subalgebra A_1^{28} obtained is not maximal in B_3 but in G_2. It holds $B_3 \supset G_2^1 \supset A_1^{28}$. The algebra G_2^1 is the single maximal S-subalgebra of B_3. Its embedding in B_3 is given by

$$f_{ik} = \frac{1}{3}\begin{pmatrix} 1 & 2 & -1 \\ 1 & -1 & 2 \\ -2 & -1 & -1 \end{pmatrix};$$

$$f(1,-1,0) = (0,1,-1)$$
$$f(-\tfrac{1}{3},\tfrac{2}{3},-\tfrac{1}{3}) = \tfrac{1}{3}(1,-1,0)+\tfrac{2}{3}(0,0,1)$$
$$f(\tilde{E}_{(1,-1,0)}) = E_{(0,1,-1)}$$
$$f(\tilde{E}_{(-1,2,-1)}) = \frac{1}{\sqrt{3}}\, E_{(1,-1,0)} + \frac{\sqrt{2}}{\sqrt{3}} E_{(0,0,1)}$$

$$\sum_k f_{ik} f_{ik} = \frac{2}{3} j_f = \frac{6}{9}; \quad j_f = 1$$

The square of the longest root of G_2 has been set equal to 2, according to the usual convention.

The maximal simple S-subalgebras have the property that there exists a representation ϕ of the algebra of dimension equal to the dimension of the defining representation of the algebra which has no non-trivial invariant subspaces under the restriction ϕf to the subalgebra. In other words, the representation ϕ forms an irreducible representation of the S-subalgebra.

(III) Non-simple maximal S-subalgebras of B_3

The algebra B_3 does not have such a subalgebra.

Now, in turn, steps I to III have to be applied for all the new simple subalgebras obtained or step IV to VI for all the semi-simple subalgebras obtained. We consider first A_3^1.

Simple algebra A_3:

(I) All r-subalgebras of A_3^1:

These have already been obtained. The maximal r-subalgebras of A_3^1 are A_2^1 and $2A_1^1$.

(II) Maximal simple S-subalgebras of A_3^1.

There exists one simple maximal S-subalgebra of A_3^1. It is the algebra B_2^1. Its defining matrix with respect to A_3 is:

$$f_{ik} = \frac{1}{2} \begin{pmatrix} 1 & 1 & -1 & -1 \\ 1 & -1 & 1 & -1 \end{pmatrix}, \qquad \begin{aligned} f(1,-1) &= (0,1,-1,0) \\ f(0,1) &= \tfrac{1}{2}(1,-1,0,0)+\tfrac{1}{2}(0,0,1,-1) \\ f(\tilde{E}_{(1,-1)}) &= E_{(0,1,-1,0)} \\ f(\tilde{E}_{(0,1)}) &= \tfrac{1}{\sqrt{2}}E_{(1,-1,0,0)} + \\ &\quad + \tfrac{1}{\sqrt{2}}E_{(0,0,1,-1)} \end{aligned}$$

$$\sum_k f_{ik} f_{ik} = j_f = 1$$

(III) Maximal non-simple S-subalgebras of A_3

There exists one non-simple maximal S-subalgebra of A_3. It is the algebra $A_1^2 + A_1^2$ in A_3:

$$f_{ik} = \frac{1}{2}\begin{pmatrix} 1 & 1 & -1 & -1 \\ -1 & -1 & 1 & 1 \\ \hline 1 & -1 & 1 & -1 \\ -1 & 1 & -1 & 1 \end{pmatrix} \qquad \begin{aligned} f(1,-1) &= (1,0,-1,0)+(0,1,0,-1) \\ f(1,-1) &= (1,-1,0,0)+(0,0,1,-1) \\ f(\tilde{E}_+) &= E_{(1,0,-1,0)}+E_{(0,1,0,-1)} \\ f(\tilde{E}_+) &= E_{(1,-1,0,0)}+E_{(0,0,1,-1)} \end{aligned}$$

$$\sum_k f_{ik}\, f_{ik} = \frac{1}{2}\, j_f = 1; \quad j_f = 2$$

This is the spin-isotopic spin subalgebra used in Wigner's A_3. The defining representation $D^4(1/4(3,-1,-1,-1))$ goes over into the representation $D^{2\times 2}$ (1/2;1/2) under the restriction to $2A_1^2$.

Non-simple algebra $2A_1^1+A_1^2$:

As next example the maximal subalgebras of the non-simple algebra $2A_1^1 + A_1^2$ are discussed.

(IV) Maximal r-subalgebras of $2A_1^1 + A_1^2$.

The subalgebras $2A_1^1$ and $A_1^1 + A_1^2$ are maximal r-subalgebras of $2A_1^1 + A_1^2$. Their defining matrices with respect to B_3 have been given earlier.

(V) Maximal simple S-subalgebras of $2A_1^1 + A_1^2$.

There are none.

(VI) Maximal non-simple S-subalgebras of $2A_1^1 + A_1^2$.

There are two maximal non-simple S-subalgebra of the algebra $2A_1^1 + A_1^2$. This is a consequence of the rule (iii) given for the defining matrices and indices. For example, the algebra $A_1^1 + A_1^1$ contains an S-subalgebra A_1^2, with defining matrix f as a sum of the two defining vectors of the two A_1^1 algebras

$$f = \frac{1}{2}\begin{pmatrix} 1 & -1 & 0 \\ -1 & 1 & 0 \end{pmatrix} + \frac{1}{2}\begin{pmatrix} 1 & 1 & 0 \\ -1 & -1 & 0 \end{pmatrix} = \begin{pmatrix} 1 & 0 & 0 \\ -1 & 0 & 0 \end{pmatrix}$$

Thus, the defining matrix for the non-simple maximal S-subalgebra $A_1^2 + A_1^2$ of the algebra $2A_1^1 + A_1^2$ is given as

$$f_{ik} = \begin{pmatrix} 1 & 0 & 0 \\ -1 & 0 & 0 \\ \hline 0 & 0 & 1 \\ 0 & 0 & -1 \end{pmatrix}, \quad j_f = 2$$

We have, however, met already the algebra $2A_1^2$, as a subalgebra of A_3 and indeed the two subalgebras $2A_1^2$ are the <u>same</u> subalgebra of B_3. Thus we have the relationship

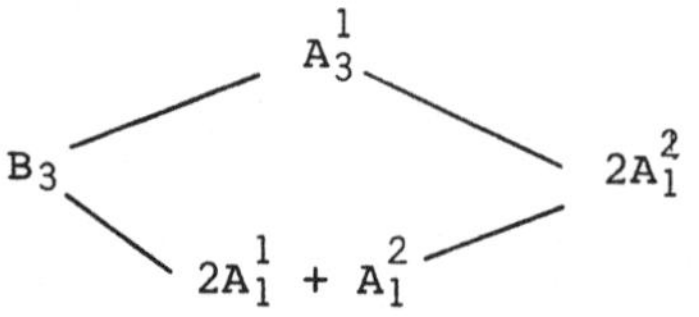

Many more such interrelationships among the subalgebras of B_3 will be found.

The defining matrices appear to be different and to contradict the statements made above. This apparent contradiction is, however, easily resolved by recognizing that we have embedded the Cartan subspace in a space with one more dimension for the case of the algebra A_3. Acting with the above defining matrix of $A_1^2 + A_1^2$ in A_3 upon the defining matrix of A_3^1 in B_3 one obtains the same defining matrix, up to equivalence, as for the subalgebra $A_1^2 + A_1^2$ of $2A_1^1 + A_1^2$ in B_3.

The other maximal non-simple subalgebra is $A_1^1 + A_1^3$. Its defining matrix in B_3 is

$$f = \frac{1}{2}\begin{pmatrix} 1 & -1 & 0 \\ -1 & -1 & 0 \\ \hline 1 & 1 & -2 \\ -1 & -1 & 2 \end{pmatrix}; \quad \sum_k f_{ik}\, f_{ik} = \frac{1}{2}\, j_f = \begin{cases} 1/2 \\ 3/2 \end{cases}$$

Subalgebra $A_1^1 + A_1^1$:

(IV) Maximal r-subalgebra: A_1^1 (in B_3); $f = \frac{1}{2}\begin{pmatrix} 1 & -1 & 0 \\ -1 & 1 & 0 \end{pmatrix}$

(V) Maximal simple S-subalgebra:

The algebra A_1^2 is a maximal simple subalgebra of $2A_1^1$. Its defining matrix has been given above and is

$$f = \frac{1}{2}\begin{pmatrix} 1 & -1 & 0 \\ -1 & 1 & 0 \end{pmatrix} + \frac{1}{2}\begin{pmatrix} +1 & +1 & 0 \\ -1 & +1 & 0 \end{pmatrix} = \begin{pmatrix} 1 & 0 & 0 \\ -1 & 0 & 0 \end{pmatrix}$$

with the index as the sum of the two indices of the two A_1 algebras, namely

$$j_f = 1 + 1 = 2$$

Of course, the index j_f could have been calculated from the defining matrix as usual.

(VI) None.

Simple algebra B_2:

(I) Maximal r-subalgebra: $2A_1^1$

(II) Maximal simple S-subalgebra: A_1^{10}
The defining matrix with respect to B_2 is given as:

$$f = \begin{pmatrix} 2, & 1 \\ -2, & -1 \end{pmatrix}, \quad \begin{aligned} f(1,-1) &= (4,2) = 4(1,-1) + 6(0,1) \\ f(\tilde{E}_+) &= 2E_{(1,-1)} + \sqrt{6}\, E_{(0,1)} \end{aligned}$$

$$\sum_k f_{ik} f_{ik} = \frac{1}{2} j_f = 5 ; \qquad j_f = 10$$

Here f is the defining matrix of A_1^{10} in B_2. The defining matrix of A_1^{10} in B_3 is then obtained as the matrix product of the defining matrices of A_1^{10} in B_2 and of B_2^1 in B_3,

$$f = \begin{pmatrix} 2 & 1 \\ -2 & -1 \end{pmatrix} \begin{pmatrix} 0 & 1 & 0 \\ 0 & 0 & 1 \end{pmatrix} = \begin{pmatrix} 0 & 2 & 1 \\ 0 & -2 & -1 \end{pmatrix} ; \quad j_f = 10.1 = 10$$

(III) None.

Simple algebra A_2:

(I) A_1^1

(II) A_1^4: $$f = \begin{pmatrix} 1 & 0 & -1 \\ -1 & 0 & 1 \end{pmatrix}, \quad \begin{aligned} f(1,-1) &= (2,0,-2) \\ &= 2(1,-1,0) + 2(0,1,-1) \\ f(\tilde{E}_+) &= \sqrt{2}\, E_{(1,-1,0)} + \sqrt{2}\, E_{(0,1,-1)} \end{aligned}$$

$$\sum_k f_{ik} f_{ik} = \frac{1}{2} j_f = 2 ; \qquad j_f = 4$$

(III) None.

Simple algebra G_2:

$$\pi = \{e_2 - e_3, \frac{1}{3}(e_1 - 2e_2 + e_3)\}$$

(I) There are two maximal r-subalgebras, A_2^1 and $A_1^1 + A_1^3$.
The defining matrix of A_2^1 in G_2 is

$$f_{ik} = \frac{1}{3} \begin{pmatrix} 2 & -1 & -1 \\ -1 & 2 & -1 \\ -1 & -1 & 2 \end{pmatrix} \qquad \begin{aligned} f(1,-1,0) &= (1,-1,0) \\ f(0,1,-1) &= (0,1,-1) \end{aligned}$$

$$\sum_k f_{ik} f_{ik} = \frac{2}{3} j_f = \frac{2}{3} ; \quad j_f = 1$$

The defining matrix of $A_1^1 + A_1^3$ in G_2 is

$$f = \frac{1}{2}\begin{pmatrix} 1 & -1 & 0 \\ -1 & 1 & 0 \\ \hline 1 & 1 & -2 \\ -1 & -1 & 2 \end{pmatrix} \qquad \begin{matrix} f(1,-1) = (1,-1,0) \\ f(1,-1) = (1,1,-2) \end{matrix}$$

$$\sum_k f_{ik}\, f_{ik} = \frac{1}{2}\, j_f = \begin{cases} 1/2 \\ 3/2 \end{cases}$$

(II) Maximal simple S-subalgebra: A_1^{28} (in G_2)

$$f = \begin{pmatrix} 2 & 1 & -3 \\ -2 & -1 & 3 \end{pmatrix}$$

$$\sum_k f_{ik}\, f_{ik} = \frac{1}{2}\, j_f = 14 \,,$$

The defining matrix of A_1^{28} in B_3 is ($A_1^{28} \subset G_2^1 \subset B_3$):

$$f = \begin{pmatrix} 2 & 1 & -3 \\ -2 & -1 & 3 \end{pmatrix} \cdot \frac{1}{3}\begin{pmatrix} 1 & 2 & -1 \\ 1 & -1 & 2 \\ -2 & -1 & -1 \end{pmatrix} = \begin{pmatrix} 3 & 2 & 1 \\ -3 & -2 & -1 \end{pmatrix}$$

(III) None.

The remaining subalgebras are simple to handle and will be treated in summary:

Maximal subalgebras of

$A_1^2 + A_1^2$: A_1^2, A_1^4

$A_1^1 + A_1^2$: A_1^1, A_1^2, A_1^3

$A_1^1 + A_1^3$: A_1^1, A_1^3, A_1^4

The defining matrices of these subalgebras are easily found.

The diagram given in Figure 1 depicts graphically the classification of the semisimple subalgebras of the algebra B_3 as well as the inclusion relations among the subalgebras.

As was mentioned in the introduction some of the algebras which appear in this diagram are applied in physics. The algebra B_3 is used in the shell model of atomic physics and its orbital angular momentum subalgebra is A_1^{28}. It is interesting to note that there is no other chain from B_3 to A_1^{28} except through G_2^1. The physically relevant algebra for Wigner's theory of supermultiplets in nuclear physics is the subalgebra $A_1^2 + A_1^2$ of A_3^1 which is one of the two subalgebras of type $A_1 + A_1$. These two subalgebras are distinct through the branching properties of the representations of A_3^1. Finally, the algebra A_2 contains two subalgebras of type A_1. It is the subalgebra A_1^4 which is used in the Elliott model of nuclear physics while A_1^1 is the isotopic spin subalgebra of par-

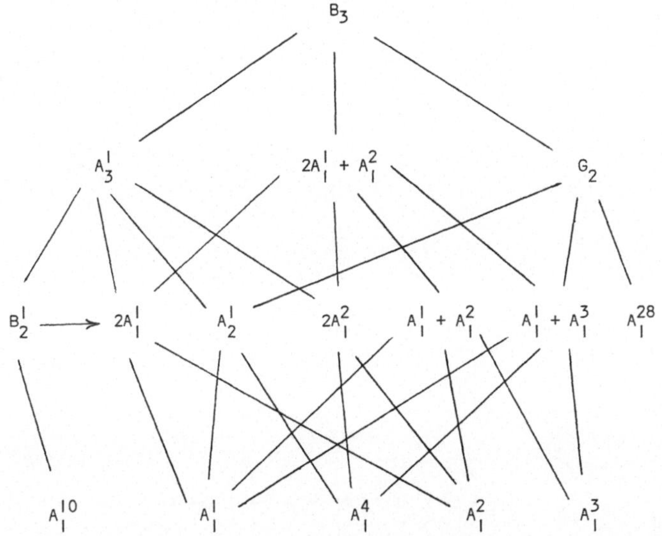

Fig. 1. Chains of subalgebras of B_3 and inclusion relations.

ticle physics.

REFERENCES

1. E. Dynkin, Am. Math. Soc. Transl. Ser. 2, 6, 111 (1957).
2. E. Dynkin, Am. Math. Soc. Transl. Ser. 2, 6, 245 (1957).
3. A.I. Malcev, Am. Math. Soc. Transl. No. 33 (1950).
4. M. Lorente and B. Gruber, Journ. Math. Phys. (Oct. 1972).

EXTERNAL (KINEMATICAL) AND INTERNAL (DYNAMICAL) CONFORMAL SYMMETRY AND DISCRETE MASS SPECTRUM

A.O. Barut
International Centre for Theoretical Physics,
Trieste, Italy
and University of Colorado,
Boulder, Colo. U.S.A.

ABSTRACT. A framework is presented in which both the space-time conformed transformations and the dynamical conformal transformations on internal co-ordinates of the relativistic two-body problem occur. Mass spectrum is discussed.

1. INTRODUCTION

There are two distinct ways in which the conformal group has been used in particle physics (aside from theories in curved space and general relativity):

(1) as a kinematical space-time symmetry group. Here the conformal group contains the Poincaré group and generalizes the relativistic invariance by dilatations and special conformal transformations.

(2) as a dynamical group acting on the internal co-ordinates of a quantum system at rest. Here the generators have an entirely different interpretation, and the only connection with the Poincaré group consists in the identification of the spin parts of both groups.

The purpose of this investigation is to propose a larger framework in which both kinematics and dynamics naturally occur together, and to derive a relativistic mass quantization principle.

2. CONFORMAL TRANSFORMATIONS ON EXTERNAL CO-ORDINATES

2.1 What is the conformal group?

The fundamental role played by the Poincaré group in relativistic quantum theory is well known. In fact, the very concept of a relativistic system is based on the representations of the Poincaré group characterized by the invariants mass and spin. The group of conformal transformations contains in addition to the transformations of relativity

$$P : x'_\mu = \Lambda_\mu{}^\nu x_\nu + a_\mu , \tag{2.1}$$

also the dilatations

$$D : x'_\mu = \rho x_\mu \tag{2.2}$$

and the special conformal transformations

$$\begin{aligned} C : x'_\mu &= \sigma^{-2}(x)\ [x_\mu + c_\mu x^2] , \\ \sigma^2\ (x) &= 1 + 2c_\mu x^\mu + c^2 x^2 . \end{aligned} \tag{2.3}$$

The transformations (2.3) can be obtained from inversions I: $x'_\mu = k(x_\mu/x^2)$ and translations T: $x'_\mu = x_\mu + a_\mu$, in the form ITI.

The transformations (1.3) are actually only well-defined in a compactified Minkowski space, or in the six-dimensional space (see Subsection 2.3).

The conformal group is the largest group preserving the light cone: $x'^2 = x^2/\sigma^2(x)$. Hence $x^2 = 0$ implies $x'^2 = 0$. It is also the smallest semi-simple group containing the Poincaré group.

2.3.1 Mathematical properties. We list a few important mathematical results on the conformal group that we shall need.

(i) The conformal group of the Minkowski space, the groups SO(4,2) and SU(2,2) are locally isomorphic:

$$C_M \overset{loc}{\simeq} SU(2,2) \overset{loc}{\simeq} SO(4,2) . \tag{2.4}$$

(ii) The generators of the transformations (2.1) - (2.3), P_μ, $M_{\mu\nu}$, D and K_μ form a basis of the 15-dimensional Lie algebra with the commutation relations

$$\begin{aligned} &[M_{\mu\nu}, P_\lambda] = -(g_{\mu\lambda}P_\nu - g_{\nu\lambda}P_\mu), \\ &[M_{\mu\nu}, M_{\sigma\rho}] = g_{\mu\rho}M_{\nu\sigma} + g_{\nu\sigma}M_{\mu\rho} - g_{\mu\sigma}M_{\nu\rho} - g_{\nu\rho}M_{\mu\sigma}, \\ &[P_\mu, D] = P_\mu, \\ &[M_{\mu\nu}, D] = 0 , \quad [P_\mu, P_\nu] = 0 , \quad [K_\mu, K_\nu] = 0, \\ &[P_\mu, K_\nu] = 2(g_{\mu\nu}D - M_{\mu\nu}) , \\ &[M_{\mu\nu}, K_\lambda] = -(g_{\mu\nu}K_\nu - g_{\nu\lambda}K_\mu) , \\ &[K_\mu, D] = -K_\mu . \end{aligned} \tag{2.5}$$

(iii) The local isomorphisms SU(2,2) ≃ SO(4,2) associates to

4 x 4 complex matrices $U \in SU(2,2)$, 6 x 6 real matrices $0 \in SO(4,2)$ by

$$O^A{}_B = \frac{1}{2}\,\mathrm{Tr}(U\Sigma_B U^\dagger \Sigma^A) \;, \qquad (2.6)$$

where Σ_A, A = 1 ... 6, are the analogue of Pauli matrices in six-dimensional space. Equation (2.6) is the basis of spinor calculus for the conformal group, counterpart of the SL(2,C)-spinor calculus formulae: $L^\nu_\mu = \frac{1}{2}\,\mathrm{Tr}(A\sigma_\mu\, A^+\, \tilde{\sigma}^\nu)$, $A \in SL(2,C)$, $L \in SO(3,1)$.

2.2 Conformal invariance of massless wave equations

The history of the conformal group goes back to the observation that free Maxwell equations are invariant under special conformal transformations and dilatations [1]. The notion of invariance here is a bit general than the form invariance. The wave operator $\Box$ is not form-invariant under special conformal transformations but goes over into

$$\Box \to \sigma^2(x)\Box \qquad (2.7)$$

Hence, only on the space of solutions of the wave equation $\Box\,\phi = 0$, we have conformal invariance. In other words, the space of solutions of the wave equation form a representation space of the conformal group. For this spin-0 massless equation the infinitesimal generators are given by

$$\begin{aligned}
&\text{Translations} && P_\mu = \partial_\mu \;; \\
&\text{Lorents transformations} && M_{\mu\nu} = x_\mu\partial_\nu - x_\nu\partial_\mu \;; \\
&\text{Dilatations} && D = x^\nu\partial_\nu \;; \\
&\text{Special conformal transformations} && K_\mu = 2x_\mu x^\nu\partial_\nu - x^2\partial_\mu \;.
\end{aligned} \qquad (2.8)$$

and we obtain

$$\begin{aligned}
&[D, \Box\,] = \lambda\Box \\
&[K_\mu, \Box\,] = -4x_\mu\Box
\end{aligned} \qquad (2.9)$$

Consequently, it follows that if ϕ_0 is a solution of $\Box\,\phi = 0$, then $K_\mu\phi_0$ and $D\phi_0$ are also solutions.

In a similar way one shows that massless wave operators for all spin values, $\gamma^\mu\partial_\mu$, etc. with the form of generators suitably generalized from (2.8), have similar properties as (2.9). Hence the solutions of the corresponding wave equations provide representations of the conformal group [2]. The special role of m = 0 wave equations can also be seen algebraically by the fact that $P_\mu P^\mu$ does not commute with D or K_μ, unless trivially for zero eigenvalue. For example, for the case given in (2.8) one finds

$$e^{ic_\nu K^\nu} P_\mu P^\mu e^{-ic_\nu K^\nu} = \sigma^2(x) P_\mu P^\mu . \quad (2.10)$$

In an irreducible representation of the conformal group the spectrum of $P_\mu P^\mu$ (which is an element of the enveloping algebra of the Lie algebra) consists either of the single point 0, or the half-real lines $-\infty < m^2 < 0$, $0 < m^2 < \infty$. A physical interpretation of this point will be given in the next section.

Because the solutions of the $m = 0$ wave equations provide us with representations both of the conformal group and of the Poincaré group P, it follows that the relevant representations of SO(4,2) remain irreducible for the subgroup P. Physically this means that no new quantum numbers are introduced by the larger symmetry.

2.3 Physical interpretation. Six-dimensional world

There is no general agreement on the physical interpretation of conformal transformations.

It is natural to interpret the dilatations D, Equation (2.2), as changes of the unit of measurements of space-time intervals. The special conformal transformations have been sometimes interpreted as transformations to accelerated frames. But this interpretation meets with difficulties [3], although formally or accidentally it has that form. A consistent interpretation of special conformal transformations seems to be to view them as space-time-dependent changes of scales. Such a change of scale is not usually carried out by experimentalists in the laboratory, but we may do it in the theoretical laboratory (passive sense), and Nature may do it*: proper units at proper space-time points. In fact, the idea is that if we contemplate a space-time dependent changes of units, the description of Nature will be simple: we may establish equivalence between events which otherwise will look completely unrelated. This is really what a larger symmetry group should establish. For example, by proper scale changes we may map a Kepler orbit of one energy into another orbit of different energy.

With this interpretation we introduce two new co-ordinates κ and λ. κ tells us the unit that we have chosen and $\lambda \equiv \kappa x^2$ something about the change of units from point to point. We introduce dimensionless space-time co-ordinates $\eta^\mu = \kappa x^\mu$ and take the 6-dimensional space $(\eta^\mu, \kappa, \lambda)$ as the physical space. The conformal group turns out to be linearly represented in the 6-space. For example, the special conformal transformations have the form

$$C : \begin{aligned} \eta'_\mu &= \eta_\mu + c_\mu \lambda \\ \kappa' &= 2c_\nu \eta^\nu + \kappa + c^2\lambda \\ \lambda' &= \lambda \end{aligned} \quad (2.11)$$

* Contemplate, for example, a slowly varying gravitational field surrounding the system.

Proof: From the first Equation of (2.11) and definition of λ:

$$\kappa' x'_\mu = \kappa(x_\mu + c_\mu x^2)$$

The second equation gives

$$\kappa' = \kappa(1 + 2c_\nu x^\nu + c^2 x^2)$$

Hence

$$x'_\mu = \frac{\kappa}{\kappa'}(x_\mu + c_\mu x^2) = \sigma^{-2}(x)(x_\mu + c_\mu x^2) ,$$

which is Equation (2.3). QED.

Further we have

$$\eta^\mu \eta_\nu - \kappa\lambda = 0 \tag{2.12}$$

which is the equation of a cone in the 6-dimensional space.

Thus both physically and mathematically we are led to the 6-dimensional space, where the conformal transformations are well defined globally as linear transformations. Another method of giving a mathematical meaning to transformations (2.3) would be to compactify the Minkowski space, i.e. making out of the infinity a definite manifold, a point, line or a cone. Then one could precisely specify which points are mapped at infinity to where. The generators of SO(4,2), as differential operators in the 6-dimensional space are

$$L_{ab} = \eta_a \partial_b - \eta_b \partial_a \quad . \tag{2.13}$$

Changing variables into $\kappa = -\ell_0(\eta_4 + \eta_6)$, $\lambda = -(1/\ell_0)(\eta_4 - \eta_6)$, where ℓ_0 is a fundamental length, and then into $x_\mu = (1/\kappa)\eta_\mu$ and eliminating λ in favour of $s = \eta_\mu \eta^\mu - \kappa\lambda$ one obtains for κ = const. precisely the form (2.8).

2.4 Conformally invariant massive equations

We have seen that the conformal transformations acting on space-time co-ordinates are symmetry formations only for zero mass particles and that for a single massless particle this is a trivial extension of the Poincaré symmetry in the sense that no new quantum numbers are introduced. Can we get non-trivial results from conformal symmetry?

According to the physical interpretation of the conformal transformations that we have adopted, namely as space-time dependent scale changes, we should also transform other basic physical quantities with dimensions, like mass, in the equations. If we allow m^2 to transform like $P_\mu P^\mu$ (cf. Equation (2.10)) then massive wave equations, e.g. $(\Box + m^2)\,\phi(x) = 0$, are all invariant under conformal transformations. The change of $m^2 \to \sigma^2(x) m^2$ is simply interpreted as the measurement of mass in different units. That is why mass has a continuous spectrum. It is then possible to introduce a dimensionless fixed mass for each particle (conformally invariant mass) and attribute the continuous values of $P_\mu P^\mu$ to its dimension [4].

It may be argued that the change of scale of everything cannot give any new physical information. However, the requirement of invariance under (special) conformal transformations is much more than changing the units of physical quantities; it allows to relate seemlingly different situations, as discussed in Subsection 2.3. As an example, we discuss massive, spin $\frac{1}{2}$, conformally invariant wave equations [5].

The smallest linear spin $\frac{1}{2}$ wave equations in 6-dimensional space, η_a, a = 1 ... 6, (cf. Subsection 2.3) must use an 8-dimensional spin space and is of the form

$$(i\beta^a\partial_a - m)\Psi(\eta) = 0 \quad , \tag{2.14}$$

where β's are 8 x 8 matrices and m is the dimensionless conformally invariant mass.

The standard description in the Minkowski space is obtained if we take the projection of this equation on the hypersurface κ = const. We then get two subspaces of solutions. In one subspace, letting $\psi = \binom{u_1}{u_2}$ we have the usual Dirac equation

$$(\gamma^\mu P_\mu - m)u_1 = 0 \text{ and } u_2 = \gamma^5 u^1 \quad , \tag{2.15}$$

But in the second subspace, letting $\psi = \binom{v_1}{v_2}$, we get

$$(\gamma^\mu P_\mu - mn\,\frac{(1-2n\theta-\theta^2)}{(1+\theta^2)})\; v_1(p) = \frac{2m\theta}{1+\theta^2}\,(1-\gamma^5)v_1(p) \tag{2.16}$$

where n = ±1 is the sign of energy and θ is a parameter related to the p^λ component of the momentum. Thus we have an apparent parity-violating interaction. It is natural to associate the first subspace with the electron states and the second with the muon states. Similarly, the single massless equation accounts for the four known neutrino states. The simplest conformally invariant spinor equation provides therefore a complete unified description of leptons and their quantum numbers.

3. CONFORMAL TRANSFORMATIONS ON INTERNAL CO-ORDINATES

In this application the representations of the conformal group are used to classify the states of a quantum system in its rest frame. The Poincaré group is not involved as a subgroup; the physical interpretations of the generators are different.

(a) The simplest example is provided by the Dirac equation

$$(\gamma^\mu P_\mu - m)\psi = 0 \quad . \tag{3.1}$$

The algebra of the Dirac matrices gives 15 independent matrices which are the elements of the Lie algebra of the group SO(4,2) in a four-dimensional representation:

$$L_{ab} = \frac{1}{2} i \gamma_a \gamma_b ; \quad a < b; \quad a,b = 1 \dots 6$$
$$\gamma_a = (\gamma_1, \gamma_2, \gamma_3, -\gamma_5, \gamma_0, -iI) \tag{3.2}$$

These generators are interpreted as follows: L_{ij} = spin $\vec{J}$, L_{i4} = "Lenz vector" $\vec{A}$, L_{i5} = Lorentz boot operators $\vec{M}$, electromagnetic current operator $L_{\mu 6} = \gamma_\mu$, and the dilatation operator L_{45}. In the rest frame of the Dirac particle Equation (3.1) becomes

$$(\gamma_0 M - m)\psi(0) = 0 \tag{3.3}$$

Hence the four-dimensional representation classifies all the rest states: two spin-$\frac{1}{2}$ particles with $n = \pm 1$, n = eigenvalue of γ_0.

The Poincaré group comes into play via the wave Equation (3.1); the latter allows us to construct induced representations of the Poincaré group from the four-dimensional representation of SO(4,2).

(b) The second example is the two-body bound system with electromagnetic interactions. We have here operators similar to Dirac matrices but in an infinite-dimensional representation of SO(4,2)

$$\vec{J}\ , \quad \vec{A}\ , \quad \vec{M}\ , \quad \Gamma_\mu\ , \ D\ .$$

The relativistic composite system is described by an infinite-component wave equation

$$(J^\mu P_\mu - M)\psi = 0\ , \tag{3.4}$$

where

$$J_\mu = \alpha_1 \Gamma_\mu + \alpha_2 P_\mu + \alpha_3 P_\mu \Gamma_4$$
$$M = \beta \Gamma_4 + \gamma\ ,$$

with minimal coupling, and describes the composite system as though it were an elementary system endowed with internal degrees of freedom.

In order to give an idea how this description is obtained, consider the Schrödinger equation in the rest frame in terms of relative co-ordinates

$$(\frac{1}{2m} P^2 - \frac{\alpha}{r} - E)\psi = 0 \tag{3.5}$$

We multiply this equation by r

$$(\frac{1}{2m} rp^2 - \alpha - Er)\psi = 0 \tag{3.6}$$

Observe now that the following three operators

$$L_{56} = \frac{1}{2}(rp^2 + r) \equiv \Gamma_0$$

$$L_{46} = \frac{1}{2}(rp^2 - r) \equiv \Gamma_4 \qquad (3.7)$$

$$L_{46} = \vec{r}\cdot\vec{p} - i \quad = D$$

close to the Lie algebra of SO(2,1).

This observation allows us to write the Schrödinger equation linear in the generators of SO(2,1)

$$[(\frac{1}{2m} - E)\Gamma_0 + (\frac{1}{2m} + E)\Gamma_4 - \alpha]\psi = 0 \qquad (3.8)$$

This equation can easily be solved using the Lie algebra (3.7). A covariant generalization of this equation leads to Equation (3.4).

We give here the representation of the remaining generators of SO(4,2) in addition to (3.7):

$$\vec{J} = \vec{r} \times \vec{p}$$

$$\vec{A} = \frac{1}{2}\vec{r}p^2 - \vec{p}(\vec{r}\cdot\vec{p}) - \frac{1}{2}\vec{r}$$

$$\vec{M} = \frac{1}{2}\vec{r}p^2 - \vec{p}(\vec{r}\cdot\vec{p}) + \frac{1}{2}\vec{r} \qquad (3.9)$$

$$\vec{\Gamma} = r\vec{p}$$

The Lie algebra SO(4,2) constitutes a dynamical algebra for the system in the sense that a single representation of the algebra accounts for all states of the systems with their multiplicities, and that electromagnetic interactions are linear in the Lie algebra generators.

It is remarkable that the representations (3.7) - (3.9) are obtained from the conformal algebra in the momentum space (x_μ and p_μ interchanged in Equation (2.8)), by putting $\underline{x^2 = 0}$, i.e. $x_0 = r$, $x_i = r_i$, $\partial/\partial x^0 = 0$, $\partial/\partial x^i = p_i$. This fact will be used in Section 4.

The generalizations of (3.7) - (3.9) to include spin or magnetic charge are also known [6]. The formalism has been applied extensively to relativistic treatment of atoms [7] as well as hadrons [8].

4. THE CONNECTION BETWEEN THE EXTERNAL AND INTERNAL CONFORMAL ALGEBRAS. DISCRETE MASS SPECTRUM [9]

We start from the (relativistic) two-body problem. Let η_1^a and η_2^a be the dimensionless co-ordinates of the (spinless) particles in the six-dimensional space and define centre of mass and relative co-ordinates by

$$Y \equiv w_1\eta^{(1)} + w_2\eta^{(2)} \quad , \quad \eta \equiv \eta^{(1)} - \eta^{(2)} \tag{4.1}$$

and the corresponding conjugate variables by

$$Q \equiv q^{(1)} + q^{(2)} \quad , \quad q = w_2 q^{(1)} - w_1 q^{(2)} \tag{4.2}$$

The generators of the conformal group are then

$$\bar{L}_{ab} = L_{ab}(Y,Q) + \ell_{ab}(\eta,q) \tag{4.3}$$

We now impose the conditions

$$Y^a Y_a = 0 \ ; \quad \eta_a^{(i)}\eta^{(i)a} = 1, \ i = 1,2 \tag{4.4}$$

This implies ($w_1 \neq 0$, $w_2 \neq 0$) in the Minkowski frame ($K_1 \neq 0$, $K_2 \neq 0$)

$$(x_1 - x_2)^2 = 0 \ .$$

This condition is also evident from the Fokker-Tetrode-Schwarzschild action principle which is essentially conformally invariant, and can be interpreted as the propagation of signals with the velocity of light. It is remarkable that the conformal invariance leads to this condition.

We now pass from Y^a, η^a to the Minkowski-space co-ordinates X^μ, x_μ and the conjugate P_μ, p_μ and write Equation (4.3) in terms of the co-ordinates

$$\begin{aligned} \bar{L}_{\mu\nu} &= L_{\mu\nu} + \ell_{\mu\nu} \\ \bar{P}_\mu &= P_\mu + p_\mu \\ \bar{K}_\mu &= K_\mu + k_\mu \\ \bar{D} &= D + d \end{aligned} \tag{4.6}$$

Transforming (4.6) with $S = e^{iux^\mu p_\mu}$ we obtain

$$\begin{aligned} L'_{\mu\nu} &= S^{-1}\bar{L}_{\mu\nu}S = L_{\mu\nu} + \ell_{\mu\nu} \\ P'_\mu &= P_\mu \\ K'_\mu &= K_\mu + k_\mu + 2u(X_\mu d - x^\nu \ell_{\nu\mu}) \\ D' &= D + d \end{aligned} \tag{4.7}$$

The Casimir operator for the algebra (4.6) or (4.7) gives

$$\begin{aligned} Q^2 &= \frac{1}{2}\bar{L}_{ab}\bar{L}^{ab} \\ &= Q^2_{ext} + Q^2_{int} - \frac{1}{2}(p_\mu k^\mu + k^\mu p_\mu) + 2i(2-u)d + k_\mu P^\mu \end{aligned} \tag{4.8}$$

For the internal algebra we use the representation (3.7) - (3.9), because of the condition $x^2 = 0$ (cf. (4.5)). For this representation $Q_{int} = -3$. Evaluating (4.8) in the rest frame and factoring out r we obtain

$$\frac{1}{2r} Q^2 = -r\, \pi_r^2 - \frac{c}{r} + \frac{1}{2} M \qquad (4.9)$$

The operator $(1/r^2)Q^2$ has a simple discrete spectrum. We solve the eigenvalue equation

$$(\frac{1}{2} Q^2 + r^2\lambda^2)\Psi = 0 \qquad (4.10)$$

or

$$(r\, \pi_r^2 - \frac{c}{r} - \frac{M}{2} - \lambda^2 r)\Psi = 0$$

In algebraic form

$$[(1 + \lambda^2)\Gamma_0 + (1 - \lambda^2)\Gamma_4 - \frac{1}{2} M]\Psi = 0 \qquad (4.11)$$

Letting $\Psi = e^{i\theta D}\phi$ and suitably choosing θ (tilting operation) we have

$$(2\lambda\ \Gamma_0' - \frac{1}{2} M)\psi = 0 \qquad (4.12)$$

The spectrum of Γ_0' in the discrete series of representation of SO(2,1) is

$$\Gamma_0' : s + \frac{1}{2} + [(j + \frac{1}{2})^2 - c^2]^{\frac{1}{2}}, \quad s = 0, 1, 2, \ldots \qquad (4.13)$$

Hence the result is the linear mass spectrum

$$M = 4\lambda\ [s + \frac{1}{2} + [(j + \frac{1}{2})^2 - c^2]^{\frac{1}{2}}], \qquad s + 0, 1, 2, \ldots \text{ for } c^2 < (j + \frac{1}{2})^2 \qquad (4.14)$$

It is remarkable that the six-dimensional framework leads us to the condition (4.5) and further to an infinite-dimensional wave Equation (4.11) and a mass spectrum (4.14). Our framework automatically incorporates something like an 1/r potential and gives us the bound-state spectrum.

REFERENCES

1. H. Bateman, Proc. London Math. Soc. 8, 228, 469 (1910).
E. Cunningham, Proc. London Math. Soc. 10, 77 (1910).
2. P.A.M. Dirac, Ann. Math. 37, 419 (1936).
L. Gross, J. Math. Phys. 5, 687 (1964).
For a more recent discussion see R. Kotecký and J. Niederle,

Ann. Phys. (NY) (1973).
3. H. Kastrup, Acta Phys. Austr., Suppl. V, 452 (1968).
4. A.O. Barut and R. Haugen, Ann. Phys. (NY) 71, 519 (1972).
5. A.O. Barut and R. Haugen, Conformally Invariant Massive and Massless Spinor Equations, Pts I and II, Nuovo Cimento A (1973).
6. A.O. Barut and G. Bornzin, J. Math. Phys. (NY) 12, 841 (1971).
7. A.O. Barut and W. Rasmussen, The H-Atom as a Relativistic Elementary Particle, Pts I and II, J. Phys. B 6, 1695 and 1713 (1973).
8. For example, A.O. Barut, Acta Phys. Austr., Suppl. (1973) and references therein.
9. In this section we follow some recent work, A.O. Barut and G. Bornzin, J. Math. Phys. (1974).

NON-LINEAR PROBLEMS IN TRANSPORT THEORY

Paul Zweifel
Virginia Polytechnic Institute and
State University,
Blacksburg, Virginia

INTRODUCTION

This material was presented in a series of four lectures to the NATO Advanced Study Institute on "Applications of Group Theory to Non-Linear Problems" held in Istanbul, Turkey, August 8-11, 1972. In conformity with the spirit of a summer school, very little of the material is new. Only the discussion in Lecture 2, of the "generalized spectrum" has not already been published elsewhere.

The author expresses his appreciation to the Kernforschungszentrum Karlsruhe for allowing him to participate in the school, and to Prof. Asim Barut, director of the school, for the invitation to attend.

1. A NON-LINEAR TRANSPORT EQUATION

Perhaps the most familiar example, in physics, of a non-linear theory is the kinetic theory of gases. Since the kinetic behavior of the gas molecules is determined by collisions whose frequency is proportional to the square of the gas density, it is clear why the theory is non-linear. The Boltzmann equation is the standard mechanism through which gas kinetics is described, and, as is well known, this equation is indeed non-linear. Admittedly, the standard way to "solve" this equation involves linearization, usually accomplished by expanding the gas distribution function in a series about some equilibrium distribution, and ignoring all except linear terms in "small quantities", a small quantity being defined as the difference between the equilibrium and the actual distribution. Incidentally, the existence of an equilibrium distribution is guaranteed by the H-theorem.

It is possible to think of physical situations described by a linear Boltzmann equation. A famous example is the "foreign gas" problem. In this problem, a very dilute sample of some active gas (i.e., it makes lots of collisions) is introduced into a background of an inert gas (one which makes few collisions) and one tries to determine what happens to the foreign gas. An example is the case of neutron diffusion in a nuclear reactor, in which the neutrons ($\rho \lesssim 10^9 cm^{-3}$) play the role of the dilute gas while the moderator ($\rho \sim 10^{23} cm^{-3}$) is the host. Because of the low neutron density, neutron-neutron collisions can legitimately be ignored and, for the same reason, the moderator never departs from its equilibrium distribution. Loosely speaking, any moderator atom "zapped" by a neutron is very unlikely to be "rezapped" before it has made several collisions with other moderator atoms, and re-entered the equilibrium distribution from which the collision with the neutron removed it.

One important situation in which the equations describing neutron transport are truly non-linear, however, should be noted. In a nuclear reactor there is generation of heat in an amount proportional to the neutron flux. This heat, in turn, effects the equilibrium moderator distribution. In a stationary situation (constant heat production and removal) the situation is still linear. If, however, the neutron density is changing, as in the case of reactor startup, shutdown, etc., there is a feedback mechanism through the moderator temperature, and one sees that the reactor kinetics equations are non-linear. Although the linearization process described earlier frequently works for this situation, reliable stability analysis requires non-linear effects to be taken into account. Considerable effort has been expended in the study of the non-linear problems of reactor dynamics, involving rather sophisticated mathematics. Since group theoretical methods, however, have not yet been applied, we shall not delve further into a study of these problems.

Besides neutrons, photons in a stellar atmosphere can also be considered an example of a foreign gas problem which, also because of feedback effects, is highly non-linear. I should like to discuss this problem in some detail, and I begin by defining notation: Denote by ω the photon angular frequency (i.e., $E = \hbar\,\omega$) and by $\psi_\omega(x,\mu)$ the photon angular density. That is

$$2\pi\ \psi_\omega(x,\ \mu)dx\ d\mu$$

represents the number of photons of frequency ω between x and x+dx with x-component of velocity between μ and $\mu + d\mu$ (we set $c = 1$). We also are assuming azimuthal and plane symmetry.)*

Photons may "scatter" from atoms in the stellar atmosphere

---

* The assumption that a star may be described by one dimension, x, is valid because of the large radius of curvature compared with distances of interest within the atmosphere.

or be "absorbed". In this context, "scatter" means scattering without change of frequency, i.e., elastic scattering. Processes which change the photon frequency are considered to be absorptions followed by re-emission. We define a scattering mean free path $\lambda_{s\omega}(x)$ and an absorption mean free path $\lambda_{a\omega}(x)$ and define absorption and scattering coefficients k and σ:

$$\lambda_{a\omega}(x) = \frac{1}{\rho(x)k_\omega} ,$$

$$\lambda_{s\omega}(x) = \frac{1}{\rho(x)\sigma_\omega} .$$

We have assumed, in writing these expressions, that the relative composition of the atmosphere is constant, so that the spatial dependence of the mean free paths depends only upon the density, $\rho(x)$. We also define a scattering frequency $f(\underset{\sim}{v}'\cdot\underset{\sim}{v})$, which describes the probability that a photon, incident upon an atom with velocity $\underset{\sim}{v}'$ will scatter into velocity $\underset{\sim}{v}$. Finally, if we define $S_\omega(x)$ to be the source of (re-emitted) photons - the original source is in the center of the star, not in its atmosphere - we can write the following Boltzmann equation for $\psi_\omega(x)$:

$$\mu \frac{\partial\psi_\omega(x, \mu)}{\partial x} + \rho(x)[k_\omega + \sigma_\omega]\psi_\omega(x, \mu)$$
$$= \rho(x)\sigma_\omega \int_1^1 \psi_\omega(x, \mu') f(\underset{\sim}{v}'\cdot\underset{\sim}{v})d\underset{\sim}{v}' + S_\omega(x) .$$

It is conventional to make two changes of variable. First, the optical thickness z is defined by

$$z = \int_0^x \rho(x')dx' .$$

Second, the energy density

$$I_\omega(x, \mu) = \hbar\omega \; \psi_\omega(x, \mu)$$

is introduced. Then the transport equation takes the form

$$\mu \frac{\partial I_\omega}{\partial z} (z, \mu) + (k_\omega + \sigma_\omega)I_\omega(z, \ \ )$$
$$= \sigma_\omega \int I_\omega(z, \mu') f(\underset{\sim}{v}'\cdot\underset{\sim}{v})d\underset{\sim}{v}' + \frac{S_\omega \hbar_\omega}{\rho} .$$

The standard way to treat this equation is to introduce the assumption of "Local Thermodynamic Equilibrium" (LTE). In other words, it is assumed that every point of the stellar atmosphere can be characterized by a local temperature, T(z). In thermodynamic "equilibrium" one has detailed balance between emission and absorption, i.e.

$$\hbar\omega \; S_\omega(z) = \rho(z)k_\omega \; B_\omega \; (T(z))$$

where $B_\omega(T(z))$ is the Planck distribution:

$$B_\omega(T(z)) = \frac{\hbar\omega^3}{2\pi^2}(e^{\hbar\omega/kT}-1)^{-1} .$$

The condition of L.T.E. is a high density approximation, the basic physical assumption being that after an atom has absorbed a photon, it makes sufficiently many collisions before reemission to reenter the equilibrium distribution. Thus, emitted photons always appear in the equilibrium, i.e., the Planck distribution.

Thus, the radiant energy transport equation becomes

$$\mu \frac{\partial I_\omega}{\partial z}(z, \mu) + (k_\omega + \sigma_\omega)I_\omega$$
$$= \sigma_\omega \int I_\omega(z, \mu')f(\underset{\sim}{v}'\cdot\underset{\sim}{v})dv' + k_\omega B_\omega(T(z)) .$$

The problem is to solve for the temperature as a function of optical depth and also for the emergent angular distribution of photons. Before discussing the solution, we introduce the so-called Schwarzschild condition of radiative equilibrium. This requires that the temperature distribution be time-independent, and that all heat transport be by radiation. In other words, the star is in a steady state neither heating nor cooling. This, in turn, implies that the net energy transport across any plane perpendicular to the z axis must be constant

$$\frac{\partial F(z)}{\partial z} = 0$$

where $F(z)$, the so-called net flux, is proportional to the energy current density:

$$F(z) = \frac{1}{\pi}\int_{-1}^{1} d\mu \int_0^\infty d\omega\, \mu I_\mu(z, \mu) .$$

We now integrate the transport equation over $d\underset{\sim}{v}$ and $d\omega$. By virtue of the Schwartzschild condition, the first term vanishes. Also, the scattering frequency is a probability distribution, i.e.,

$$\int f(\underset{\sim}{v}'\cdot\underset{\sim}{v})d\underset{\sim}{v}' = 1 .$$

Thus, we obtain

$$\int d\omega\, 4\pi(k_\omega + \sigma_\omega)J_\omega = \int d\omega[4\pi\sigma_\omega J_\omega + 4\pi\, k_\omega B_\omega(T(x))] ,$$

where we have introduced a further notation, J_ω, called the average intensity

$$J_\omega = \frac{1}{2}\int_{-1}^{1} d\mu I_\omega(z, \mu) .$$

Thus, we find, indepent of σ_ω

$$\int_0^\infty d\omega k_\omega J_\omega = \int_0^\infty d\omega k_\omega B_\omega(T(z)) \ .$$

We can consider this equation, along with the transport equation, simultaneous equations for the unknowns $I_\omega(z, \mu)$ and $T(z)$. This set is, we see, highly nonlinear and, in fact, solutions have been found for only rather special cases. Even numerical solutions are very difficult. For example, an interative procedure suggests itself. A temperature distribution, $T(z)$ is assumed. Then the transport equation can be solved for I_ω, a new temperature distribution calculated from the Schwartzschild condition, and the procedure iterated. C.E. Siewert (unpublished) has carried their procedure out for a rather simple physical model and found that convergence rate to be so slow as to make the method essentially worthless. More sophisticated numerical techniques have been developed and, in some cases, have proved useful. However, I should like to consider one simple model which is available to analytical solution. This is the so-called "grey" atmosphere:

$$k_\omega = k = \text{constant}$$

$$\sigma_\omega = 0 \ .$$

We can now integrate the transport equation over $d\omega$ obtaining ($I \equiv \int_0^\infty I_\omega \, d\omega$):

$$\mu \frac{\partial I(z, \mu)}{\partial z} + kI(z, \mu) = k \int_0^\infty B_\omega(T(z))d\omega$$

$$= k\alpha T^4(z) \ ,$$

where α is the Stefan-Boltzmann constant divided by 2π. Furthermore, the Schwarzschild condition reduces to

$$\alpha T^4(z) = \frac{1}{2} \int_{-1}^{1} d\mu \; I(z, \mu) \ .$$

Thus, the following procedure is convenient. Solve the transport equation

$$\mu \frac{\partial I(z, \mu)}{\partial z} + kI(z, \mu) = \frac{k}{2} \int_{-1}^{1} d\mu \; I(z, \mu)$$

for I, then deduce the temperature from the Schwarzschild equation above, relating T. and J. This particular problem has a name: the grey Milne Problem.

2. GENERAL PROPERTIES OF THE SOLUTION

We now discuss the solution of the transport equation. First, a word about boundary condition. We have a physical situation like that shown in Figure 1.

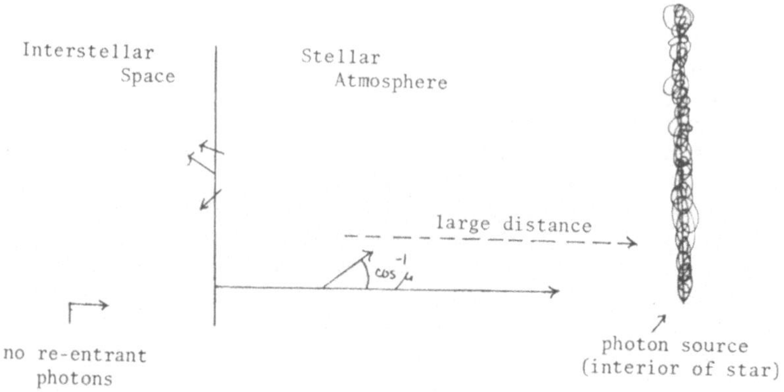

Clearly,

$$I(0, \mu) = 0 \, , \quad 0 < \mu \leq 1 \, .$$

At "infinity", i.e., the center of the star, the energy density diverges. We assume it approaches infinity less than exponentially, however; that is, for every positive number,

$$\underset{z \to \infty}{L} \; e^{-\varepsilon z} I(z, \mu) = 0 \, .$$

Returning now to the equation for $I(z, \mu)$, we note that the translation group in one dimension is an invariance group for the operator B:

$$BI = \mu \frac{\partial I}{\partial z} + kI(z, \mu) - \frac{k}{2} \int_{-1}^{1} d\mu I(z, \mu) \, .$$

This suggests that the solutions should form the bases for irreducible representations of the translation group. Thus, take

$$I = \zeta_{\nu}(\mu) e^{-x/\nu}$$

We obtain for $\zeta_\nu(\mu)$ the equation

$$\left(k - \frac{\mu}{\nu}\right) \zeta_{\nu}(\mu) = \frac{k}{2} \int_{-1}^{1} d\mu \zeta_{\nu}(\mu) \, ,$$

or

$$\left(1 - \frac{\mu}{\nu}\right) \zeta_{\nu}(\mu) = \frac{1}{2} \int_{-1}^{1} d\mu \, \zeta_{\nu}(\mu')$$

where, without loss of generality, we have set $k = 1$.

Our general program will be to seek eigensolutions of this equation; to expand the solution to the Milne problem in terms of these eigensolutions, and to determine the expansion coefficients from the boundary conditions enumerated above. The above equation can be cast in canonical form

$$O\zeta_\nu = \frac{1}{\nu}\zeta_\nu$$

with

$$Of = \frac{1}{\mu}f - \frac{1}{\mu}\int_{-1}^{1} f(\mu')d\mu'$$

and our first problem is to determine the spectrum of operator O. This operator O has a number of obnoxious properties, since it is

1) unbounded
2) non self-adjoint ($O \neq O^*$)
3) and, in fact, not normal ($OO^* \neq O^*O$).

The existence of the adjoint requires, incidentally, that O be densely defined. We assign

Problem 1. The operator O is densely defined on $L_2(-1,1)$.

The study of the spectrum $\sigma(O)$ is less convenient than the study of the generalized spectrum $\sigma^g(O)$ as considered for example by Kuščer and Vidav (J. Math. Anal. Appl. 25, 80, 1969). The theory of generalized spectrum may be sketched as follows. Consider a linear transformation $T; B \to B$, where B is a Banach space, with T unbounded and $T^* \neq T$ (naturally we require $\overline{D(T)} = B$, where $D(T)$ is the domain of T). Suppose further that T can be decomposed as follows:

$$T = RS ,$$

where R^{-1} and S are bounded. Then the eigenvalue equation for O may be recast in the form

$$S\,\zeta_\nu - \frac{1}{\nu}R^{-1}\zeta_\nu = 0 .$$

Then, the generalized point spectrum $\sigma_p^g(O)$ is defined by

$$\sigma_p^g(O) = \{\frac{1}{\nu}: \omega = 0 \in \sigma_p(S - \frac{1}{\nu}R^{-1})\} .$$

Similarly, the continuous spectrum is defined by

$$\sigma_c^g(O) = \{\frac{1}{\nu}: \omega = 0 \in \sigma_c(S - \frac{1}{\nu}R^{-1})\} .$$

The following results are left as exercises:

Problem 2. $\sigma(O) \subseteq \sigma^g(O)$.

Problem 3. Let $\frac{1}{\nu} \in \sigma_p(O)$ with eigenvector ζ_ν. Then $\frac{1}{\nu} \in \sigma_p^g(O)$ with generalized eigenvector ζ_ν, and conversely.

Problem 4. $\sigma_c(O) \subseteq \sigma_c^g(O)$

Proof: If $\frac{1}{\nu} \varepsilon \sigma_c(0)$ and $\varepsilon > 0$, then there exists a unit vector f_ε and some vector h such that

$$(0 - \frac{1}{\nu})\, f_\varepsilon = h \tag{16}$$

and

$$\|h\| < \varepsilon / \|R^{-1}\| \,. \tag{17}$$

Then

$$\|(S - \frac{1}{\nu} R^{-1}) f_\varepsilon\| = \|R^{-1}(0 - \frac{1}{\nu})\, f_\varepsilon\| = \|R^{-1}h\| \leqslant \|R^{-1}\|\ \|h\| < \varepsilon \,. \tag{18}$$

So $(S - \frac{1}{\nu} R^{-1})^{-1}$ is unbounded.

Finally, we must show that if $(0 - \frac{1}{\nu})^{-1}$ is densely defined then $(S - \frac{1}{\nu} R^{-1})^{-1}$ is also densely defined. Assume by way of contradiction that $\overline{D((0 - 1/\nu)^{-1})} = \overline{R(0 - 1/\nu)} = X$, but that $\overline{R(S - \lambda R^{-1})} \subsetneq X$. Since $\overline{D(R)} = X$ there exists at least one vector, call it x, in $D(R) \setminus \overline{R(S - 1/\nu\ R^{-1})}$. Furthermore, $D(R) \cap R(S - \frac{1}{\nu} R^{-1}) \neq \phi$ since $R(S - \frac{1}{\nu} R^{-1}) = 0 - \lambda$ is densely defined. Thus there exists a positive number m such that $\|x - y\| > m$ for every vector $y \,\varepsilon\, D(R) \cap R(S - \frac{1}{\nu} R^{-1})$. Now choose $\varepsilon > 0$. Since $\overline{R(0 - 1/\nu)} = X$ there exists a vector $z \,\varepsilon\, R(0 - \lambda)$ such that $\|Rx - z\| \leq \varepsilon$. Furthermore, there exists $z_o \,\varepsilon\, D(0 - 1/\nu)$ such that $z = (0 - \frac{1}{\nu}) z_o$. We define $z_1 = (S - \lambda R^{-1}) z_o \,\varepsilon\, D(R) \cap R(S - \frac{1}{\nu} R^{-1})$. It follows that $\varepsilon > \|Rx - z\| = \|Rx - Rz_1\| = \|R\| \cdot \|x - z_1\| \geq \|R\|\, m$. Hence we conclude that R is not bounded below, which contradicts the fact that A is bounded above.

This theorem is very useful. It means that one can carry out the usually simpler calculation of σ_c^g instead of σ_c; each element of σ_c^g must then be checked to determine if it is indeed in σ_c, but at least this "checking" need not be carried out for the entire complex plane.

Problem 5. Let

$$S\zeta_\nu = \frac{1}{\nu} R^{-1} \zeta_\nu$$

and

$$S^*\zeta_{\nu'} = \frac{1}{\nu'} R^{-1*}\zeta_\nu$$

Then ζ_ν and $\zeta_{\nu'}$ are orthogonal in the sense that either $\nu = \nu'$ or

$$(\zeta_\nu, R^{-1}\zeta_{\nu'}) = 0$$

This may be referred to as a generalized orthogonality relation. There is also a generalized Ritz variational principle which, since we do not use it, we do not write down.

The above problems indicate that the generalized spectrum and the spectrum may not coincide.

Problem 6. Let 0 be the transport operator

$$0 = \frac{1}{\mu} - \frac{1}{2\mu}\int_{-1}^{1} d\mu' \quad .$$

Then $\sigma_c(0) = \sigma_c^g(0)$. Detailed proofs of the results, problems 1-8 are to be published elsewhere.

In the subsequent analysis, we shall calculate the generalized spectrum of the reduced transport operator.

We thus seek values of ν such that

$$(1 - \frac{\mu}{\nu})\zeta_\nu(\mu) = \frac{1}{2}\int_{-1}^{1} d\mu\zeta_\nu(\mu) \quad .$$

First, assume $\mu \notin [-1,1]$ so that the factor $(1 - \frac{\mu}{\nu})$ is invertible. A simple calculation indicates that the generalized eigenvalues ν (or $\frac{1}{\nu}$) obey the equation

$$\Lambda\ (\nu) = 1 - \frac{1}{\nu}\int_{-1}^{1}\frac{d\mu}{1-\mu/\nu} = 0 \quad .$$

This equation has the (degenerate) solution $\nu = \pm\infty$, i.e., $1/\nu = 0$. Thus, the one-dimensional representation $(e^{-z/\nu})$ will not suffice. Consider then the two-dimensional representations. We know that an Abelian group has only one-dimensional unitary irreducible representations. But what about non-unitary representations? Consider the matrices

$$T(a) = \begin{bmatrix} 1 & a \\ 0 & 1 \end{bmatrix} e^{-a/\nu} \quad .$$

Clearly

$$T(a)T(b) = T(a+b) \quad ,$$

So the T(a) are a two-dimensional, non-unitary, irreducible representation of the translation group.

Let us introduce a little more generality, although we do not need it in our present context. Let $\phi_1(z)$, $\phi_2(z)$,... $\phi_n(z)$ be a set of basis vectors for an irreducible n-dimensional representation. Then to each translation of distance we can associate the operator T(a) such that

$$T(a)\ \phi_i = \phi_i(z + a) \quad .$$

Since the ϕ_i are basis vectors, we know that

$$T(a)\phi_i = \sum_{j=1}^{n} \phi_j(z)T_{ij}(a),\ i = 1,\ldots n$$

where the matrix $T_{ij}(a)$ is the representation matrix. This set of equations can be solved in two ways. The most general way is to introduce the infinitesimal generator, thus obtaining differential equations for the ϕ_j (see Case and Zweifel, Linear Transport Theory, Addision-Wesley, p 290ff). A more direct way is to use the representation matrix directly. In the two dimensional case, for example, we obtain immediately

$$\phi_1(z + a) = [\phi_1(z) + a\phi_2(z)]e^{-a/\nu}$$

$$\phi_2(z + a) = e^{-a/\nu}\phi_2(z) \ .$$

Thus

$$\phi_2(z) = e^{-z/\nu}$$

and

$$\phi_1(z) = ze^{-z/\nu}$$

(The three-dimensional representation matrices are

$$\begin{bmatrix} 1 & a & \frac{a^2}{2} \\ 0 & 1 & a \\ 0 & 0 & 1 \end{bmatrix} e^{-z/\nu}$$

and one easily works out

$$\phi_1(z) = e^{-z/\nu}$$

$$\phi_2(z) = ze^{-z/\nu}$$

$$\phi_3(z) = \frac{z^2}{Z} e^{-z\nu}$$

and so forth.

Problem 7. Obtain the non-unitary representation matrices of the translation group in two and three space dimensions.

In our particular case, we have determined $\nu = \infty$. Thus we can conclude that there are two eigensolutions of the transport equation which must be linear combination of ϕ_1 and ϕ_2. That is

$$I = a_1(\mu) + a_2(\mu)z \ .$$

Inserting into the transport equation, we find

$$\mu a_2(\mu) + a_1(\mu) + a_2 z = \frac{1}{2}\int_{-1}^{1} a_1(\mu)d\mu + \frac{z}{2}\int_{-1}^{1} a_2(\mu')d\mu' \ .$$

Equating powers of z, we find

$$a_2(\mu) = \frac{z}{2}\int_{-1}^{1} a_2(\mu')d\mu' \quad ,$$

i.e., $a_2(\mu)$ = constant,
and

$$a_1(\mu) + \mu a_2(\mu) = \frac{1}{2}\int_{-1}^{1} a_1(\mu')d\mu' \quad .$$

Suppose we choose $a_2(\mu) = 0$. Then

$$\int_{-1}^{1} a(\mu')d' \neq 0 \text{ (otherwise } I = 0) \; .$$

Thus we can normalize $\int_{-1}^{1} a_1(\mu')d\mu' = 1$. This implies $a_1 = 1/2$, or we have one solution

$$I_1(z,\mu) = 1/2 \; .$$

If $a_2(\mu) \neq 0$, normalize so that

$$\int_{-1}^{1} a_2(\mu')\, \mu' = 1 \; .$$

Then

$$a_2 = 1/2 \; .$$

Thus

$$a_1(\mu) = -\frac{\mu}{2} + C \; .$$

Where C is another constant.

Thus

$$I_2 = \frac{1}{2}(z - \mu) \; ,$$

where we have made the simple choice C = 0.

The generalized continuous spectrum turns out to be the interval $\nu \in [-1,1]$ on the real line (i.e., $\frac{1}{\nu} \in [1,\infty) \cup [-1,-\infty)$). This follows immediately from applying the Weyl Theorem to the generalized operator $\mu 0 = 1 - \frac{\mu}{\nu} - \frac{c}{2}\int_{-1}^{1} d\mu'$ and noting that the integral term is a compact perturbation (this is an example of the frequent simplicity obtainable from the generalized spectrum). Having thus treated the reduced transport operator, we now know the eigensolutions of the transport equation

$$\mu \frac{\partial I}{\partial z} + I = \frac{1}{2}\int_{-1}^{1} I(z,\mu')d\mu' \; .$$

Summarizing:

The transport equation has two eigenvalues, $\nu = \pm\infty$ with corresponding eigenvectors

$$I_1 = \frac{1}{2}$$

$$I_2 = \frac{1}{2}(z - \mu)$$

and a continuous spectrum $\nu \in [-1,1]$.

In the next lecture we will use these results to solve the Milne problem.

3. SOLUTION OF THE MILNE PROBLEM

We now seek solutions of the equation

$$BI = \frac{\partial I}{\partial z} + I - \frac{1}{2}\int_{-1}^{1} I(z,\mu')d\mu' = 0$$

$$I(0,\mu) = 0 \, , \quad 0 < \mu \leq 1$$

$$I \to \infty \quad \text{as} \quad z \to \infty \, , \quad \text{but } e^{-\varepsilon}I \to 0 \forall \varepsilon > 0 \, .$$

We have separated variables according to

$$I_\nu(\mu) = e^{-z/\nu}\zeta_\nu(\mu)$$

with

$$0\,\zeta_\nu = \frac{1}{\mu}\zeta_\nu - \frac{1}{2\mu}\int_{-1}^{1}\zeta_\nu(\mu')d\mu' = \frac{1}{\nu}\zeta_\nu \, .$$

The generalized spectrum consists of two eigenvalues, $\nu = \pm\infty$ (i.e., $\frac{1}{\nu} = 0$ is a doubly-degenerate eigenvalue) with associated eigenvectors $I_1(z, \mu) = 1/2$; $I_2(z, \mu) = 1/2(z-\mu)$. We recall the dispersion relation

$$\Lambda(z) = 1 - \frac{z}{2}\int_{-1}^{1}\frac{d\mu}{z-\mu} = 0$$

where Λ, as a function of the complex variable z, has the following properties

$\Lambda(z) \in A$ on the complex plane cut from -1 to +1 on the real line;

$$\Lambda(z) \to 1 \text{ as } z \to \infty \, .$$

The boundary values of $\Lambda(\mu)$ on the cut obey

$$\Lambda^{\pm}(\mu) = \lambda(\mu) \pm \pi\frac{i\mu}{2} \, ,$$

where

$$\lambda(\mu) = 1 - \frac{\mu}{2} P \int_{1}^{1}\frac{d\mu'}{\mu-\mu'}$$

Problem 8. Derive the expressions for $\Lambda^{\pm}$.

Furthermore, we have a continuous spectrum $\sigma_c^g(0) = [-1,1]$. With no attempt to be rigorous (although a rigorous treatment is in fact possible) we introduce generalized functions as eigensolutions corresponding to the continuous spectrum. Writing

$$(1 - \frac{\mu}{\nu})\zeta_\nu - \frac{1}{2}\int_{-1}^{1} \zeta_\nu(\mu')d\mu' = (\nu - \mu)\zeta_\nu - \frac{\nu}{2} = 0 ,$$

where we have normalized

$$\int_{-1}^{1} \zeta_\nu(\mu')d\mu' = 1$$

we find

$$\zeta_\nu = \frac{\nu}{2} P \frac{1}{\nu-\mu} + \eta(\nu)\delta(\nu - \mu)$$

as the generalized solution for ζ_ν, as may be verified by direct substitution (noting $x\delta(x) = 0$). The "arbitrary" function $\eta(\nu)$ may be found from the normalization condition:

$$\eta(\nu) = 1 - \frac{\nu}{2} P \int_{-1}^{1} \frac{d\mu}{\nu-\mu} \equiv \lambda(\nu)$$

where $\lambda(\nu) = \frac{1}{2}(\Lambda^+ + \Lambda^-)$ has been defined above.

Now that we have a set of eigensolutions, we are equipped to expand the Milne solution in terms of them, and try to fit the expansion coefficients to the boundary condition. We proceed as follows. The solutions I_1, I_2, and $e^{-z/\nu}\zeta_\nu$ for $\nu \geq 0$ all satisfy the homogeneous transport equation, and obey the boundary condition at infinity. Then, the general solution will be some linear combination of these "eigensolution"

$$\Phi_M(z, \mu) = I_1(z, \mu) + aI_2(z, \mu) + \int_0^1 A(\nu)e^{-z/\nu}\zeta_\nu(\mu)\, d\mu ,$$

where a and $A(\nu)$ are "expansion coefficients". The boundary condition at z = 0 gives us

$$-I_1(0, \mu) = aI_2(0, \mu) + \int_0^1 A(\nu)\zeta_\nu(\mu)d\mu$$

or

$$\frac{1}{2}\mu = \frac{1}{2}a + \int_0^1 A(\nu)\zeta_\nu(\mu)d\mu .$$

Let us denote $\frac{1}{2}\mu - \frac{1}{2}a$ by $\psi(\mu)$ and try to solve the equation

$$\psi(\mu) = \int_0^1 A(\nu)\zeta_\nu(\mu)d\mu$$
$$= P\int_0^1 \frac{\nu}{2}\frac{A(\nu)d\nu}{\nu-\mu} + \lambda(\mu)A(\mu) .$$

We will seek solutions $A(\nu)$ which are Schwartz distributions ε K' (i.e., linear functionals on $K = \{f(\mu):f$ is infinitely differentiable$\}$.

The standard way to solve such a singular integral equation is to introduce the Hilbert transform (see N.I. Muskhelishvili, Singular Integral Equations, Noordhoff, Groningen, Holland, 1953). This is defined by

$$N(z) = \frac{1}{2\pi i} \int_0^1 \frac{\nu}{2} \frac{A(\nu)}{\nu - z} \, d\nu \ .$$

Then $N(z)$ has the following properties, which we know to hold for the Cauchy integrals of Schwartz distributions:

1) $N(z) \in A$ in the complex plane cut from $[0,1]$
2) $N(z) \to 0$ as $\frac{1}{|z|}$ at ∞,
3) $N^+ + N^- = \frac{1}{\pi} P \int_1^1 \frac{\nu}{2} \frac{A(\nu)}{\nu - z} \, d\nu$,

and the "inversion formula" for Hilbert transforms:

4) $N^+ - N^- = \frac{\nu}{2} A(\nu)$.

We now eliminate A from our singular integral equation, to obtain

$$\psi(\mu) = \pi i (N^+ + N^-) + \frac{1}{2} (\Lambda^+ + \Lambda^-) \frac{2}{\nu} [N^+ - N^-]$$

or, since $\pi i \nu = \Lambda^+ - \Lambda^-$,

$$\frac{\nu}{2} \psi(\mu) = \frac{1}{2} (\Lambda^+ - \Lambda^-)(N^+ + N^-) + \frac{1}{2} (\Lambda^+ + \Lambda^-)(N^+ - N^-)$$

or

$$\frac{\Lambda^+ - \Lambda^-}{2\pi i} \psi(\mu) = \Lambda^+ N^+ - \Lambda^- N^- \ .$$

If Λ and N had the same branch cuts, we would be finished, because we could view this as an equation between the boundary values of an analytic function, and could write

$$\Lambda N = \frac{1}{2\pi i} \int_{-1}^1 \frac{\Lambda^+(\mu) - \Lambda^-(\mu)}{2\pi i} \psi(\mu) \frac{d\mu}{\mu - z} \ .$$

However, Λ is cut from $[-1,1]$ and N is cut from $[0,1]$. Thus, a different procedure must be followed. Specifically, we seek a function $X(z)$ whose boundary values are in the same ratio as those of Λ, but which has the right cut, i.e., $[0,1]$. (This process is sometimes called the "Wiener-Hopf factorization of Λ".)

To clarify the procedure, we divide the Hilbert equation through by Λ^-, obtaining

$$\left(\frac{\Lambda^+}{\Lambda^-} - 1 \right) \frac{\psi(\mu)}{2\pi i} = \frac{\Lambda^+}{\Lambda^-} N^+ - N^- \ .$$

We now introduce a function $X(z)$ with a branch cut from 0 to +1, such that

$$\frac{X^+(z)}{X^-(z)} = \frac{\Lambda^+(z)}{\Lambda^-(z)} \quad , \quad z \in [0,1] \ .$$

Assuming momentarily that such a function can be found, our Hilbert equation becomes

$$N^+X^+ - N^-X^- = \gamma(\mu)\psi(\mu) \ ,$$

where we have introduced the abbreviation

$$\gamma(\mu) = \frac{1}{2\pi i}\left[X^+(\mu) - X^-(\mu)\right] \ .$$

Then, since N and X have the same branch cuts, a solution to the Hilbert equation is

$$N(z) = \frac{1}{X(z)}\,\frac{1}{2\pi i}\int_0^1 \frac{\gamma(\mu)\psi(\mu)}{\mu - z}\,d\mu \ .$$

From the form of this solution, it is seen that $X(z)$ must be analytic and non-vanishing in the cut plane. For example, if $X(z_0) = 0$ for some z_0 in the cut plane, $N(z)$ would have a pole at z_0 which, according to property 1) above ascribed to $N(z)$ is not allowed. A possible choice for $X(z)$ is the function $X_0(z)$ defined by

$$X_0(z) = \exp\left[\frac{1}{2\pi i}\int_0^1 \frac{d\mu'}{\mu' - z}\ln\frac{\Lambda^+(\mu')}{\Lambda^-(\mu')}\right] \ .$$

Since $\Lambda^+(\mu) = \overline{\Lambda^-(\mu)}$ this can be written

$$X_0(z) = \exp\left\{\frac{1}{2\pi i}\int_0^1 \frac{d\mu'}{\mu' - z}\,\theta(\mu')\right\} \ ,$$

$$\theta(\mu) = \arg \Lambda^+(\mu) \ .$$

The function $X_0(z)$ is clearly analytic and non-vanishing in the cut plane, except perhaps at the endpoints 0,1 (and, of course, obeys the ratio condition). We assign as

Problem 9. $X_0(z) \to$ Const, $z \to 0$
$X_0(z) \to (1 - z)$, $z \to 1$.

Thus $X_0(z)$ has a zero at $z = 1$, so that $N(z)$ has a pole at $z = 1$. But $z = 1$ must be a branch point of $N(z)$, not a pole. However, the function

$$X(z) = \frac{X_0(z)}{1 - z}$$

satisfies the ratio condition, and meets the appropriate analyticity conditions for $N(z)$.

We finally investigate the behavior of $N(z)$ as $z \to \infty$. Since

$$\int_0^1 \frac{\gamma(\mu)\psi(\mu)}{\mu - z}\,d\mu \to \frac{1}{z}$$

for large z, and since $X(z) \to z$, it appears that $N(z) \to$ Const as $z \to \infty$, rather than $\frac{1}{z}$, as property 2) requires. We "fix up" this behavior by bringing in the discrete expansion coefficient a. In particular, using

$$\frac{1}{\mu - z} = -\frac{1}{2}\left[1 + \frac{\mu}{z} + \ldots\right]$$

we see that if

$$\int_0^1 \gamma(\mu)\psi(\mu)d\mu = 0 \ ,$$

then N(z) will, after all, have the right asymptotic behavior at infinity. Recalling that

$$\psi(\mu) = \frac{1}{2}(\mu - a)$$

this requirement fixes the value of a:

$$a = \frac{\int_0^1 \mu\gamma(\mu)d\mu}{\int_0^1 \gamma(\mu)d\mu}$$

The solution is now, in principle, obtained (numerical evaluation will be described in lecture 4). We have

$$\psi_M(z, \mu) = \frac{1}{2}(z - \mu) + \frac{a}{2} + \int_0^1 A(\nu)\zeta_\nu(\mu)e^{-z/\nu}d\nu \ ,$$

where, we recall, $A(\nu)$ must be calculated from N(z) by

$$A(\nu) = \frac{2}{\nu}\left[N^+(\nu) - N^-(\nu)\right] \ .$$

The <u>average intensity</u> is

$$J(z) = \frac{1}{2}\int_1^1 d\mu\psi_M(z,\mu)$$

$$= \frac{z}{2} + \frac{a}{2} + \frac{1}{2}\int_0^1 A(\nu)e^{-z/\nu}d\nu \ .$$

(we have used the fact that ζ_ν is normalized as

$$\int_{-1}^1 \zeta_\nu(\mu)d\mu = 1 \ .)$$

The asymptotic solution, far from the boundary at $z = 0$, is given by

$$J_{as}(z) = \frac{1}{2}(z + a) \ .$$

The <u>extrapolated endpoint</u>, z_0, is the distance beyond the boundary at which the asymptotic distribution extrapolation to zero. We see

$$z_0 = a \ .$$

The temperature distribution is, of course, proportional to the fourth root of $J(z)$, as we have seen in lecture 1. Finally, the "law of darkening" is the name applied to the energent angular distribution:

$$\Psi_{em}(0,\mu) = \frac{1}{2}(z-\mu) + \frac{a}{2} + \int_0^1 A(\nu)\frac{c\nu}{2}\frac{d\nu}{\nu-\mu}, \quad \mu < 0,$$

where we have taken advantage of the fact that $\zeta_\nu(\mu)$ is regular for $\mu < 0$, $\nu > 0$.

4. EXPLICIT EVALUATION OF THE MILNE PROBLEM SOLUTION

A more or less formal solution to the Milne problem was obtained in Lecture 3. In today's concluding lecture we give some insight into how this formal solution may be converted into practical (i.e., numerical) results. Since time is so short, we cannot be comprehensive, by any means, and refer the interested reader to the previously cited work, Linear Transport Theory (Case and Zweifel) for further details.

The basic idea is to try to express all results in terms of two transcendental functions, nearly the X-function, introduced in Lecture 3, and the function

$$N(\nu) = \nu\Lambda^+(\nu)\,\Lambda^-(\nu)$$

$$= \nu/g(c, \nu) .$$

These functions are widely tabulated (the X-function, for example, in Linear Transport Theory, Appendix L, and $g(c, \nu)$ in Introduction to the Theory of Neutron Diffusion by Case, de Hoffmann and Placzek [U.S. Gov't Printing Office, 1953]. Furthermore, a function closely related to $X(z)$ namely Chandrasekhar's H-function, is widely tabulated in its own right and the various expansion of interest in the Milne problem can just as easily be expressed in terms of H as X.

We began by proving a number of identities.

Case's Identity A.

$$X(z) = \int_0^1 \frac{\gamma(\mu)}{\mu-z}d\mu .$$

(The function $\gamma(\mu)$, introduced in the previous lecture, was defined as

$$\gamma(\mu) = \frac{1}{2}[X^+(\mu) - X^-(\mu)] .)$$

Proof: From Cauchy's theorem, we can write

$$X(z) = \frac{1}{2\pi i}\oint_{C_1+C_2}\frac{X(z')}{z'-z}dz' ,$$

where C_1 is a contour enclosing the branch cut [0,1], while C_2 is a contour at infinity ($X(z)$, we recall, was analytic in the cut plane.) However, $X(z) \to \frac{1}{z}$ as $z \to \infty$. Thus, the integral over C_2 vanishes, and we have

$$\begin{aligned}X(z) &= \frac{1}{2\pi i} \int_{C_1} \frac{X(z')}{z'-z} dz' \\ &= \frac{1}{2\pi i} \int_0^1 \frac{X^+(z')dz'}{z'-z} + \int_1^0 \frac{X^-(z')dz'}{z'-z} \\ &= \frac{1}{2\pi i} \int_0^1 \frac{X^+(z')-X(z')}{z'-z} dz' \\ &= \int_0^1 \frac{\gamma(\mu)}{\mu-z} d\mu \ .\end{aligned}$$

Case's Identity B. $X(z)X(-z) = 3\Lambda(z)$.

Proof: Consider

$$R(z) = \frac{\Lambda(z)}{X(z)X(-z)}$$

and calculate

$$\begin{aligned}R^+(\mu) - R^-(\mu) &= \frac{\Lambda^+(\mu)}{X^+(\mu)X(-\mu)} - \frac{\Lambda^-(\mu)}{X^-(\mu)X(-\mu)} , \quad \mu > 0 \\ &= 0\end{aligned}$$

($X(-\mu)$ is continuous for $\mu > 0$). Similarly, for $\mu < 0$, $X(\mu)$ is continuous, and we again calculate

$$R^+(\mu) - R^-(\mu) = 0, \quad \mu < 0 \ .$$

Thus $R(z)$ is an entire function and hence, by Liouville's Theorem, it is a constant. We evaluate it at infinity

$$\begin{aligned}\underset{z\to\infty}{L} R(z) &= \underset{z\to\infty}{L} \frac{\Lambda(z)}{X(z)X(-z)} \\ &= \underset{z\to\infty}{L} \frac{\Lambda(z)}{-1/z^2}\end{aligned}$$

Since $X(z) \sim \frac{1}{1-z}$ at infinity. Furthermore

$$\begin{aligned}\Lambda(z) &= 1 - \frac{z}{2} \int_{-1}^1 \frac{d\mu}{z-\mu} \\ &= 1 - \frac{1}{2} \int_{-1}^1 \frac{d\mu}{1-\mu/z} \\ &= 1 - \frac{1}{2} \int_{-1}^1 d\mu \left(1 + \frac{\mu}{z} + \frac{\mu^2}{z^2} + \dots \right)\end{aligned}$$

$$\sim -\frac{1/3}{z^2} \text{ for large } z.$$

Thus

$$R(z) \to 1/3 \text{ as } z \to \infty \text{ or}$$

$$R(z) = \frac{\Lambda(z)}{X(z)X(-z)} = 1/3 \ .$$

Case's Identity C.

This identity is really a non-linear, non-singular integral equation which can be solved iteratively for numerical evaluation of the X-function. Since this a conference in non-linear problems of physics, it is perhaps appropriate to mention that non-linear equations are used extensively in transport theiry, Identity C being only one special example. (See S. Chandrasekhar, Radiative Transfer, Dover Publications, New York, 1966).

We begin with identity A

$$X(z) = \int_0^1 \frac{\gamma(\mu')}{\mu'-z} d\mu'$$

and note that

$$\gamma(\mu) = \frac{1}{2}[X^+ - X^-]$$

$$= \frac{1}{2\pi i} X^- \left[\frac{X^+}{X^-} - 1\right]$$

$$= \frac{1}{2\pi i} X^- \left[\frac{\Lambda^+}{\Lambda^-} - 1\right]$$

$$= \frac{1}{2\pi i} X^- \left[\frac{\Lambda^+}{\Lambda^-} - 1\right]$$

$$= \frac{1}{2\pi i} \frac{X^-}{\Lambda^-} [\Lambda^+ - \Lambda^-]$$

$$= \frac{\mu}{2} \frac{X^-}{\Lambda^-} = \frac{\mu}{2} \frac{X^+}{\Lambda^+}$$

since $\Lambda^+ - \Lambda^- = \pi i \mu$. Thus

$$X(z) = \int_0^1 \frac{\mu}{2} \frac{X^-}{\Lambda^-} \frac{d\mu}{\mu - z} \ .$$

Now, from Identity B

$$\frac{X^-}{\Lambda^-} = \frac{3}{X(-\mu)} \ ,$$

so

$$X(z) = \int_0^1 \frac{\mu}{2} \frac{3}{X(-\mu)} \frac{d\mu}{\mu - z}$$

or, changing variables

$$X(z) = \frac{3}{2} \int_{-1}^{0} \frac{\mu d\mu}{X(\mu)(\mu+z)} \quad .$$

This is the first form of the non-linear equation for $X(z)$. Note that if $X(\mu)$ is known on the interval $[-1,0]$, it is known everywhere.

It is customary to subtract $X(0)$ from both sides of the integral equation. From Identity B

$$X(0) = \sqrt{3}\Lambda(0)$$

$$= \sqrt{3}$$

Since $\Lambda(0) = 1$.
Thus

$$X(z) - X(0) = \frac{3}{2} \int_{-1}^{0} \frac{\mu d\mu}{X(\mu')} \left[\frac{1}{\mu+z} - \frac{1}{\mu}\right]$$

or

$$X(z) = \sqrt{3} - \frac{3z}{2} \int_{-1}^{0} \frac{d\mu}{X(\mu')(\mu'+z)}$$

The Chandrasekhar H-function incidentally is related to X through

$$H(z) = \frac{\sqrt{3}}{X(z)} \quad .$$

The final step in the analyses is to express the solution to the Milne problem in terms of $X(\mu)$. First

$$\gamma(\mu) = \frac{\mu}{2} \frac{X^-(\mu)}{\Lambda^-(\mu)} \quad .$$

From Identity B

$$\frac{X^-(\mu)}{\Lambda^-(\mu)} = \frac{3}{X(-\mu)}$$

so that

$$\gamma(\mu) = \frac{3\mu}{2} \frac{1}{X(-\mu)} \quad .$$

Thus the discrete coefficient

$$a = \frac{\int_0^1 \gamma(\mu)\mu d\mu}{\int_0^1 \gamma(\mu) d\mu}$$

becomes

$$a = \frac{\int_0^1 \frac{d\mu}{X(-\mu)}}{\int_0^1 \frac{\mu}{X(-\mu)}\, d\mu}$$

The expression for the continuum coefficient, A(ν) and the law of darkening, require somewhat more analysis which may be found, for example, in the books referred to previously, by Case and Zweifel or by Chandrasekhar. One finds

$$A(\nu) = -\frac{1}{9}\frac{X(-\nu)}{N(\nu)}\frac{1}{\int_0^1 \frac{\mu d\mu}{X(-\mu)}}$$

$$\psi_{em}(0,\mu) = \frac{1}{3X(\mu)}\left(\int_0^1 \frac{\mu d\mu}{X(-\mu)}\right)^{-1}, \quad \mu < 0$$

Thus we find the famous result, that the law of darkening is given by the H-function.

These four lectures have represented only the barest introduction to the subject of radiative transfer. In particular, the important topic of orthogonality relations was not discussed at all. In practice, the generalized orthogonality relations (discovered by I. Kuščer in 1963) among the eigensolutions to the transport equation permit rapid and convenient evaluation of all quantities of interest as, for example, the law of darkening above. The work has also been extended to cases in which the radiation field is a vector, rather than a scalar (case of polarized light) the non-conservative case (atmosphere not in equilibrium) etc. Also, many of these concepts and methods have been applied to other areas of physics - neutron transport, gas dynamics, electron discharge, plasma oscillations, etc. I hope, however, that the superficial introduction which I have given in these lectures might give the student an ideal of the field, and make it possible for him to proceed further on his own.